THERMAL FATIGUE
OF METALS

MECHANICAL ENGINEERING

A Series of Textbooks and Reference Books

Editor: L.L. FAULKNER   Columbus Division, Battelle Memorial Institute, and Department of Mechanical Engineering, The Ohio State University, Columbus, Ohio

Associate Editor: S.B. MENKES   Department of Mechanical Engineering, The City College of the City University of New York, New York

1. Spring Designer's Handbook, *by Harold Carlson*
2. Computer-Aided Graphics and Design, *by Daniel L. Ryan*
3. Lubrication Fundamentals, *by J. George Wills*
4. Solar Engineering for Domestic Buildings, *by William A. Himmelman*
5. Applied Engineering Mechanics: Statics and Dynamics, *by G. Boothroyd and C. Poli*
6. Centrifugal Pump Clinic, *by Igor J. Karassik*
7. Computer-Aided Kinetics for Machine Design, *by Daniel L. Ryan*
8. Plastics Products Design Handbook, Part A: Materials and Components; Part B: Processes and Design for Processes, *edited by Edward Miller*
9. Turbomachinery: Basic Theory and Applications, *by Earl Logan, Jr.*
10. Vibrations of Shells and Plates, *by Werner Soedel*
11. Flat and Corrugated Diaphragm Design Handbook, *by Mario Di Giovanni*
12. Practical Stress Analysis in Engineering Design, *by Alexander Blake*
13. An Introduction to the Design and Behavior of Bolted Joints, *by John H. Bickford*
14. Optimal Engineering Design: Principles and Applications, *by James N. Siddall*
15. Spring Manufacturing Handbook, *by Harold Carlson*
16. Industrial Noise Control: Fundamentals and Applications, *edited by Lewis H. Bell*
17. Gears and Their Vibration: A Basic Approach to Understanding Gear Noise, *by J. Derek Smith*

Additional Volumes in Preparation

Mechanical Engineering Software

HERMAL FATIGUE OF METALS

ANDRZEJ WEROŃSKI
TADEUSZ HEJWOWSKI

Technical University of Lublin
Lublin, Poland

CRC Press
Taylor & Francis Group
Boca Raton London New York

CRC Press is an imprint of the
Taylor & Francis Group, an **informa** business

CRC Press
Taylor & Francis Group
6000 Broken Sound Parkway NW, Suite 300
Boca Raton, FL 33487-2742

First issued in paperback 2019

© 1991 by Taylor & Francis Group, LLC
CRC Press is an imprint of Taylor & Francis Group, an Informa business

No claim to original U.S. Government works

ISBN-13: 978-0-8247-7726-5 (hbk)
ISBN-13: 978-0-367-40295-2 (pbk)

Library of Congress Cataloging-in-Publication Data

Weroński, Andrzej
    Thermal fatigue of metals / Andrzej Weroński, Tadeusz Hejwowski
      p.   cm. -- (Mechanical engineering; 74)
    Includes bibliographical references and index.
    ISBN 13: 978-0-8247-7726-5
    1. Metals--Thermal fatigue. 2. Thermal stresses. I. Hejwowski, Tadeusz. II. Title. III. Series: Mechanical engineering (Marcel Dekker, Inc.); 74.
TA460.W46   1991
620.1'61--dc20   91-20558
                       CIP

Visit the Taylor & Francis Web site at
http://www.taylorandfrancis.com

and the CRC Press Web site at
http://www.crcpress.com

Preface

Systematic investigation into the nature of thermal fatigue began in the early 1950s, the impetus coming principally from the continually increasing working temperatures employed in industry and a growing need for greater efficiency and reliability. The commonly accepted definition of thermal fatigue is that it is a complex of phenomena appearing in material exposed to cyclically varying temperatures in the presence or absence of external load. The process of thermal fatigue of a particular component is usually considered in association with mechanical fatigue, creep, corrosion, and similar factors. These phenomena undoubtedly contribute to the failure, as thermal cycling causes alternating strains to occur and the average temperature of the cycle is usually high enough for both corrosion and creep to become significant factors. All these topics have been studied extensively by researchers, but unfortunately, attempts to approximate thermal fatigue by considering each of these phenomena in isolation, then summing up the damages introduced, usually produce erroneous results.

The next difficulty encountered is that in considering thermal fatigue in practice, the exact form of the loading cycle of a particular component is very rarely known—what we know, in fact, is the approximate range of thermal variations at its surface. These should, in turn, be translated into thermal variations below the surface and thermal stresses and strains there. Usually, considerable experimental effort is needed to obtain true temperature profiles. Calculations of thermal stresses and strains require detailed knowledge of material properties, and these change with an increasing number of cycles and can be only roughly approximated.

This book brings together applied and analytical aspects of these phenomena. Attention is focused on the practical aspects and the book is devoted to both experimentalists who interpret and generate thermal fatigue data and users whose principal concern is to find a remedy to thermal fatigue problems. Accordingly, the book is intended to serve for both instruction and reference.

This book consists of separate chapters on mechanical fatigue, creep, heat-resistant materials, and thermal fatigue. The phenomena are presented in the order in which they attracted the notice of technicians. The advantage of this sequence is that we approach thermal fatigue gradually, meeting en route phenomena some or all of which are always present in any practical case in which thermal fatigue might occur. Also shown in the book is how to make practical use of the information provided on thermal fatigue. The example utilized is that of a mold for centrifugal casting working under extreme thermal conditions. The following steps are taken to mitigate the effects of thermal fatigue:

Defining the problem
Specifying the working conditions
Designing a test procedure to simulate the crucial factors
Running tests to determine the effect of thermal cycles on material and to rank the
    candidate materials
Finally, improving the mold design

Thermal fatigue arises in a greater variety of situations than can be discussed in detail herein. The wide domain in which thermal fatigue may make a significant or total contribution to failure has three boundaries:

1.   The area in which mechanical stress cycling is severe and thermal stressing is moderate
2.   The thermal shock region, in which temperature change and rate of change is so great that failure occurs within one or a very few cycles
3.   The region of creep failure, characterized by constant mechanical loading and relatively high temperatures during very long thermal cycles

Within these bounds there are situations normally understood as thermal fatigue. The discussion covers:

1.   End products subject to thermal fatigue in operation: for example, internal combustion engines, where the benefits of thermal fatigue design and manufacture appear as increased reliability and efficiency
2.   Manufacturing tools subject to thermal fatigue, which must be minimized in the interest of production costs. The tools considered include ingot molds, equipment for heat treatment, mill rolls, and forging equipment.

Chapter 8, the heart of the book, comprises a discussion of thermal fatigue in particular industrial components in reference to their working conditions. Examples are provided of a variety of machinery in which thermal fatigue problems are common and must be overcome. This book was written to assist in satisfying that imperative of modern engineering.

We wish to express our gratitude to the authors and publishers who kindly granted us permissions to cite their works, but first of all we would like to thank George Munns, who read most of the manuscript and made many helpful comments.

Andrzej Weroński
Tadeusz Hejwowski

Contents

3. Creep-Resisting Materials   81

4. Thermal Fatigue   108

5. Experimental Methods   136

6. Lifetime Predictions   161

7. Investigations of the Structure and Properties of Metals in the Course of Thermal Fatigue   173

Contents

THERMAL FATIGUE
OF METALS

THERMAL FATIGUE OF METALS

Mechanical Fatigue

1.1 INTRODUCTION

The commonly accepted definition of mechanical fatigue is that it is a complex of phenomena caused by progressive, cumulatively increasing damage to material accumulated in each of successive loading cycles. It is noteworthy that fracture is produced by stresses that are only a fraction of the static strength and that the number of loading cycles encountered in service by an element can be as high as 10^8. Some authorities maintain that about 80 to 90% of all failures ending in fracture are connected with fatigue. Fatigue failure caused by material imperfections is rather rare; the majority of accidents have, as a sole or contributory cause, a design or maintenance fault. Since Wöhler's investigation of the causes of a series of catastrophic failures to railway axles, summarized in his 1870 article, scores of thousands of publications devoted to mechanical fatigue have been published, but the knowledge gathered has been neither cohesive nor comprehensive. Furthermore, it seems unrealistic to reach a converse opinion when considering the variety of factors that influence the fatigue life and the difficulties in describing the state of real solid matter.

Mechanical fatigue is a particular case of thermal fatigue. Both have the same source of damage: repeatedly applied stresses of different origin, caused by either external load or temperature transients. This intuitively deducted identity is only sometimes found to be valid; its particular aspects are discussed later. In many situations encountered in industry, a device such as a forging die is loaded simultaneously by external forces and thermal stresses. Moreover, to discuss thermal fatigue it is convenient to accept the descriptive methods developed in earlier studies.

The process of fatigue failure can be divided into the following successive stages:

Formation of a microcrack nucleus
Propagation (growth) of the fatigue crack
Final rupture

These three stages are discussed in separate sections of this chapter.

1.2 FATIGUE TESTS

Any attempt to create a model of crack nucleation must be well supported by experimental data, and therefore some findings should be obtained first. Mechanical fatigue experiments are carried out in a range of temperatures from almost absolute zero to approaching the melting point of the specimen, in various media, often in vacuum or laboratory air and with a coexisting factor such as exposure to neutron flux while loading with fluctuating force. Despite these differences, all such experiments can be classified as follows:

1. Those in which the external cyclic or fluctuating load amplitude is held constant throughout the entire test run, whether by bending, push-pull, torsion, rotating cantilever, or a more "sophisticated" method. Such a setup is called *load controlled*. An example is the classical experiment of Wöhler.
2. Those in which the amplitude of load is not a crucial factor and thus is allowed to change during the test. Interest focuses solely on the strain amplitude. Experiments in this group are called *strain controlled*. The method is realized in practice by attaching a suitable extensometer to the specimen to measure deformation and using it as a controlling device. During testing the stress amplitude is allowed to vary while the strain amplitude is held constant.

Thermal stresses are usually strain controlled; for example, an ingot mold wall is exposed to thermal shock when poured with molten metal but is restrained from deformation by cooler portions of the wall material. By analogy, pressure loads are usually load controlled. Between these two extremes are a vast number of real loading situations that are mixed and more complicated in nature.

The external load is usually applied repeatedly with a chosen frequency while maintaining constant conditions. However, in a few situations a random variety of loads are utilized: to study the response of carriage springs, for example. This load spectrum can either be recorded during use of the device and reproduced in laboratory tests, or can be generated by a suitable device, usually a computer.

The load spectrum is characterized by the following quantities:

Medium stress, the arithmetic mean of maximal and minimal stresses in the cycle
$[\sigma_{av} = (\sigma_{min} + \sigma_{max})/2]$

Stress amplitude, half the difference between the maximal and minimal values of stress in the cycle

Stress ratio, the ratio of maximal value to minimal value of stress in the cycle
$(R = \sigma_{max}/\sigma_{min})$

The information concerning loading conditions becomes complete on specifying the cycle shape. Both cycle shape and frequency of cyclic loading are especially important in tests carried out at elevated temperature or in aggressive media.

The results obtained in the load-controlled tests are usually plotted in terms of the stress versus the logarithm of the number of stress cycles imposed. A batch of

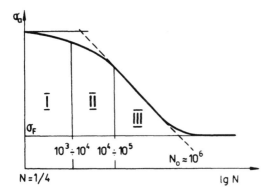

Figure 1.1 Diagram of Wöhler plot.

identical specimens is tested in practice under the same conditions, and the average number of stress cycles withstood is treated together with the cycle amplitude as coordinates of the point displayed in the plot. It is worth noting that the results obtained within one batch of specimens can differ by as much as one order of magnitude (e.g., Yokobori, 1954), and the curve drawn to represent the entire set of results is, in fact, the curve of best fit. A plot made in such a way is commonly referred to as a *Wöhler plot*. The leftmost point on the plot (see Fig. 1.1) corresponds to one quarter of the cycle and is usually in reasonable accord with the results of a static tensile test. As shown, the curve can be approximated by three dashed-line segments. Hück (1981) published a procedure to calculate the inclination of the middle segment and the limiting magnitude of stress, below which the indefinite life of the tested specimen is expected, being the ordinate of the rightmost segment and called the *fatigue strength* or *fatigue limit*. This procedure was developed on the basis of statistical considerations, over 600 Wöhler plots appearing in the literature having been taken into account for this purpose. The input data are the static strength or hardness, yield strength, quantities characterizing the size and shape of the specimen, cyclic stressing conditions, and the thermal treatment given to the specimen. Of course, such a description of the Wöhler plot is only approximate—to be used until experimental data regarding the material in question are provided. The portion of the plot (Fig. 1.1) marked "I" is called a *quasi-static range*, as the elongations of fatigued specimens reached just before the fracture are comparable with those found in static tests. Area II represents *low-cycle fatigue*, and at such stress amplitudes, yielding cannot be neglected, plastic deformation being a substantial part of the total strain range. The stress amplitudes and corresponding numbers of cycles relating to yielding that is restricted to the zone just ahead of the crack tip constitute region III, called a *high-cycle fatigue region*.

The presence of elastoplastic deformations causes hysteresis loops to appear in a plot of stress imposed on the specimen versus resultant strain (see Fig. 1.2). The area encircled by the loop is a measure of the energy supplied to the specimen.

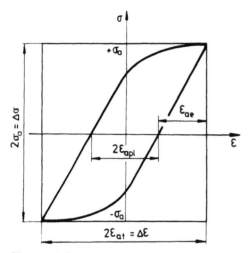

Figure 1.2 Hysteresis loop. σ_a, amplitude of cyclic stress; ε_{ae}, amplitude of elastic strain; ε_{apl}, amplitude of plastic strain; ε_{at}, total strain amplitude.

Hysteresis loops can be recorded in both strain- and load-controlled experiments. The experimental data obtained show that the hysteresis loop usually changes during early stress cycles to stabilize before one-third to one-half of the fatigue life is spent and achieves a saturation state characterized by constant values of stress amplitude or strain. The termination of fatigue life is accompanied by a loss of stability. The behavior of materials under test is illustrated schematically in Fig. 1.3. The upper curves were obtained in strain-controlled experiments, the lower ones in load-controlled experiments. The instances shown in Fig. 1.3a and d are referred to as *cyclic hardening*, those in Fig. 1.3b and e as *cyclic softening*, specimens c and f having revealed stability.

It should be stressed, however, that under more severe conditions (e.g., at slightly elevated temperatures), a stable hysteresis loop cannot be observed at all. Investigating the cyclic response of AISI 316 stainless steel, Jaske and Frey (1982) found that the saturation state was achieved in room-temperature tests and that a small alteration in test temperature (i.e., to 427 K) resulted in an ever-changing hysteresis loop up to 10^5 cycles.

The following relationship, determined experimentally by Manson (1953) and Coffin (1954), is an alternative way of describing the fatigue process:

$$N_f^K \, \Delta\varepsilon_{pl} = C$$

where

N_f = number of stress cycles to fracture

$\Delta\varepsilon_{pl}$ = plastic strain range

K, C = coefficents dependent on specimen material and test conditions

In the double-logarithmic coordinate system the plot of the foregoing equation is a straight line, the slope of which is defined by a K coefficient ranging from 0.4 to 0.8 but generally near 0.5. The C coefficient is usually found from the ultimate strain in the tensile test ($N_f = \frac{1}{4}$). Moreover, what is especially interesting, the relationship was proved valid in thermal fatigue tests. Attempts to replace the thermal fatigue test by a general fatigue tests, which provides a thermal strain range by imposing sufficiently large stress amplitudes, are of limited validity. In many instances even an application of stresses at the upper temperature of the thermal cycle provided very optimistic results—longer specimen endurance times than those obtained in a thermal fatigue test (e.g., Coffin, 1958).

The extent of the cyclic hardening increases with the stacking fault energy. The hypotheses explaining this phenomenon grew on the basis of dislocation consider-ations, formerly restricted to static tests. The hypothesis of Feltner (1965) explaining cyclic hardening applies to face-centered cubic (fcc) metals. It involves the creation of dislocation loops as a feature of cyclic hardening and the forcing of screw dislocations to circumvent the debris obstacles by cross-slip, which results in the formation of additional dislocation loops. In the saturation state, the number of dislocation loops remains constant, with the loops performing flip-flop motions between the two stable locations, enabling the reversibility of cyclic deformations.

High-cycle fatigue is relatively poorly explored. Tests on steels are usually terminated at about 10^6 cycles; continuation is very time consuming and expensive. It compels us to apply higher cycle frequencies, which in turn creates problems in maintaining constant ranges of stress or strains. For that reason, the fatigue strength is related to the number of cycles defined, usually being 1×10^6 for steels and only

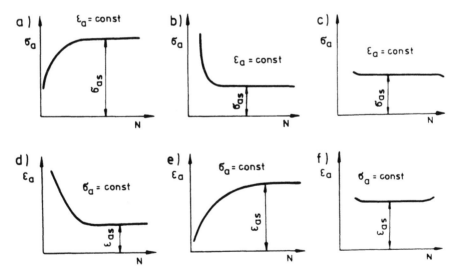

Figure 1.3 Behavior of various materials in fatigue tests in strain (a–c) and in load-con-trolled (d–f) experiments. (From Kocańda, 1985.)

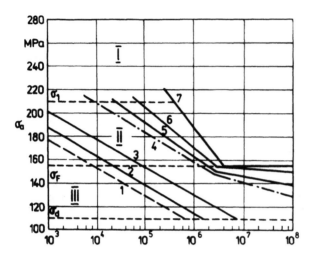

Figure 1.4 Wöhler plot for Armco iron.

exceptionally as high as 10^7 cycles. The value of the fatigue strength is found in tests involving reversed symmetrical cycles and more rarely, fluctuating cycles. The choice of cycle number limits is connected to the fact that in moderate test conditions, the knee of the S/N plot is observed below about 10^5 to 10^6, and the curve runs parallel to the axis of the abscissa. In more severe conditions, the horizontal portion of the plot can be absent altogether.

There are two possible design concepts:

1. A fail-safe design, in which the cyclic stresses occurring in the element are, by proper design, kept within the stress range that provides long life,
2. A design that assumes the existence of flaws in the applied component, and their gradual extension, described below.

1.3 MICRO- AND MACROSCOPIC PATTERNS OF FATIGUE

Investigations of fatigue failures almost invariably produce the same result: A fatigue crack nucleates at the surface of a loaded element and gradually grows to a critical length above which the crack is no longer stable, and propagates with subsonic velocity, causing final rupture of the element. This suggests that the state of the free surface of the element describes the damage accumulation within the fatigued element, and that changes in the microstructure found on the surface are crucial. Kocańda (1961a, b) investigated Armco iron in bending, interrupting the test several times to find the sequence of microstructural changes accompanying the fatigue damage. The fatigue limit was 155 MPa, the yield strength was 210 MPa, the amplitudes of stresses imposed on the specimens were in the range 80 to 240 MPa, and the corresponding numbers of cycles ranged from 10^3 to 10^8. Details of microstructural changes were studied on carbon replicas by means of transmission microscopy in addition to examination under a metallographic microscope. Figure 1.4 demonstrates a Wöhler plot (marked 7) obtained by experiment and the zones of characteristic microstructural changes. For stresses exceeding σ_1 (region I) the observed changes were very distinct, contrary to the very low stresses below σ_d, where the microstructure remained intact. In region III of stress amplitudes the microstructural changes, however, visible, did not lead to rupture, unlike the situation in region II. The inclined, numbered segments introduce a classification of the characteristic phases of fatigue.

1. In stress amplitude region I, distinct plastic deformation is observed after a few fatigue cycles. Narrow slip bands are present which are not grouped in any particular direction but rather, are randomly oriented to the acting stresses. Cracks are observed at the boundaries of heavily deformed grains and within slip bands. A rapid extension of cracks, passing grain boundaries, is observed between lines 6 and 7.
2. In region II, slip bands appear between lines 1 and 2, growing to form distinct bands between lines 2 and 3. Line 3 corresponds to blocking slip bands by grain boundaries. Between lines 4 and 5, slips induced in adjacent grains, when joined, constitute the microcrack nuclei clearly visible between lines 5 and 6; lines 6 and 7 mark the region of crack extension from an initial size of 0.1 to 0.5 mm until final rupture.
3. The pattern observed in region III is similar to the one described just above. The lower amplitude of acting stresses reduces the number of grains with developed slips and extends their incubation period. The microcracks revealed are blocked considerably longer on grain boundaries.

The pattern presented seems sketchy—the chemical composition of tested material enables precise interpretation of the phenomena observed, at the expense of oversimplification. As commercial alloys usually have a multiphasic structure, they are much more difficult to study, and the pattern would need to be broadly supplemented. Armco iron, which has a very low content of alloying additives, was shown to be sensitive to interstitials (e.g., N and C), which shift the Wöhler plot relative to coordinates, leaving the shape and pattern of changes intact. The presence of a fatigue limit is controlled by grain size, and for fine-grained steels is often observed. It therefore is interesting to note the experimental results obtained on commercial materials. For example, Maiya (1977) revealed short, one-grain-diameter-size (0.1 mm) cracks in stainless steel tested in air at 866 K in a low-cycle regime after 74 to 91% of fatigue endurance.

Irreversible microstructural changes, which threaten specimen integrity, appear in a very early stage of the fatigue proces. Thorough annealing, repeated after some number of cycles, may be efficient for a low-cycle situation when repeated after, say, 0.1% of fatigue endurance.

Nisitani and Takao (1981) investigated two grades of low-carbon steels (0.13 and 0.17 C): annealed α-brass and aged Al alloy. Bending tests were carried out at ambient temperature in air, and interrupted after a defined number of cycles in order to take replicas. It was found that in aged Al alloy, slip bands were confined to a small portion of the grain, the surrounding matrix remaining almost intact. In the course of fatigue progress, these bands developed and were transformed into a crack nucleus. In contrast, in the other materials the region of accumulating damage was far greater, reaching grain size. The repetition of stresses produced new slip lines, with the region of heavy slips subsequently transformed into a crack. It was also revealed that in low-carbon steels tested below the fatigue limit, microcrack nuclei were formed, but their slow growth was soon terminated.

The slips observed in Armco iron had a medium width of 100 lattice constants and the surface irregularity was on the order of 1000 lattice constants (Kocańda, 1961b). Slips proceed along definite crystallographic planes and axes. Two situations are frequently encountered: lamellar slips occurring in parallel planes and turbulent slips where mutually inclined slip planes are operating. The wavy appearance of slip lines points either to the presence of obstacles on the path of moving dislocations or to the presence of cross-slip. The latter is favored by a high value of staking fault energy. Microirregularities of the surface, even amounting to several micrometers, are products of slips and microcracks as well as of grain deformations. The pattern of extrusions and intrusions, rarely observed in its pure form, is depicted in Fig. 1.5. Extrusion is a squeezed-out flake of metal, the height of which is about 1 μm, and the neighboring extrusions are separated by 1 to 10 μm for copper. Intrusion, which does not necessarily accompany the former, is a fine crevice between extrusions. It is intuitively obvious that squeezing metal flakes out of the surface produces a void underneath. The risk of seizing up its walls by adhesion forces is reduced by the scale of the phenomena described (i.e., translation of relatively high volumes of material), and therefore it is a tempting entry point for crack nucleation hypotheses.

Lozinski and Romanov (1965) investigated commercial iron containing 0.03% C.

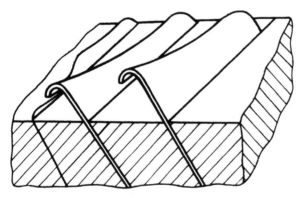

Figure 1.5 Pattern of extrusions and intrusions.

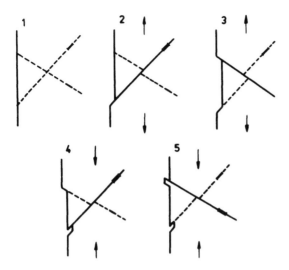

Figure 1.6 Sequence of slip movements producing extrusion and intrusion. (From Cottrell and Hall, 1957.)

Tests were done in high vacuum, with repeated bending, at temperatures of 293 to 1073 K. The sets of extrusions accompanied by intrusions were found only at elevated temperatures (above 573 K). The microrelief areas observed on the surface were often of complicated form, because in each grain (composed of subgrains) there is a set of various slip systems. The given test conditions favored some possibilities, rendering others negligible. An important achievement was the experimentally proved presence of voids below extrusions.

Cottrell and Hull (1957) proposed a model of extrusion and intrusion formation based on the coexistence of two slip systems and the sequence of slip motions (Fig. 1.6). It can be seen that the two intersecting slip systems are actuated in sequence and that during the tension half-cycle an intrusion is formed, as opposed to an extrusion during the compressive phase.

The model introduced by Mott (1958) is more independent of microstructure than that of Cottrell and Hall. In this model, extrusion results from the motion of a screw dislocation (the conservative and the cross-slip) having its extremities attached to the free surface, along the path $ABCD$, and the cavity lying underneath it, along $A'B'C'D'$ (Fig. 1.7). The cavity involved in the model can originate as a result of dislocation interaction following the mechanism described by Fujita (1958).

The models of extrusion and intrusion formation described cannot be treated as a complete explanation. Microcracks are not necessarily preceded by microrelief on the surface. Models introduced below were constructed mostly on the ground of static tests on pure materials, and therefore any conclusions regarding fatigue tests and commercial alloys should be drawn with care.

The hypothesis of Fujita (1958) involves two parallel rows of dissimilar edge

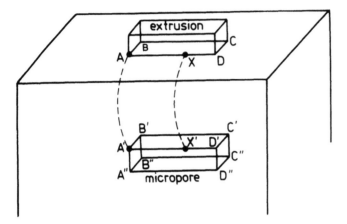

Figure 1.7 Dislocation movement in a crystal during extrusion formation. (From Mott, 1958.)

dislocations piling up against each other. If the distance separating these two slip planes is on the order of 10^{-7}cm, the annihilation of dislocation pairs produces the cavity of nb length (where n is the number of annihilated dislocations and b is the Burgers vector). The cavity is expected go grow because of the excess dislocations it captures. The model is supported by the findings of Pangborn et al. (1981). The measured density of excess dislocations in the fatigued specimen of Al2043-T3 alloy was several times higher at the surface than in the core. The surface density increased rapidly to 15% of the fatigue life, then grew at a moderate rate to 90% of the fatigue life, when it accelerated again, reaching some critical value at failure. In light of this experiment, assumptions of this hypothesis seem pertinent.

Stroh (1954) developed a suggestion, encountered earlier, that the concentration of stresses produced by a piled-up group of dislocations is sufficient for microcrack to occur. This is illustrated in Fig. 1.8. The plane of maximum stresses is at $\Theta = 70.5°$ to the slip plane.

Averbach (1965) based his model on the assumption that the stress concentration produced by the piled-up slip band of the critical width can locally exceed the fracture stress. The pattern of the model is shown in Fig. 1.9. A slip band of the width p_c (in iron of 10µm grain size, it is of 0.2µm) formed under the external tensile stress and cannot propagate into the B-grain, giving rise to the microcrack formation. In turn, in the C-grain, the narrower crack is formed to relieve shear displacement. Microcracks of this type were observed in mild steel within Lüders bands at 77K. Similarly, the microcrack can be nucleated by blocking the twin by the other or by the grain boundary, which was also proved in experiments.

Models involving the piling up of dislocations at grain boundaries are useless in crystalline material. The model of Cottrell (1958) overcomes this restriction,

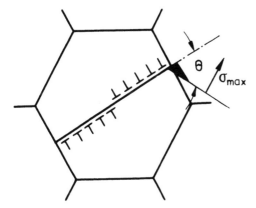

Figure 1.8   Model of Stroh (1954).

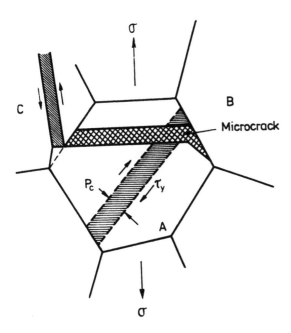

Figure 1.9   Model of Averbach (1965).

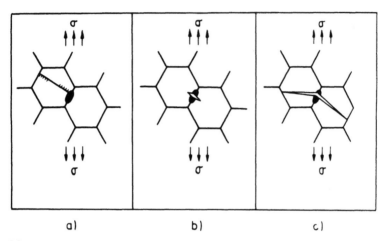

Figure 1.10 Phases of crack nucleation and propagation: (a) pileup of stresses impinging grain-boundary carbide; (b) cracking of carbide; (c) advancing crack encounters grain boundary. (From Saario et al. 1984.)

assuming that the two intersecting slip systems and the microcrack are a result of dislocation coalescence, which was also proved by the experiment.

Smith's (1968) model introduced the grain-boundary carbide. The blocking of twins or pileup of dislocations at the carbide particle produces a crack of its size propagating into the matrix (see Fig. 1.10). The model introduces the effect of grain size (through the number of dislocations) and the influence of temperature (through the change in the dominating deformation mode—twinning at low temperatures and slip at ambient or elevated temperature). Subsequent growth of the crack is considered to occur according to Orowan's model.

In commercial alloys, precipitations, nonmetallic inclusions, flaws, and inter-phase boundaries are sites of crack nucleations. For example, in steels of pearlitic structure, nuclei were found at pearlite–ferrite boundaries and also in cementite lamellas. The models described are applicable at moderate temperatures, whereas at elevated temperatures the models relevant to creep become valid.

1.4 CRACK PROPAGATION

According to Griffith's (1920) theory, the cause of the large discrepancy (tens of times) between the theoretically derived and experimentally found values of the cohesive strength is the presence of minute slits. These cause a pileup of stresses in the adjacent regions to a value exceeding the loading capacity of the material. The resulting cracks then develop in an avalanche-like manner under a sufficiently large external load at a subsonic velocity. The theory held true in the case of classical experimentation with stressing glass rods where the size of revealed microslits was comparable with that calculated from the results of tensile testing. However, in metals, unlike glass, which is brittle in the common sense of the word, plastic deformation is always present, appearing either in the macroscale as an elongation or in the microscale proved by x-ray back-reflection photographs. Moreover, according to Griffith's theory, the expected slit size is unusually large in plastic materials, to account for their mechanical properties (e.g., of millimeter order for zinc). For this reason it should be expected that the crack has its beginning in the form of a very fine nucleus growing gradually under the external load and reducing the volume of sound material that carries the load, subsequently leading to rupture.

The appearance of the fracture surface is a consequence of material properties and the geometry of the element made of that material, test conditions, material heterogeneities, and an increase in stress with a reduction in the load-carrying portion of the cross-sectional area, and rarely, inaccuracies introduced by the testing rig that promote differences in the fracture surface appearances of adjacent areas. The fracture surface appearances produced in static tests are commonly classified as follows (Maciejny, 1973):

1. *Brittle fracture*, where cracks run along cleavage planes (referred to as brittle, transcrystalline fracture), or *intercrystalline fracture*, when the grain interior has higher strength than its boundaries.
2. *Ductile fracture* composed of a series of concavities and convexities forming a honeycomb-like pattern. It is further divided into *fibrous fracture*, with the fracture surface perpendicular to the direction of maximal tension stress, and *shear fracture*, when this surface is inclined at 45°.

It was found experimentally that most materials follow the pattern sketched by Forsyth (1970), comprising three stages:

Stage 1: called *slip plane cracking*; occurs in planes closely aligned with the direction of maximal shear stress. This stage appears in pure metals, whereas in commercial alloys it does not necessarily occur.

Stage 2: directly following stage 1, in which the crack develops perpendicularly to the maximal tensile stress.

Stage 3: covering the moment of final fracture.

The first two stages are illustrated in Fig. 1.11 for an axially loaded specimen. Investigating NiCrMo cast iron, Mang and Wallace (1968) found that for lives on the order of 10^5 stress cycles, only 10% is spent in stage 2, a few cycles in stage 3, and the remaining major part of the life was spent in producing a minute crack 0.1 mm in size, which developed further at later stages. The appearance of the final fracture surface is usually similar to that depicted in Fig. 1.12. The locus at which the crack started is marked with the number (1); the adjacent zone (2) is usually small with a fine-grained appearance and a smooth surface. The steps numbered (3) occur when a few microcracks laying in mutually parallel planes develop simulta- neously, ad the principal crack results from a coalescence of some of them. The region marked (4) is called a *real fracture area*; the concentric lines encircling the

page14   **Chapter 1**

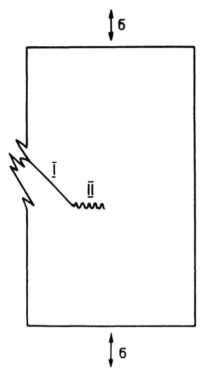

Figure 1.11 Stages of crack propagation. (From Hutchinson and Ryder, 1980a.)

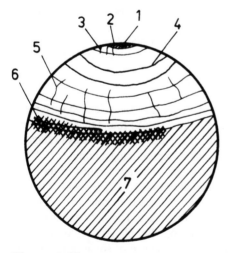

Figure 1.12 Appearance of fracture surface. (From Kocańda and Szala, 1985.)

fracture origin, called *beach marks*, are probably due to a momentary arrest of the crack. The principal crack is accompanied by secondary cracks (5); zone (6) is intermediate and adjoins the final rupture surface (7), which is closely similar to the fracture surfaces produced in static tests. The degree of plastic deformation can be deduced from the fracture's appearance, a useful experimental technique being heat tinting, where the clean fracture is heated in air until it forms an oxide film. The thickness of the oxide layer formed depends on the mode of fracture, local deformation, and also proved useful in identifying the fracture origin (Madeyski, 1984d).

Detailed examinations performed on the shadowed replicas of the real fracture area by means of transmission electron microscopy demonstrate that it has a fine structure consisting of a pattern of striations, each believed to correspond to one stress cycle. There is a definite relation between the fatigue striation spacing and the stress amplitude, and therefore, by examining the fracture surface in postmortem analysis and then by comparing it with the results of laboratory tests, the stress amplitude and number of cycles spent in crack growth can be deduced (Madeyski and Albertin, 1978). A comprehensive review of fractographic methods used in failure analysis appears in Madeyski (1984a,b,c).

1.5   PHYSICAL DESCRIPTION OF FATIGUE CRACK GROWTH

In the Griffith's (1920) theory based on the assumption of preexisting cracks, the necessity of crack shape knowledge was circumvented by considering the crack in terms of energy. According to that, crack growth produces a drop of elastic deformation energy, but simultaneously, it increases the surface energy, which is due to creation of the new surface. Calculations lead to the conclusion that a crack of initial $2a$ length grows spontaneously after the stress reaches some critical value. The appropriate condition for a two-dimensional crack (slit) in the plate subject to tensile stress acting perpendicularly to the crack surfaces, defined for plane stress state, is

$$\sigma = \left[\frac{2\alpha E}{\pi a (1 - v^2)} \right]^{1/2}$$

where

σ = critical value of stress, causing spontaneous growth of the crack initially having a length of $2a$

E = Young's modulus

a = surface energy per unit of thickness

v = Poisson's ratio

Plastic deformation can be treated as an equivalent of the surface energy and incorporated into the expression above by replacing the α value by the sum of α

and the plastic deformation energy, which makes it possible to extend the validity of the equation to cover the case of small-scale yielding, which is confined to the narrow region adjacent to the crack walls (Orowan, 1948).

An alternative approach, calculations of the piled-up stress magnitude, was developed by Orowan (1934), who assumed the radius of crack tip curvature to be approximately equal to one interatomic spacing, the critical value of the stress being convergent with the experimental data. Subsequent investigators developed Griffith's criterion, extending its validity for further instances. For example, Margolin (1984) introduced the compressive stress and friction forces.

The simplest case of loading is illustrated in Fig. 1.13. The center crack, of $2a$ length in the infinitely wide plate, is opened by tensile stress σ perpendicular to the crack line. The stresses acting on the small element at the advancing edge are (Irwin et al., 1958):

$$\sigma_x = \sigma \left(\frac{a}{2r}\right)^{1/2} \cos\frac{\theta}{2}\left(1 - \sin\frac{\theta}{2}\sin\frac{3\theta}{2}\right)$$

$$\sigma_y = \sigma \left(\frac{a}{2r}\right)^{1/2} \cos\frac{\theta}{2}\left(1 + \sin\frac{\theta}{2}\sin\frac{3\theta}{2}\right)$$

$$\tau_{xy} = \sigma \left(\frac{a}{2r}\right)^{1/2} \sin\frac{\theta}{2}\cos\frac{\theta}{2}\cos\frac{3\theta}{2}$$

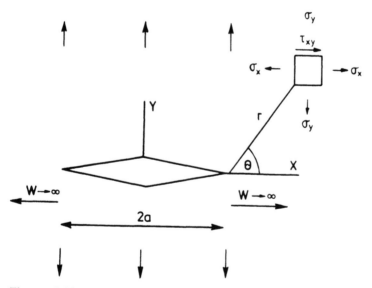

Figure 1.13 Crack in infinitely wide plate. (From Hutchinson and Ryder, 1980b.)

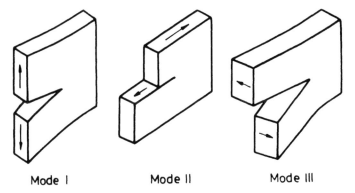

Mode I   Mode II   Mode III

Figure 1.14   Modes of crack opening. (From Hutchinson and Ryder, 1980b.)

For plane stress: $\sigma_2 = 0$ in thin sheet
For plane strain: $\sigma_2 = v(\sigma_x + \sigma_y)$ in thick plate

In general, the three basic crack opening modes are considered (Fig. 1.14). Correspondingly, the stresses at the advancing edge are

$$\sigma_{ik} = \frac{K_n}{\sqrt{2\pi r}} F_{ik}^n(\theta)$$

where

ik = stress component

n = subscript denoting the mode of crack opening (n = I, II, or III)

$F_{ik}^n(\theta)$ = functions depending solely on the angle

K_n = stress intensity factors, $K_z = \sigma(\pi a)^{1/2} M_k$, where M_k is a correction factor

The values of stress intensity factors play the role of a universal material constant, the maximal allowable stress in given stress conditions, and the geometry is related to the K_n value by the right value of the M_k factor. Values of the stress intensity factor for some geometries were tabulated by Sih and MacDonald (1974). The crack grows rapidly when the stress intensity factors reach the critical values called *fracture toughness* (for model we use the notation K_{Ic}). The typical values of the K_{Ic} are given in Table 1.1.

The radii of the plastic zones advancing ahead of the crack tip for a model crack are (McClintock and Irwin, 1965)

Table 1.1 Typical values of KIc

Material description	Yield strength (MPa)	Fracture toughness, K_{Ic} (MPa $\cdot$ m$^{1/2}$)
Aluminum 2014-T651	393	24.2
Aluminum 7075-T651	544	29.5
Titanium 6Al-4V	965	87.9
A517-T1 steel	758	194.5
5Ni-Cr-Mo-V steel	1027	306.6
17-7 PH steel	1151	53.3
12Ni-5Cr-3Mo steel	1282	256.0
AISI 4330 V mod steel	1310	60.4
18Ni (200) maraging steel	1324	117.6
AISI 4340 steel	1586	59.3
18Ni (250) A-538 steel	1696	95.6
18Ni (250) maraging steel	1786	75.2
18Ni (300) maraging steel	1965	56.9

Source: Blake (1982).

$$r_{p1} = \frac{K_I^2}{2\pi\sigma_y^2} \cos\frac{\theta}{2}\left(1 + 3\sin^2\frac{\theta}{2}\right) \qquad \text{for plane stress}$$

$$r_{p2} = \frac{K_I^2}{2\pi\sigma_y^2} \cos\frac{\theta}{2}\left[1 + 3\sin^2\frac{\theta}{2} - 4v(1-v)\right] \qquad \text{for plane strain}$$

where σ_y is the yield stress of the material. The shape of plastic zones for both cases is illustrated in Fig. 1.15. It is noteworthy that the knowledge of the K_I value is sufficient to estimate the size of the plastic zone, but its presence creates the necessity to correct the equations for the stress intensity factors for plasticity, and the commonly accepted way is to adopt the apparent crack length as the sum of crack length and plastic zone radius instead of crack length alone.

The other phenomenon connected with plastic deformation is the crack closure; that is, the crack is open for only a part of the stress cycle (Elber, 1970). For this reason, instead of the stress intensity factor range calculated as a difference between factors corresponding to the maximal and minimal values of stress in the cycle, the effective range of stress intensity factor is used. The lower value of stress taken into account is the opening stress.

Dugdale (1960) considered wedge-shaped plastic zones formed ahead of crack

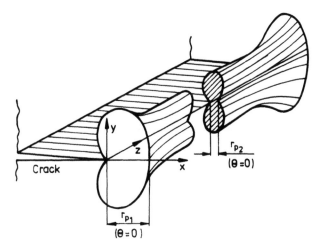

Figure 1.15 Shape of plastic zones. (From Kocańda, 1985.)

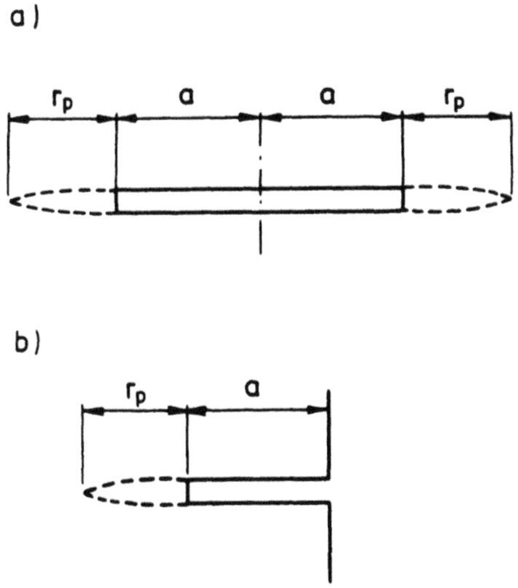

a)

b)

Figure 1.16 Plastic zones ahead of crack in thin sheet: (a) center crack; (b) edge crack. (From Dugdale, 1960.)

tips in thin sheets (see Fig. 1.16). The theoretically derived ratio of the plastic zone size to the apparent crack length corresponded well with the measured values, satisfying the equation

$$\frac{r_p}{r_p + a} = 2 \sin^2\left(\frac{\pi}{4} \frac{\sigma}{\sigma_y}\right)$$

The model of the plastic zone was justified by the dislocation analogy of the crack (the continuous distribution of edge dislocations was substituted for the crack) and extended on the shear stress by Bilby and co-workers (1963). Such an enriched model is called the BCS model after the names of the authors. Waku et al. (1983) proved the applicability of this model to describe the size of plastic zones and demonstrated the transition from plane stress to plane strain for a real, finite-geometry specimen when investigating deeper and deeper layers.

Lardner (1968) further developed the BCS model to cover the case of fatigue loading and obtained the following formula for the crack growth rate:

$$\frac{da}{dN} \sim \Delta K^2$$

where da/dN is the crack advance per stress cycle and ΔK is the stress intensity factor range. Weertman (1966) also anticipated, on the basis of the BCS theory, crack growth rate dependence on the fourth power of ΔK.

A very convincing model for describing crack advance was proposed by Tomkins (1968), the assumption being the existence of one slip band for stage I of cracking and the presence of two bands emanating at $\pm 45°$ from the crack tip for stage II (see Fig. 1.17). The process of plastic deformation is irreversible to a large degree under varying stress and produces a rippled pattern on the fracture surface.

Moreover, plastic deformation is confined to a small region; consequently, the BCS model is applicable. The coefficients of the Manson–Coffin relation calculated by means of this theory corresponded well with the experimental data. The derived equation for small stresses is

$$\frac{da}{dN} = A \, \Delta\sigma^3 \, \sigma_{av} \, a$$

where

       A = material constant

       σ_{av} = medium stress in the cycle

       $\Delta\sigma$ = stress amplitude

       a = actual crack length

Subsequent modifications of the model (Tomkins, 1971) and (Wareing et al., 1973) extended the equation on the high stresses

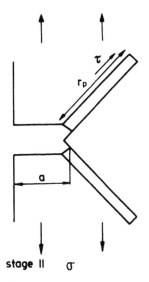

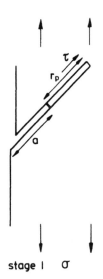

Figure 1.17 Model proposed by Tomkins.

$$\frac{da}{dN} = (\Delta\varepsilon_p + b\,\Delta\varepsilon_e)r_p$$

where

$b \approx 1/6$

$\Delta\varepsilon_e$ = range of elastic strain

$\Delta\varepsilon_p$ = range of plastic strain

Homma and Makazawa (1978) considered crack growth on the basis of the experimentally proven fact that growth is not a continuous process but rather, intermittent. Momentary crack arrests, known as *dormant* periods, are spent to accumulate the damage ahead of the temporarily stopped crack. The following relationship, well satisfied by experimental data, was found:

$$\frac{da}{dN} = C_1 C_2 (K_{\text{max}})^2 (K_a)^{\alpha(2-n)}$$

where

C_1 = constant relating the size of the plastic zone with the length of elementary crack advance, $da = C_1 r_p$

C_2 = constant connected with the yield stress, Young's modulus, and $\alpha(2-n)$, where α is the exponent in the Manson–Coffin law defined for the material considered and n is the exponent of work hardening relating true stress with true strain ($\sigma_e = F\varepsilon_e^n$, where F is a plastic coefficient); the α and n values range from 1.0 to 4.0 and 0 to 1.0, respectively

K_{max}, K_a = maximum and the amplitude of the stress intensity factor, respectively

The model presented relates the exponent in the fatigue crack growth rate equation with the material properties.

The equation proposed by Forman et al. (1967) is often used:

$$\frac{da}{dN} = \frac{C(\Delta K)^m}{(1-R)K_c - \Delta K}$$

where R is the stress ratio, the equation being written for plane stress conditions. The applicability was proved for aluminum 2024-T3 alloy for R between 0 and -1. For plane strain conditions the equation should be rewritten in the following form:

$$\frac{da}{dN} = \frac{C(\Delta K)^m}{[(1-R)K_c - \Delta K]^{1/2}}$$

Paris and Erdogan (1963) introduced the practice of validating crack propagation laws on the basis of a small number of data assuming that the range of stress intensity factor describes stress field at the crack tip and that its relation with crack growth rate is of a power form:

$$\frac{da}{dN} = C(\Delta K)^m$$

With regard to the growth of cracks in aluminum alloys, they propose

$$\frac{da}{dN} = C(\Delta K)^4$$

where C and m are treated as material constants, is called *Paris's law*. However, depending on the test conditions and the grade of material, they take various values; for example, for aluminum 7079-T6 alloy the exponents were 2 to 2.7 and 7 for stress intensity factor ranges of 6.2 to 14 and 14 to 22 MPa·m$^{1/2}$, respectively. The typical logarithmic plot of the crack propagation rate versus the stress intensity factor range is shown in Fig. 1.18. Paris's equation is seen to be valid for low stress values. This division into crack propagation rate ranges is often applied. The additionally marked ΔK_{th} value is a threshold amplitude of the stress intensity factor

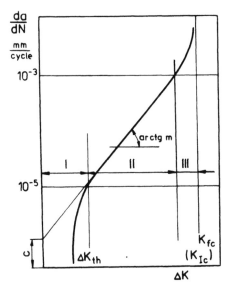

Figure 1.18 Crack propagation rate as a function of stress intensity factor range.

range below which cracks do not propagate; the K_{fc} value that corresponds in practice to the K_{Ic} value is called the *dynamic fracture toughness*.

Young's modulus normalizes the plot to some degree, and the data obtained for different materials and presented in the coordinates da/dN and $\Delta K/E$ are spread in a narrow region, thus enabling a curve to be drawn reasonably well fitted.

The coefficients in Paris's law depend on the geometry of the specimen being tested and that of a machined notch, as any testing of crack growth rate needs a localized crack. According to Wolkov and Berzov (1982), such differences in low-alloy 22 KWD steel (0.23% C, 0.6% Mn, 0.26% Cr) are of one order of magnitude for C and several percent for the m exponent. Consequently, any conclusions based on results should be treated cautiously.

It is interesting to note the following facts that illustrate the extent of plasticity effects. In the aluminum 2219-T581 alloy, the closure stress of grain size cracks exceeds 0.2 of the yield stress and is the function of the distance between the crack tip and the grain boundary (Morris et al., 1981). Most published data regarding fatigue were obtained in experiments where the average stress of the cycle was either zero or tensile. The results presented by Saal (1972) were gathered under quite different conditions, and the notched specimens were subject to the fluctuating compressive stress. The presence of a notch caused a pileup of stresses, with the resultant yielding at the notch bottom under a sufficiently large external load. For entirely compressive cycles ($R = 0$) crack arrest was observed at the boundary of the plastic zone, the crack growth being dependent on the magnitude of the stress applied. For cycles comprising tensile parts, cracks grow beyond the plastic zone at a rate that depends on the value of the tensile stress.

The practical limit of applicability of linear fracture mechanics—namely, the K-factor concept—is $\sigma \approx 0.7\sigma_y$ (σ_y is the yield stress). The appearance of a distinct plastic zone at the crack tip contradicts the previously assumed ideal elasticity and introduces an error of order, say 20%. The correction of results based on the apparent crack length is an inconvenient tool, and therefore the two new parameters are introduced to correlate with the crack growth rate: the crack tip opening displacement (CTOD) and the J-integral.

The first parameter, crack tip opening displacement (δ) is introduced in Fig. 1.19. The choice of the place to measure the distance of crack walls is difficult in the case of crack tip blunting and can lead to some uncertainties. The location depicted, just at the crack tip, is mostly commonly used but is not compulsory. It follows from purely geometric considerations that ½ CTOD = Δa, the elementary crack advance. In contrast, Tanaka and co-workers (1984) found that the contribution of the CTOD to Δa in SUS 304 steel tested in a fully reversed push-pull mode depends on the range of the K. factor and decreases five times with decreasing crack growth rate from 5×10^{-5} to 3×10^{-6} m/cycle. The crack growth rate dependence on the CTOD (δ) value measured 250 μm behind the crack tip was

$$\frac{da}{dn} = C(\text{CTOD})^p$$

where

$C = 1.9 \times 10^4, p = 2.26$ for 0.04% C steel
$C = 2.25 \times 10^4, p = 2.24$ for SUS 304 steel

Relations were satisfied in the range of the crack growth rates 10^{-9} to 5×10^{-5} m/cycle.

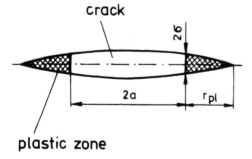

Figure 1.19 Crack opening displacement.

The CTOD (δ) value and the K factor are interconnected with each other by the following equation:

$$\delta \approx \frac{K^2}{E \cdot Re} \qquad \text{for plane stress}$$

$$\delta \approx \frac{1}{2} \frac{K^2}{E \cdot Re} \qquad \text{for plane strain}$$

In the case of stresses exceeding the yield strength, the energetic state into the vicinity of the advancing crack is to be described by means of the J integral. Considering a homogeneous body of material containing a crack and subject to the two-dimension deformation field, it can be found that the following integral is path independent (Rice, 1967):

$$J = \int_{\Gamma} \left[W(\varepsilon)\, dy - \mathbf{T}\, \frac{\partial \mathbf{u}}{\partial x}\, \partial s \right]$$

The Γ path is traversed in the counterclockwise direction (see Fig. 1.20), $\mathbf{T}$ is the traction vector on Γ ($\mathbf{T} = \sigma_{ij} n_j$), $\mathbf{u}$ the displacement vector, s the distance along the path of integration, and $W(\varepsilon)$ the density of the strain energy (i.e., the energy per unit volume, $W = \sigma_{ij}\varepsilon_{ij}/2$). On choosing a particular, circular path with the r radius, we get

$$J = r \int_{-\pi}^{+\pi} \left[W(r,\varphi)\, \cos\varphi - \mathbf{T}(r,\varphi)\, \frac{\partial \mathbf{u}(r,\varphi)}{\partial x} \right] d\varphi$$

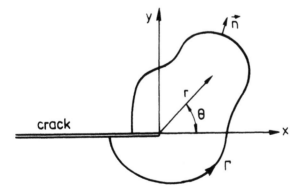

Figure 1.20 Coordinates for description of crack tip deformation field and typical contour Γ. (From Rice and Rosengren, 1968.)

In the elastic case,

$$J = \frac{K^2}{E} \qquad \text{plane stress}$$

$$J = \frac{(1 - v^2)K^2}{E} \quad \text{plane strain}$$

Wilson and Chrepko (1983) reported on application of the J integral to crack branching. The calculated K factors were dependent on the number and orientation of branches. Branching was found to alleviate the stress severity.

The concept of the J integral has frequently been modified. Dowling (1977) introduced the cyclic J integral, ΔJ. Taira et al. (1979) adopted this concept to describe shallow cracks and specimens of rectangular cross section:

$$\Delta J = \frac{(\Delta K*)^2}{E} + \frac{U*}{2Bb}$$

where

$\Delta K* =$ effective range of the K factor

$E =$ Young's modulus

$U* =$ area encircled by the hysteresis loop [formed in the system $\Delta P*$ (effective range of stresses) versus crack center opening displacement (CTOD)]

$B =$ specimen thickness

$b =$ ligament length (specimen width minus crack depth)

It was found that for AISI 316 steel tested at 923 K in the push-pull mode the following relation was valid:

$$\frac{da}{dn}\left(\frac{mm}{cycle}\right) = 1.18 \times 10^{-4}(\Delta J)^{1.32}$$

1.6 FATIGUE-LIFE ESTIMATION

The criterion of fatigue failure is a widely understood dysfunction of the component considered. It can be either a permanent deformation exceeding specified limits, a deterioration of required properties, or in the worst case, rupture. In practice, the imposed stress can very rarely be characterized by one set of values (e.g., frequency, stress amplitude, and load ratio), since the load spectrum usually is stochastic in nature. Considering such spectra, it can be established that there are:

Different frequencies and different cycle shapes and amplitudes
Uncertainties in the method of cycle counting, simplifying the picture; the small stress excursions are imposed on the major stress changes

The most widely accepted approach is the rain flow method (Dowling, 1972).

Laboratory tests indicate that approximation of the real spectrum by a set of loading blocks, each comprising unified cycles that have the same amplitudes and frequencies within the block, is of limited use and that the sequence of blocks is also important.

Fatigue-life estimation can be performed on the basis of the Manson–Coffin equation or an assumption of damage accumulation. According to the theory formulated initially by Palmgren (1924) and reintroduced by Miner (1945), the failure condition for a multistress history is

$$\sum_{i=1}^{k} \frac{n_i}{N_i} = 1$$

where

   k = number of stress levels

   n_i = number of cycles imposed at the σ_i stress

   N_i = number of stress cycles needed to produce failure at the σ_i stress

However, it was found that this theory overestimates the fatigue life by factors of 2 to 3½ for the random spectra. Moreover, it was found that for spectra comprising largely different stress amplitudes, the summation results often depart from 1, sometimes severalfold. Therefore, the theory was subsequently modified to produce more conservative fatigue-life predictions. Grover (1959) proposed dividing the fatigue process into the stages of crack initiation and crack propagation, and using linear summation in these stages separately. Tests performed on specimens of 2024-T4 aluminum alloy, prestrained before testing, gave summation values in the range 0.36 to 1.50 for complicated stress histories (Dowling, 1972). The tests were essentially limited to an analysis of crack progress.

Feltner and Morrow (1961) postulated a hysteresis energy criterion of fatigue failure. The plastic strain energy is accumulated in small portions over the life of the specimen until a critical value is achieved. This critical value was associated with the area under the true stress–true strain curve in the static test. Kliman and Bilý (1984) proved that the assumption of the stable hysteresis energy is true for 0.4% C steel in tests carried out in air, at ambient temperature, with a loading frequency of 1 to 15 Hz and in the strain range 1×10^{-4} to 5×10^{-3}. Of course, in more severe conditions (e.g., at high temperatures), hysteresis loop energy can vary throughout the test. The supposition of the frequency independence is also rough: For high-strength 0.86% C steel the increase in fatigue limit when the frequency is raised from 50 Hz to 10^4 Hz is almost twofold; at higher, ultrasonic frequencies, the fatigue limit decreases.

Halford (1966) demonstrated for metals and commercial alloys that the energy necessary to rupture a specimen increases with the one-third power of the fatigue life; for 500,000 cycles this energy is 100 times higher than that for fracture in monotonic tension. Obviously, this makes the concept of Feltner and Morrow applicable to the limited number of cases.

Considering the mechanical energy supplied to the fatigued specimen, it is worth noting that its major part is manifested as heat and is removed from the specimen by conductive, convective, and radiative heat transfer. The amount of heat evolved can be a few times higher than that to melt the specimen. The temperature distribution over the specimen surface has local maxima, which shows that the mechanical deformation is also nonhomogeneously distributed over the specimen surface. In their study on a notched mild steel specimen tested in bending in a high cycle regime at room temperature at 30 Hz, Charles and co-workers (1978) reported a temperature increase of 2 K at the notch bottom immediately before origination of the visible fatigue crack. Subsequent temperature increase was obviously higher, reaching approximately 20 K at the termination of the test.

Another, quite recent approach is the damage tolerance concept, developed to utilize components above their safe design life. In practice, only a minor percentage of components have developed a detectable crack before their retirement. In this approach, it is assumed that any flaws exist in a component at the time of manufacture but that their sizes are just below the detection limit of the NDI equipment. It is assumed further that these flaws are located in highly stressed parts of the component and grow at a rate governed by the stress prevailing in their loci. The component can be utilized until the risk of rapid, unstable crack growth becomes severe, provided that it is checked at regular specified intervals to protect against any excursion from experimentally derived crack growth formulas. Implementation of this approach introduces the probability of flaw detection or the probability of occurrence of flaws of a certain size (Koul et al., 1985).

1.7 EFFECT OF MICROSTRUCTURE

It happens most often that fatigue failure occupies some period of time, and therefore one can envisage the fact that the structure gradually changes and the structure corresponding to the termination of life differs considerably from that accompanying crack nucleation. Apart from that, in austenitic steels localized martensitic transformation is often encountered in the stress field ahead of the crack tip, which increases the specific volume of metal in the zone of crack propagation, either accelerating or retarding it, depending on the stress intensity range ΔK or the test temperature. The subject of this section is the actual microstructure ahead of the crack tip. It also happens that seemingly identical coupons of material supplied by other producers display quite different lives, and the detailed analysis shows that it is often caused by different severity of inclusions and pollution with tramp elements. The former can be illustrated by the work of Hebsur et al. (1980). The commonly employed AISI 4340 steel was tested in two forms: (1) conventionally cast and forged rods, and (2) electroslag refined [i.e., the steel coupon was arc remelted under the cover of the synthetic slag mixture containing $CaF_2/CaO/Al_2O_3(60:20:20)$].

Table 1.2 Chemical Analysis (wt %) of AISI 4340 Steel Before and After ESR

	C	Si	Mn	P	S	Cr	Mo	Ni
Unrefined	0.45	0.34	0.45	0.042	0.025	0.99	0.24	1.34
ESR ingot								
Top	0.44	0.30	0.47	0.040	0.004	1.01	0.24	1.34
Bottom	0.46	0.28	0.42	0.040	0.004	1.00	0.23	1.36

Source: Hebsur et al. (1980).

The high basicity of slag together with the high temperature gave effective desulfurization. Large inclusions were removed and the fine ones were randomly redistributed. The chemical analysis of the as-received and refined materials is given in Table 1.2. Distribution of the nonmetallic inclusions is shown in Fig. 1.21. The rate of fatigue crack propagation is depicted in Fig. 1.22. The improvement in fatigue performance and static tensile properties of the refined material were

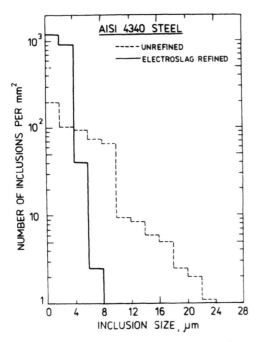

Figure 1.21 Number of inclusions per mm^2 versus their size for AISI 4340 steel before and after ESR. (From Hebsur et al., 1980.)

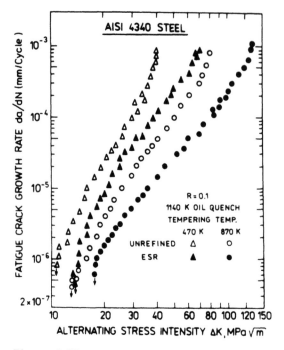

Figure 1.22 Crack propagation rates at room temperature measured on as-received and refined materials, heat treated in two different ways. (From Hebsur et al., 1980.)

attributed chiefly to the lower inclusion content and also, however to a minor extent, to the reduction in sulfur content.

The influence of microdiscontinuities on the mechanical properties of cast steels has been reviewed by Mitchell (1977). This term was used to cover inclusions, gas porosities, shrinkages, and so on. As inclusions generally are complex compounds of metallic alloying elements with oxygen, carbon, phosphorus, sulfur, and silicon, they differ considerably in shape and size. Generally, silicates are spherical in shape, and sulfides can undergo drastic morphological changes depending on the amount of aluminum that is added as a deoxidizing agent. The size of inclusions also depends on the solidification rate; inclusions encountered in large ingots can reach a considerable size. The number of gas porosities can be reduced by agitation of the partially solid melt, thus allowing additional sites to form dendrites as the gases evolved from solidifying melt are trapped at solid–liquid interfaces. Wrought steels in which internal porosities are welded shut are known to have better properties than those of as-cast material. However, working enlarges inclusions in the longitudinal direction with respect to the working direction and causes fatigue life to be appreciably lower in the transverse direction. The influence of all the discontinuities—their shape, size, and distribution—should be considered in terms of the local

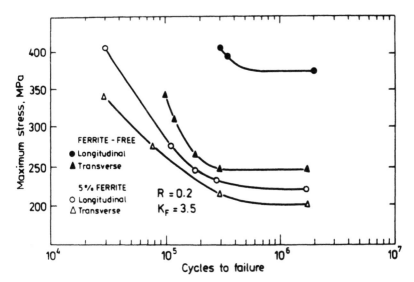

Figure 1.23 Notched fatigue strength of AISI 431 stainless steel. (From Scott, 1969.)

stress concentration. Microstructural banding, also dangerous to material durability, is usually produced by a local chemical segregation which stabilizes a phase not normally present in the alloy. The severity of the banding can be seen in Fig. 1.23. Stainless steel was heat treated to 1240 to 1380 MPa and the notched fatigue strength for both ferrite-free specimens and for those containing 5% ferrite is displayed in Fig. 1.23. The distinct reduction in fatigue performance was caused by the pileup of stresses at the order of phases.

ESR is a powerful tool but only one of several measures available. Clean steel technology has been a focal point of researchers for years. The methods listed below (after Chandler, 1982) are competitive with ESR:

Quelle-basic oxygen process (Q-BOP)
Ladle refining (LR)
Vacuum induction melting (VIM)
Vacuum degassing (VD), in some cases in conjunction with electromagnetic or argon stirring
Argon oxygen decarburization (AOD)
Vacuum arc remelting (VAR)
Vacuum arc double-electrode remelting (VADER)
Electron-beam remelting (EBR)

Practically, it is possible to obtain steels containing less than 20 ppm of hydrogen, 0.01% sulfur, and 0.01% phosphorus. Improvement in the pouring phase can also be achieved through the use of bottom pouring, with ceramic filters to remove slag and inclusions from the stream of molten metal. The method of HIP (hot isostatic pressing) involving simultaneous application of pressure and temperature is very promising.

The application of modern technologies does not encounter a technological barrier, the only restriction being the quality of the product. For example, low pressures become necessary when started with reactive elements such as Al, Ti, or Cb to the nickel- and nickel/iron-based alloys to form the strengthening phases [i.e., $(\gamma')Ni_3(Al,Ti)$ and $(\gamma'')Ni_3Cb$] to enhance high-temperature properties. Other additives (i.e., Cr, Mo, W, V, and Ta) combine with carbon to control grain size and reinforce grain boundaries. Elements such as zirconium, boron, and hafnium migrate to grain boundaries, enhancing creep rupture properties. The affinity for oxygen and nitrogen of such alloying additives compels us to make use of vacuum metallurgy. The first commercial heats of VIM-produced ingot were obtained in 1952. Since then the capacity of VIM furnaces has increased from 5 kg to 15 tons (Boesch et al., 1982). However, most VIM superalloy ingots possess coarse grains, and therefore a second production stage (VAR) was introduced in the early 1960s. In this method, the VIM-produced ingot is treated as a consumable electrode and arc remelted into a water-cooled crucible. The electric arc is applied between the crucible and the consumable electrode. Further grain refinement can be obtained by the VADER method, in which two consumable electrodes made of VIM ingots are mounted horizontally and melting arcs are established between them. This method can be utilized as an alternative to powder processing.

The most widely used disk alloy, IN-718, was produced in the United States by the double vacuum process VIM-VAR, which produces ingots of an acceptable quality. The demand for improved elevated-temperature properties prompted us to study the capability of the ESR method. The segregation of alloying additives in the alloy IN-718 takes the form of *freckles*, that is, chains of randomly oriented Nb-rich grains and *white spots*, Nb-lean grains, both of which have a detrimental effect on the properties of the product. The former are present in VAR and in ESR-treated ingots, the latter in VAR ingots. The formation of freckles is connected with the flow of a solute-rich interdendritic liquid during solidification due to contraction, gravity, and electromagnetic forces. White spots seem to be caused by segregation. Yu et al. (1986) demonstrated that thorough control of ESR production variables can provide an ingot of acceptable quality.

In the case of cast iron the problem is also the segregation of elements. In nodular iron, the boundary between the graphite spherule and the matrix is the site of the segregation (e.g., Franklin and Stark, 1984). The crucial influence on fatigue properties and impact strength of Si-enhanced regions around graphite spherules is well known, and improvement can be achieved by the thermocycling treatment. Unwanted precipitation of Si takes place during the cooling of metal and thus the thermal cycles applied to eliminate this should involve rapid cooling. For example, Fiediukin (1984) utilized eight thermal heating cycles to temperatures 30 to 50 K below A_{c1} at a rate of 30 to 40 K/min, followed by rapid cooling. The resultant impact strength was five times higher than that for as-cast material, and the fatigue limit increased from 90 MPa to 230 MPa.

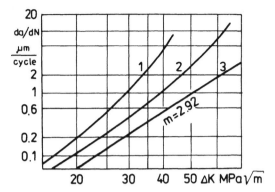

Figure 1.24 Effect of structure on crack growth rate. 1, Lamellar pearlite; 2, pearlite of mixed shapes; 3, globular pearlite. (Adapted from Heald et al., 1972.)

The more advantageous shape of cementite (i.e., globular) can be obtained by the treatment involving four to six cycles of heating at a rate of 30 to 40 K/min to the temperatures 30 to 40 K above A_{c1}, followed by air cooling to 30 to 50 K below A_{r1}, followed by water quenching. Brieusov and co-workers (1973) reported an increase in the fatigue limit from 165 MPa to 200 MPa after similar thermal treatment, whereas thermocycling involving cycles above A_{c3} gave a better result (i.e., 230 MPa), due to the more profitable shape of cementite.

The effect of pearlite shape is also remarkable in steels as shown in Fig. 1.24. The tested material was carbon steel (1% C, 1.35% Mn, 0.5% Cr, 0.5% W). The higher rate of crack propagation for the pearlitic lamellar structure was caused by the cleavage cracking of cementite lamellas. Only the third curve was in reasonable agreement with Paris's equation ($m = 2.92$).

The influence of thermal treatment was also investigated by Eifler et al. (1981). Steel of grade 42Cr-Mo4, which contains 0.44% C, 0.32% S, 0.73% Mn, 0.014% P, 0.028% S, 1.11% Cr, 0.22% Mo, and 0.11% Ni, was subjected to four different thermal treatments:

1. Normalizing at 1203 K for 3 h in argon, furnace cooled, the resultant ultimate tensile strength being 550 MPa.
2. Quenching and tempering (austenitizing at 1023 K for 20 min in a salt bath, oil quenching followed by tempering at 923 K for 16 h), the resultant ultimate tensile strength being 730 MPa.
3. Quenching as above and tempering at 923 K for 2 h, the resultant ultimate tensile strength being 865 MPa.
4. Austenitizing at 1123 K for 3 h in a muffle furnace, oil quenching, and tempering at 843 K for 4 h, the resultant ultimate tensile strength being 1110 MPa.

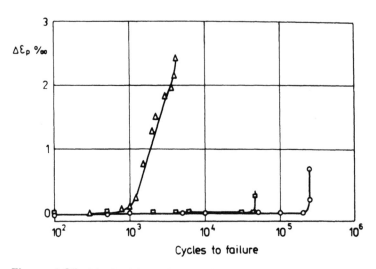

Figure 1.25 Plastic deformation amplitude as a function of cycle number for three batches of specimens: Δ, UTS = 730 MPa; $\square$, UTS = 865 MPa; $\bigcirc$, UTS = 1110 MPa.

The tests performed were of push-pull type at room temperature under load control. The measured strain amplitudes are shown in Fig. 1.25 for the three different thermal treatments. The results observed are rationalized in terms of differences in the microstructure. For the most resistant quenched and tempered steel, fine carbides are separated by minute spacings; for other thermal treatments these spacings are bigger, which facilitates the dislocation motion. In the normalized specimen, for which the data were omitted in Fig. 1.25, very low fatigue endurance was connected with a great ease of displacement in the normalized ferrite.

Yamaguchi and Kanazawa (1983) studied the effect of thermal aging on the low cycle fatigue of SUS 321 stainless steel (0.07% C, 0.6% Si, 1.84% Mn, 0.024% P, 0.014% S, 10.0% Ni, 18.95% Cr, 0.45% Ti). Push-pull fatigue tests were carried out under total strain control in air at temperatures of 873 and 973 K. The strain waveshape was triangular, fully reversed, and the strain rates were 0.4 and 40% per minute. The specimens were in three states, resulting from thermal treatments applied to simulate metallurgical changes during in-service exposure:

1. Solution annealed at 1473 K of 0.5 h, water quenched
2. As state 1 plus aging at 1173 K for 1 h, slowly cooled
3. As state 1 plus aging at 1023 K for 24 h, slowly cooled

In specimens subjected to the first type of thermal treatment, no precipitates were observed; in the second group of specimens, very fine TiC precipitates were seen; and in the specimens subjected to the third variant of thermal treatment, large $M_{23}C_6$ precipitates were found near the grain boundaries. The results of fatigue experiments can be summarized as follows (see also Fig. 1.26):

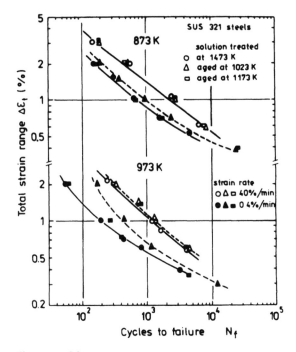

Figure 1.26 Relation between total strain range and fatigue life.

1. In tests where a high strain rate was involved, the cracks were transgranular with a well-defined striated pattern. Crack propagation rates remained almost independent of the thermal treatment.
2. In tests involving low strain rates, cracks were intergranular; the solution-annealed specimen displayed the shortest life, whereas specimens aged at 1023 K had the longest life, which is due to the presence of carbides at the grain boundaries.

Neate (1985) investigated 1 Cr-Mo-V steel in long-dwell displacement-controlled cycling at 873 K under both coarse-grained and ferritic conditions. The investigation was undertaken following the discovery of extensive cracking in the valve chests of the two 500-MW steam turbines. The valve material, which was originally in a high-creep-strength coarse-grained bainitic condition, was subsequently heat-treated to obtain refined ferritic microstructure with improved toughness. Such a reheat treatment enabled the unit to be returned to service with minimum delay. Tests were of two kinds: fluctuating tensile with a 2-h hold imposed at maximum displacement and fully reversed, with a 24-h hold at maximum tensile displacement. In both groups of specimens cracks were intergranular, and in the bainitic structure steel, cracks initiated first and developed considerably faster.

A distinct effect of thermal treatment on fracture toughness is also apparent in titanium alloys. Yoder and co-workers (1977) tested Ti-6Al-4V and Ti-6Al-6V-2Sn, both of which are used in weight-critical aerospace applications. The best results were provided by β-annealing, which for the Ti-6Al-4V alloy reduced the fatigue crack growth rate by an order of magnitude that was connected with the change in cracking mode (multiple cracking). In the Ti-6Al-6V-2Sn alloy it lowered crack growth rate by a factor of 2. A second type of heat treatment, recrystallization annealing, resulted in negligible reductions in fatigue growth rates over lower values of ΔK, and for higher ΔK values it produced substantial increases.

The influence of grain size on tensile properties is well established (the Hall–Petch relation). Mannan et al. (1983) have proved the validity of this relation up to 1023 K in type 316 stainless steel. Yamaguchi and Kanazawa (1980) tested the same steel together with other austenitic stainless steels for fatigue at elevated temperatures (873 K and 973 K). The cycles were either triangular of trapezoidal, with the included 30-min hold-time period on the tension side. Some results are plotted in Figs. 1.27 and 1.28. The grain size number was measured in accordance with Japan Industrial Standard (JIS) G0551. Details of the tests are given in the figures. The results can be summarized as follows:

1. For fast cycles (strain rate 6.7×10^{-3} s^{-1}) the fracture mode was always transgranular and the fatigue life was independent of the type of steel and of grain size.

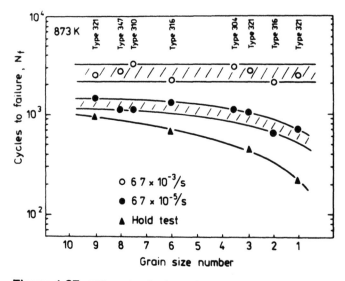

Figure 1.27 Effect of grain size on fatigue performance at 873 K (total strain amplitude 1%). (Reprinted with permission from ASM International.)

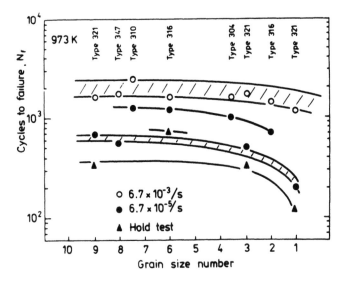

Figure 1.28 Effect of grain size on fatigue performance at 973 K (total strain amplitude 1%). (Reprinted with permission from ASM International.)

2. For slow cycles imposed at 873 K, fatigue lives decreased with decreased grain size number, and the fracture mode changed from dominantly transgranular to completely intergranular, with an increasing grain size number.

3. For slow cycles at 973 K, the fracture modes of steels types 304, 316, and 310 were mixed because of recovery that took place during the tests; types 321 and 347 fractured in the intergranular mode. Recovery removes the cyclic hardening, resulting in cracks that have a mixed character. The lack of recovery enables the grain interior to harden, causing cracks to run along grain boundaries.

The question that naturally arises relates to the fatigue performance of materials having an ordered structure: monocrystals, columnar grained alloys, and composites. Composites received particular attention, as they offer possibilities of increasing further service temperatures over those withstood by conventional superalloys. The particular composites studied were directionally solidified eutectic superalloys (DSESA) and artificially fiber-reinforced superalloys (AFRSA). Lemkey and Thompson (1972) tested one of the DSESA composites, referred to as γ, $\gamma' + \delta$, and found a microstructure consisting of an aligned lamellar Ni_3Cb phase (δ) surrounded by a gamma nickel–chromium matrix (γ) containing a γ' precipitate (Ni_3Al). Interlamellar spacing was determined by the solidification rate. The alloy displayed better creep performance than that of IN-100 and columnar-grained MAR-M200. When tested in high cycle fatigue, another eutectic alloy, Ni_3Al-Ni_3Nb, revealed better performance than that of a AFRSA composite, Hastelloy X–35V/0/W (reinforced with tungsten fibers), and that in turn behaved better than conventional Hastelloy X (Alexander et al., 1969). However, what is noteworthy is that the ordered structure and the difference in coefficients of thermal expansion of particular phases create a pileup of stresses on the interphase boundary that can considerably reduce thermal fatigue life.

Monocrystals have encouraging properties because the lack of grain boundaries excludes paths of easy crack propagation, which is a great advantage in creep and thermal fatigue. Miner et al. (1986a,b) and Gabb et al. (1986) investigated fatigue properties of the crystalline René N-4 of nominal chemical composition: 3.7% Al, 4.2% Ti, 4% Ta, 0.5% Nb, 6% W, 1.5% Mo, 9% Cr, 7.5% Cr, 7.5% Co, and balanced Ni. It is seen that elements usually added to reinforce grain boundaries (i.e., C, B, Zr, and Hf) are missing. The alloy contains a high fraction of γ' (65 vol %). This phase is fairly evenly distributed in the form of cuboidal particles. Interdendritic pores occupied about 0.38% and were often sites for crack initiation. The most advantageous orientation toward acting stresses was [001] in fully reversed total strain-controlled push-pull experiments. In tests under inelastic strain control at 1033 K there was no preferred orientation; at 1253 K the [111] orientation is best.

1.8 EFFECT OF TEMPERATURE

1.8.1 Low Temperatures

At low temperatures the fatigue strength of metals is usually higher than it is at ambient temperatures. For instance, in heat-treated low-alloy steel (0.3% C, 0.7% Mn, 3.5% Ni) tested in reversed bending, the fatigue strength at 195 K was 10 times higher than it is in laboratory air. However, the factor that must be taken into consideration is the risk of brittle fracture, a very spectacular example of which was given by Pogodin-Aleksiejev (1969). The number of railway rails in Siberia replaced each month corresponds to the average monthly temperature. The 7- to 15-fold difference was attributed to the change into twinning as the dominant deformation mode.

The discovery and exploration of gas in arctic regions created a demand for higher yield strength and more resistance to the notch effect. The same properties were required of welds, parent metal, and heat-affected zone. Tests carried out by Jesseman and Smith (1975) on the candidate steels simulated the strains induced by fabrication and installation. The impact strength was measured on Charpy V-notch specimens. For two superior steels, IN-787 and high-alloy Mn-Mo-Cb, the impact strength exceeded 68 J in the temperature range 211 to 269 K, which was treated as the criterion for the test. The beneficial effect of reductions in the content of sulfur and interstitial nitrogen was expected.

An increase in nickel content is known to reduce the brittle transition temperature, and for the ASTM A 533-72a (9% Ni) steel this is below 76 K. Tobler and co-workers (1978) investigated 18% nickel maraging steel under solution-annealed conditions and found a moderate drop (from 165 MPa·m$^{1/2}$ to 83 MPa·m$^{1/2}$) in fracture toughness, corresponding to a reduction in temperature from 295 K to 4 K. The associated increase in the yield stress was from 831 MPa to 1596 MPa. Thus the fracture toughness in the steel investigated was at 4 K only 8% greater than in the less expensive 9% Ni steel.

The brittle transition is not encountered in aluminum alloys, which makes this group of materials competitive to steels for use in superconducting machinery. Tobler and Reed (1977) tested an aluminum 5083-0 alloy strengthened by solid solution additions of magnesium incorporated in the chemical specifications. It was found that the difference in fracture toughness between the value at the expected service temperature (i.e., 111 K) and that at the liquid helium temperature is almost negligible. The crack propagation rate for the range $\Delta K > 15$ MPa·m$^{1/2}$ is only slightly inferior at room temperature, whereas for lower ΔK the difference is within an order of magnitude.

1.8.2 Elevated Temperatures

At elevated temperatures the strength of materials is reduced. At 573 to 673 K, the reduction is moderate, at higher temperatures the decrease observed is much more drastic, and for common steel at a temperature of, say, 900 K, the yield strength and static strength constitute only one-half of the comparable values at room temperature. The fatigue strength is generally lower at high temperatures; however, it does now show a monotonic character, even in relatively short tests where the effect of exposure to the test environment is negligible. For example, Ueda and Matsuo (1966) tested high-speed M2 steel (C = 0.89%, Si = 0.26%, Mn = 0.32%, Cr = 4.22%, Mo = 5.25%, W = 6.67%, V = 1.80%) in rotating bending. The observed fatigue limits at 573, 673, and 773 K were 598, 392, and 392 MPa, respectively. Wellinger and Sautter (1973) investigated two cast steels of G-X22 Cr-Mo-V12 1 and GS-22 Mo4 grades and 14Mo-V63 and 13Cr-Mo44 steels. The imposed cycles had a trapezoidal shape, with hold times of 5, 20, 40, and 180 min at maximum and minimum stresses. The failure criterion involved was the first visible crack occurrence. The reported numbers of cycles withstood by tested materials were at 803 K lower in relation to room temperature: 2.6 times for the G-X22 Cr-Mo-V12 1 cast steel in tests with a 20-min holding period and a strain range amplitude of 3.5 × 10^{-3} and 2.7 times for the GS-22 Mo4 cast steel. These results can be explained in terms of the relaxation of stresses and creep. For rolled steels, the decrease observed was more moderate: 1.7 for 14Mo-V63 steel and 1.4 for 13Cr-Mo44 steel.

In general, the situations encountered in fatigue testing at elevated temperatures can be divided into *fatigue-dominated cracking*, which is transgranular, and *creep-dominated cracking*, where a crack links the already existing cavities and minute cracks located at grain boundaries, producing an intergranular macrocrack. The fracture surface appearance is usually complicated; the final crack is often one that started as a transgranular crack, then changed into the intergranular mode at higher levels of stress resulting from a reduction in the cross-section load-carrying area, enhanced by the presence of cavities at grain boundaries which further lower the loading capacity of the material. Fatigue-dominated cracking is favored by high cyclic amplitudes and frequencies and relatively low temperatures, whereas intergranular cracking is promoted by the opposite conditions. The incipient cracking connected with the presence of low-melting eutectics at grain boundaries and the effect of neutron irradiation introducing structural disorder and such products of transmutation of elements as helium and hydrogen are known to enhance the latter. The reported life reduction was as high as 35-fold for Incoloy 800 (Abo-El-Ata, 1978).

In conditions where creep is significant and cavitational damage interacts with the growing crack, the ΔK value does not correlate well with the crack growth rate. The alternative approach, introducing a net stress amplitude, is only partly applicable, due to its inherent dependence on specimen type. The modified J integrals are the most accepted concepts (Koterazawa and Mori, 1977).

Corrosion, which was neglected in the considerations above, is evidently a very marked factor. For instance, in superalloys tested in fatigue at room temperature, the crack growth rate was two to three times higher in air at normal pressure than at ultrahigh vacuum. At elevated temperatures the life of specimens loaded by a constant force can be even a few hundred times longer in vacuum.

Reuchet and Remy (1983a,b) investigated two heats of the cobalt superalloy MAR-M 509. The tests were carried out in the temperature range 293 to 1373 K in

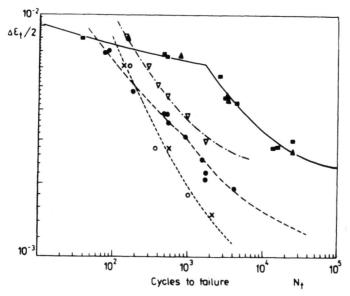

Figure 1.29 Fatigue life N_f as a function of the total strain amplitude ($\Delta\varepsilon_t/2$) at various temperatures: ——■, 292 K; ▲, 873 K. — · — ∇, 973 K. – – ●, 1173 K; O, 1273 K. – – ×, 1373 K. (From Reuchet and Mowbray, 1983 a, b.)

air with sawtooth cycling under plastic strain control. The total strain rate was held constant at 7×10^{-4} s^{-1}. The microstructure developed by the heat treatment consisted of a face-centered cubic (fcc) matrix rich in cobalt, lamellar eutectic composed of $M_{23}C_6$ and matrix, and primary MC carbides of Chinese script morphology. The results of fatigue tests are depicted in Fig. 1.29. It can be seen that at a temperature of about 873 K, there is a distinct change in material performance. Supplementary electron microscopy testing revealed carbide precipitation along dislocations with resultant hardening; less precipitation was present at higher temperatures. At room temperature, cracks nucleated along crystallographic planes. At higher stress amplitudes cracks initiated predominantly from cracked primary MC carbides. At a temperature of 873 K, the crystallographic slips played a more limited role; cracking initiated from MC carbides. Oxidation did not have a marked influence on crack initiation; the crack propagation path was transgranular. At a temperature of 973 K, oxidation became a significant factor. The extrusions of material observed on the surface of the specimen were caused by the preferential oxidation of primary MC carbides. The reduction in fatigue life at high temperatures was attributed to the interaction of fatigue and oxidation. The importance of the latter factor was demonstrated simply by running comparative tests in vacuum. The results obtained at 1173 K were comparable to those provided by tests carried out at room temperature in laboratory air.

1.9 NOTCH EFFECT

In places where either the shape or cross-sectional area is changed, a pileup of stresses takes place. This makes the relevant regions of the component crack-prone, which shows up in both impact tests and in static or fatigue tests. A survey of industrial failures caused by fatigue shows that the notch effect is responsible for almost one-third of them.

Experiments have shown that the seriousness of the notch effect depends on the geometry (i.e., sharpness) of the notch and on the material of which the component is made. Where there is more than one groove, the final effect depends on their relative position (e.g., spacing). This has been illustrated by Vishnevsky and Wallace (1968), who investigated several Ni-Cr-Mo fine-grained, Al-treated cast steels in reverse bending. Test materials were in two states: water quenched and tempered (ultimate tensile strength of 861 to 1000 MPa) and normalized and tempered (ultimate tensile strength of 206 to 620 MPa). The notches studied were 0.9 mm deep with a 60° included angle. Blunt notches have a radius at the bottom of 0.4 mm, and sharp notches, one of 0.025 mm. Fatigue testing demonstrated that the former reduced the fatigue limit by 36%, and the latter by 42 to 53%, depending on the tensile strength level of the material tested. For quenched and tempered specimens the reduction was generally higher. Detailed considerations [e.g., those of Neuber (1958)] lead to the conclusion that it is justified to characterize the notch region by the magnitude of the piled-up stresses and the value of the stress gradient. The former is quantified by the theoretical stress concentration factor

Table 1.3 Relationship Between K_t and K_f

K_t	K_f
1.35	1.05–1.3
1.60	1.1 –1.4
2.50	1.6 –2.0
3.60	2.1 –2.7
5.20	2.8 –3.4

Source: Haibach and Matschke (1981).

$$K_t = \frac{\sigma_{max}}{\sigma_{min}}$$

where σ_{max} is the maximal stress at the notch bottom and σ_{min} is the nominal stress, also referred to as the net section stress. Values of K_t are known and tabulated for a variety of notch geometries. However, sensitivity to the notch effect is an inherent feature of the material. Thum and Buchmann (1932) introduced the notch sensitivity, or q, factor:

$$q = \frac{K_f - 1}{K_t - 1}$$

where K_f is a fatigue strength reduction factor relating the actual stress increase with a reduced lifetime. The q value lies in the range 0 to 1, approaching zero for low-strength cast irons and 1 for brittle materials, but it is a well-established fact that it remains dependent on the notch geometry. Haibach and Matschke (1981) investigated CK 45 steel (0.45 to 0.52% C, 0.17 to 0.27% Si, 0.61 to 0.80% Mn) and 42Cr-Mo4 steel (0.43% C, 0.23 to 0.31% Si, 0.68 to 0.75% Mn, 0.22% Mo). The relationships that were found are presented in Table 1.3. Such a dependence was believed to be caused by the susceptibility of these steels to recovery and recrystallization, as proved in earlier works.

The notch effect can be greatly alleviated by nitriding (Kudriavcev, 1951) and by surface tempering performed using radio-frequency currents (Khodzher et al., 1987). In the latter work, smooth, quenched, 473 K-tempered specimens displayed the same fatigue life as specimens with a machined sharp notch, surface tempered to a depth below the bottom of the notch, and annealed at 473 K to relieve residual stresses.

Siebel (1955) demonstrated that the fatigue strength of notched members depends on the ratio of the stress gradient to the maximum stress calculated for the notch bottom locus:

$$x = \left(\frac{d\sigma}{dx}\right)_{\sigma_{max}} \frac{1}{\sigma_{max}}$$

In the technical literature diagrams are available that relate the x value with the ratio of the smooth specimen fatigue strength to the fatigue strength of the grooved specimen. Of course, each material and each specimen size is represented by a separate curve.

Several methods were proposed to take into account, through adequate corrections, plasticity at the notch tip. The most widely accepted is the postulate of Neuber (1961):

$$K_t = \sqrt{K_{t\sigma}K_{t\varepsilon}}$$

where

$K_{t\sigma} = \sigma_{max}/\sigma_{nom}$ (stress-dependent correction factor)

$K_{t\varepsilon} = \varepsilon_{max}/\varepsilon_{nom}$ (strain-dependent correction factor)

ε_{max} = strain at the notch bottom, corresponding to σ_{max}

ε_{nom} = strain corresponding to the nominal stress σ_{nom}

1.10 SURFACE TREATMENTS THAT IMPROVE THE SURFACE LAYER

Fatigue resistance can be increased by mechanical, thermochemical, or purely thermal treatment of the surface. The benefits arise from hardening the material in the immediate vicinity of the surface and, simultaneously, introducing compressive stress in the region. Mechanical treatments improve the smoothness of the surface, reducing or eliminating micronotches.

Fatigue failure results from the nucleation of cracks and their subsequent development. Various hypotheses regarding the nucleation and propagation of cracks were described earlier, and while none of them are entirely satisfactory for universal application, they all point to the widely observed facts that fatigue cracks almost invariably begin in the surface region, that hard materials are less susceptible than soft materials to crack nucleation, and that crack propagation is slower in a soft or plastic material than in a hard and brittle one. Hence the special benefit of appropriate surface treatments stem from their effects being localized. The initiation of microcracks in the surface zone is delayed because it is hardened, while their propagation in the adjacent material is retarded because its plasticity was unaffected by the treatment.

The compressive stress induced in the surface zone is superimposed on the stresses due to cyclic loading, so that tensile stress in the surface zone is eliminated or greatly reduced. This extends fatigue life, as compressive stress tends to prevent the creation of voids and the opening of cracks, in contrast to varying tensile stress, which is far more destructive.

Of course, when and where the applied load itself produces compressive stress, this is increased by the compressive stress produced by the surface treatment, but they are nevertheless of advantage overall. All hardening processes induce compressive stress, some for obvious mechanical reasons and in other cases because the martensitic or nitrided zone produced has a greater specific volume than that of the original material.

The compressive stress produced by roller burnishing is particularly high and may exceed 1000 MPa. That produced by nitriding is about 700 MPa, but its effect is limited, as the layer affected is relatively thin. Shot peening produces compressive stress to a depth of some tens of micrometers. Delitzia (1984) recommends a two-stage process employing low-energy shots initially to smooth the surface and to induce compressive stress in the immediately adjacent layer. The depth of compression is then increased by further peening with higher-energy shots. It is claimed that where single-stage shot peening increased the number of cycles to failure by a factor of 10, the two-stage process increased it by a factor of 20. Tests carried out by Meguid and Chee (1983) indicated complete life recovery, even when the specimens had been partially fatigued before shot peening by as much as 75% of their fatigue life. The life improvement was governed by the size of cracks formed during the fatigue test prior to shot peening. Case hardening by surface heat treatment alone or by carburizing can be beneficial, but all forms of case hardening increase the sensibility to the notch effect and so cannot be used indiscriminately (Weronski, 1984).

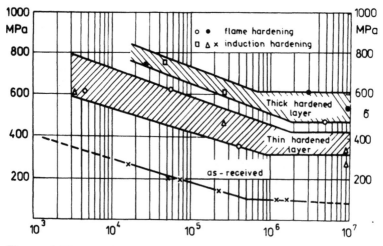

Figure 1.30 Effect of surface hardening on type 45 steel.

Table 1.4 Effect of Short-Duration Nitriding on the Bending Fatigue Strength of Specimens

Steel	Specimen	State	Bending fatigue strength (MPa)
0.4% C	Smooth	As received	240.4
		Nitrided at:	
		820 K, 2.5 h, furnace cooled	360.8
		870 K, 1.5 h, furnace cooled	367.8
		920 K, 1.0 h, furnace cooled	364.8
		870 K, 2.0 h, furnace cooled	387.5
		920 K, 2.0 h, furnace cooled	384.5
	Notched (90° notch);	As received	145.1
	included angle 0.4mm	Nitrided at:	
	deep, 0.3 mm radius of	870 K, 1.5 h, furnace cooled	178.5
	curvature	920 K, 1.0 h, furnace cooled	152.1
40H (0.4% C,	Smooth	Normalized	349.1
0.65% Mn,		Normalized, nitrided at:	
0.27% Si,		820 K, 2.75 h	407.1
0.95% Cr)		930 K, 2.0 h	399.3
40 H (0.4% C,	Smooth	Quenched and tempered	399.0
0.65% Mn,		Quenched , tempered, and nitrided	
0.27% Si,		at 870 K, 2 h	507.9
0.95% Cr)			
	Notched (90° notch);	Normalized	145.1
	included angle	Normalized, nitrided at 820 K	296.2
	0.4 mm deep, 0.3 mm		
	radius of curvature		

Source: Kudriavcev (1951).

Table 1.5   Effect of Carburizing and Quenching on Fatigue Strength of Specimens

Material	Specimen diameter (mm)	Depth of carburized layer (mm)	State of material	Fatigue strength (MPa) in bending
E2 steel (0.16% C, 0.74% Cr, 1.85% Ni)	10	—	Normalized	333.4
St2 (0.12% C)	10	—	Normalized	292.2
		1–1.2	Carburized, tempered at 470 K	502.0
St3 (0.22% C)	15	—	Normalized	207.0
		1–1.2	Carburized, tempered at:	
			445 K	431.5
			620 K	441.3
St5 (0.35% C)	15	—	Normalized	248.2
		1–1.2	Carburized, tempered at 470 K	647.2
	50	—	Normalized	178.5
		1.2–1.3	Carburized, tempered at 570 K	343.2

Source: Kotov (1961).

Any form of surface hardening employed should be applied to the entire surface of the component, and no part of the hardened surface should be removed by subsequent machining, as the border between hardened and unhardened surfaces frequently provides a site for crack nucleation. The effect of surface hardening (by flame or induction) on the fatigue strength of type 45 steel (0.45% C) is shown in Fig. 1.30. The effect of short-duration nitriding on the fatigue strength of various steels is shown in Table 1.4 and that of carburizing and hardening in Table 1.5.

1.11 CORROSION

Fatigue strength is reduced substantially by exposure to corrosive media such as moist air, seawater, or, of course, any more aggressive substances. Where, in the absence of corrosion, a Wöhler plot would consist of two straight lines intersecting at some point, the plot for the same material when exposed to corrosion consists of two straight lines intersecting at a different point, each line having a greater inclination below the horizontal than previously. Thus the fatigue limit or stress below which fatigue failure will never occur does not exist in the case of exposure to corrosion. The reduction in fatigue strength is due to the notch effect at corrosion pits and also to the continual loss of sound metal, which increases the intensity of stress in that which remains. The effects of the corrosive medium are limited to some degree if the products of corrosive attack protect the material below them. This can occur if they are not solvable in the surrounding medium and are sufficiently elastic to withstand a load cycle without spalling. However, once a crack has formed, corrosion products within it may suffer abrasion due to relative motion of the opposite surfaces during the load cycle.

Thin layers of corrosion products are generally stronger in compression than in tension, so that their protective effect is better maintained and their fatigue strength less seriously reduced if the stress cycle is mainly compressive (Sors, 1971).

In the course of continuous testing, the reduction of fatigue strength by corrosion is greater if the frequency of the load cycle is low, as for a given number of cycles, the period of exposure to corrosion is longer overall. Also, when a crack has been initiated, the portions of each cycle during which it is open and admitting the corrosive medium are longer that at higher cycling frequencies. If in-service operation is intermittent, allowing long periods of exposure between periods of load cycling, a similar effect may be expected.

The result of the simultaneous action of a corrosive medium and load cycling cannot always be predicted by superposition of the effects of the two processes proceeding in isolation. It is therefore generally regarded as a specific phenomenon and is referred to as *corrosion fatigue*. Wöhler plots for various alloys in dry air and in a salt mist are shown in Fig. 1.31.

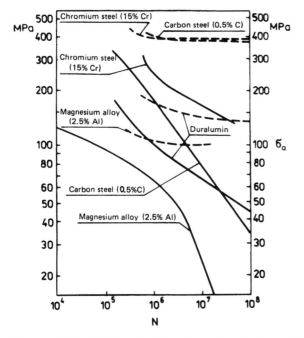

Figure 1.31 Wöhler plots for various alloys in dry air (dashed lines) and in a salt mist (solid lines). (Adapted from Wyatt and Dew-Hughes, 1978.)

1.12 PROTECTIVE COATINGS

In a noncorrosive environment the combination of a coating and a base material generally exhibits a lower fatigue strength than that of the base material alone. The use of a coating is therefore inadvisable unless the environment is such that corrosion would severely reduce fatigue life. The extent to which application of a coating reduces fatigue strength depends on the materials involved and the method of application. The reduction may be due to any of four possible mechanisms:

1. Notch effect caused by microcracks in the coating—especially apparent in the cases of brittle coatings (e.g., chromium) and flame-sprayed coatings, which possess microregions of imperfect fusion between the sprayed particles.
2. Notch effect at the boundaries of regions where adhesion between coating and base material is imperfect, possibly the result of inadequate cleaning prior to deposition.
3. The existence of tensile stress in the coating.
4. Hydrogen embrittlement due to hydrogen being trapped between the base and coating during electroplating. (The significance of this factor is not accepted by all researchers.)

Despite this, the adverse effects of zinc or cadmium coatings on steel may be considered negligible. Nickel coatings reduce the fatigue life of mild steels (C = 0.08%) by only a few percent and that of steels containing 0.6% carbon by some 30% (Sors, 1971). The effect of electro-deposited chromium on the fatigue strength of various steels is shown in Fig. 1.32, where the fatigue strengths of unprotected steels are read from the vertical axis and the reduced strengths resulting from chromium plating to the thicknesses t are read from horizontal axis. It can be seen

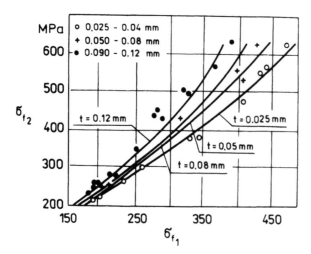

Figure 1.32 Effect of electrodeposited chromium on fatigue strength.

that the thicker the coating, the lower the fatigue strength. The adverse effects of all coatings, however applied, are more severe in the vicinity of stress risers such as holes and grooves required by the design, and it is advantageous for these to be cut after the coating has been applied.

1.13 FRETTING CORROSION

Fretting corrosion occurs at metal surfaces that are in close contact under compressive pressure but nevertheless, rub against one another. The surfaces concerned often consist of two parts which are, nominally, well fastened together (e.g., by riveting or by one being a forced fit on the other). The amount of relative motion is then very small, but an amplitude of 10^{-8} cm is sufficient to cause fretting corrosion which substantially reduces the fatigue strength of one or both components. In such cases the relative motion may be due to differences in the elastic deformation of the two parts when subjected to the same stress variation.

If the contacting surfaces are initially uncorroded, their relative motion first detaches minute particles of metal (molecular plucking). In air these oxidize rapidly, forming an abrasive that causes further wear and repeated by stripping away the oxide that forms on the freshly exposed surfaces. Wear leads to surface pitting, and the pits constitute sites for the nucleation of fatigue cracks. The phenomenon is more pronounced in soft materials than in hard ones and is accelerated if the method of clamping the surfaces together is imperfect. In the presence of water or water vapor, fretting is reduced, probably because the water lubricates and for hydrated oxides is relatively soft. The literature contains a report of fatigue life being reduced to less than a third by fretting at ill-fitting joints, and another case of fretting reducing fatigue strength by only 5%. Usually the reduction in fatigue strength is put in the region of 30%. There is general agreement that fretting corrosion can be reduced by the following measures (Sors, 1971) and (Low, 1985):

1. Protective coating of the metal surfaces
2. Damping and reducing vibration
3. Lubrication (e.g., with MoS_2)
4. Surface hardening (e.g., nitriding or shot peening) together with a provision for surfaces that are widely different with respect to finish and hardness
5. Bonding the surfaces together with an elastic material

In some cases fretting corrosion has been virtually eliminated by one or more of these methods, which may be regarded as safe, as they do not themselves cause local damage or stress concentration. In contrast, excessive tightening of a joint will not necessarily eliminate fretting and will dangerously increase local stress. In multiple bolted or riveted joints, excessive clearance between the fasteners and the holes accommodating them may encourage fretting, but the tolerances must allow for possible imperfect alignment of holes so as to permit insertion without creating scratches and stress concentrations.

1.14   SIZE AND SHAPE OF COMPONENT

A great many data regarding the fatigue strength of materials have been obtained from tests on small specimens. Tests on large specimens are much less common, as they require bigger rigs and are costly to perform. However, it is well established that the fatigue strength determined with large specimens is significantly lower than that found in similar tests (e.g., bending) with small specimens of the same material. Field experience supports the laboratory findings, and it is therefore necessary when designing larger components to apply a reduction factor to the fatigue strength found with small specimens. A correction is also necessary when the shape of the component is different from that of the test pieces.

The reduction in fatigue strength with increased size is usually attributed to practical rather than theoretical factors. Thus an increase in size:

1. Increases the area of the surface, where most cracks begin, and hence the number of potential crack initiation sites due to variation in dislocation density, surface imperfections (e.g., scratches), and so on
2. Increases the volume of material and hence the number and range of internal flaws, inclusions, and the like
3. Increases the possibility and extent of inhomogeneity of the microstructure
4. Results in a more complex state of stress within the component

In addition, it is sometimes suggested that small specimens may have better grain structure or other qualities than larger items because of the additional working the metal has received while its size was undergoing reduction.

The relative importance attached to these explanations depends on the circumstances. A designer using high-purity material melted and cast in vacuum is less concerned with the danger of inclusions than is one working with inferior material. However, there is overall agreement that large components do exhibit a lower fatigue strength than is seen in small test samples of the same material.

When the shape of a component differs from that of the specimens from which fatigue strength data have been obtained, it must again be assumed that the fatigue strength of the component will be different from, and usually less than, that shown by the specimens. Stress analysis will reveal any unfavorable stress concentrations in the new shape but beyond what can be found by calculation, some allowance must be made for the higher probability of crack initiation somewhere in the surface exposed to maximum stress if its area is greater than that of the test pieces.

REFERENCES

Abo-El-Ata, M. M. (1978). *J. Eng. Mater. Technol.*, *100*(4): 121.
Alexander, J., Shaver, R. G. and Withers, J. C. (1969). NASA Contract NASW-1779.
Averbach, B. L. (1965). *International Conference on Fracture*, Sendai, Japan.
Bilby, B. A., Cottrell, A. H. and Swinden, K. H. (1963). *Proc. R. Soc. A*, *272*: 304.

Blake, A. (1982). *Practical Stress Analysis in Engineering Design*, Marcel Dekker, New York, p. 654.
Boesch, W., Tien, J. K. and Howson, T. E. (1982). *Met. Progr.*, *122*(5): 49.
Brieusov, W. P., Fiediukin, W. K. and Pustovoit, W. K. (1972). *Progresivnyie procesy w czugunno-litiejnym proizwodstwie*, Izdatelstvo LGU, Leningrad, p. 90.
Chandler, H. (1982). *Metal Prog.*, *122*(5): 9.
Charles, J. A., Appl, F. J. and Francis, J. E. (1978). *J. Eng. Mater. Technol.*, *100*(4): 200.
Coffin, L. F. Jr. (1954). *Trans. ASME*, 76: 931.
Coffin, L. F., Jr. (1958). *SESA Proc.*, *15*: 117.
Cottrell, A. H. (1958). *Trans. AIME*, *212*: 192.
Cottrell, A. H. and Hull, D. (1957). *Proc. R. Soc. A*, *242*: 211.
Delitizia, A. T. (1984). *Manuf. Eng.*, *92*(5).
Dowling, N. E. (1972). *J. Mater.*, *7*(1): 71.
Dowling, N. E. (1977). *Crack Growing During Low Cycle Fatigue of Smooth Axial Specimens: Cyclic Stress–Strain and Plastic Deformation Aspects of Fatigue Crack Growth*, STP 637, ASTM, Philadelphia, p. 97.
Dugdale, D. S. (1960). *J. Mech. Phys. Solids*, 8: 100.
Eifler, D., Mayr, P. and Macherauch, E. (1981). *Stahl Eisen*, *101*(2): 17.
Elber, W. (1970). *Eng. Fract. Mech.*, 2: 37.
Feltner, C. E. (1965). *Philos. Mag.*, *12*: 1229.
Feltner, C. E. and Morrow, J. D. (1961). *J. Basic Eng.*, *83*(3): 15.
Fiediukin, W. K. (1984). *Metod termocykliczeskoj obrabotki metallov*, Izdatelstvo LGU, Leningrad, p. 190.
Forman, R. G., Kearney, V. E. and Engle, R. M. (1967). *J. Basic Eng. D*, *89*(9): 459.
Forsyth, P. J. E. (1970). *The Physical Basis of Metal Fatigue*, Blackie, London.
Franklin, S. E. and Stark, R. A. (1984). *3rd International Symposium on the Physical Metallurgy of Cast Iron*, Stockholm, Aug. 29–31.
Fujita, F. E. (1958). *Acta Metall.*, 6: 543.
Gabb, T. P., Gayda, J. and Miner, R. V. (1986). *Metall. Trans. A*, *17*(3): 491.
Griffith, A. A. (1920). *Philos. Trans. R. Soc. A*, *221*: 163.
Grover, H. J. (1959). *Symposium on Fatigue of Aircraft Structures*, ASTM STP 274, ASTM, Philadelphia, p. 120.
Haibach, E. and Matschke, C. (1981). *Stahl Eisen*, *101*(3): 21.
Halford, G. R. (1966). *J. Mater.*, *1*(1): 3.
Heald, P. T., Lindley, T. C. and Richards, C. E. (1972). *Mater. Sci. Eng.*, *9*(10): 235.
Hebsur, M. G., Abraham, K. P. and Prasad, Y. V. R. K. (1980). *Eng. Fract. Mech.*, *13*(4): 851.
Homma, H. and Nakazawa, H. (1978). *Eng. Fract. Mech.*, *10*(3): 539.
Hück, M. (1981). *Stahl Eisen*, *101*(24): 57.
Hutchinson, D. and Ryder, D. A. (1980a). *Engineering 220*: 410.
Hutchinson, D. and Ryder, D. A. (1980b). *Engineering 220*: 566.
Irwin, G. R., Kies, J. A. and Smith, H. L. (1958). *Proc. ASTM*, 58: 640.
Jaske, C. E. and Frey, N. D. (1982). *J. Eng. Mater. Technol.*, *104*(4): 137.
Jesseman, R. J. and Smith, R. C. (1975). *J. Eng. Ind.*, *97*(5): 408.
Khodzher, P. K., Wolkow, A. K. and Dvoyashov, K. V. (1987). *Probl. Prochn.*, *214*(4): 23.
Kliman, V. and Bilý, M. (1984). *Mater. Sci. Eng.*, *68*(7): 11.
Kocańda, S. (1961a). *Biul. Wojskowej Akad. Tech.*, *91* (Suppl.): 159.
Kocańda, S. (1961b). Über die Ermündunserscheinungen im Eisen und Kohlenstoffstahl im Bereiche von Zeitfestigkeit und unter Wechselfestigkeit, *Proc. 2nd Congress on Material Testing*, Budapest, Hungary: 68.

Kocańda, S. (1985). *Zmeczeniowe pekanie metali*, WNT, Warsaw, p. 487.

Kocańda, S. and Szala, J. (1985). *Podstawy obliczen zmeczeniowych*, PWN, Warsaw, p. 276.

Koterazawa, R. and Mori, T. (1977). *J. Eng. Mater. Technol.*, 22(4): 298.

Kotov, O. K. (1961). *Povierchnostnyje uprocznienie dietalej maszin chimikotiempieraturnymi metodami*, Maszgiz, Moscow.

Koul, A. K., Thamburaj, R., Raizenne, M. D., Wallace, W. and De Malherbe, M. C. (1985). Laboratory Technical Report LTR-ST-1537, National Aeronautical Establishment, Canada.

Kudriavcev, I. W. (1951). *Wnutrenye naprazenya kak rezerw prochnosti w maszinostroienii*, Maszgiz, Moscow.

Lardner, R. W. (1968). *Philos. Mag.*, 17: 71.

Lemkey, F. D. and Thompson, E. R. (1972) *Proc. International Conference on Composites Grown in Situ*, Lakeville, Conn., p. 105.

Low, M. B. J. (1985). *Wear*, 106: 315.

Lozinski, M. G. and Romanov, A. N. (1965). *Proc. 2nd Conference on Dimensions and Strength Calculations, Budapest.*

Maciejny, A. (1973). *Kruchosc metali*, Slask, Katowice, Poland, p. 200.

Madeyski, A. (1984a). *Met. Progr.*, 125(May): 57.

Madeyski, A. (1984b). *Met. Progr.*, 126(June): 37.

Madeyski, A. (1984c). *Met. Progr.*, 126(July): 33.

Madeyski, A. (1984d). *Met. Progr.*, 126(Aug.): 54.

Madeyski, A. and Albertin, L. (1978). Special Technical Publication 645, American Society for Testing and Materials, Philadelphia.

Maiya, P. S. (1977). *Scripta Metall.*, 11: 331.

Mang, J. F. and Wallace, J. F. (1968). *AFS Cast Met. Res. J.*, (June): 141.

Mannan, S. L., Samuel, K. G. and Rodriguez, P. (1983). *Met. Sci.*, 17(2): 637.

Manson, S. S. (1953). *Behaviour of Materials Under Conditions of Thermal Stress*, NACA TN-2933.

Margolin, I. G. (1984). *Eng. Fract. Mech.*, 19(3): 539.

McClintock, F. A. and Irwin, G. R. (1965). STP 381, ASTM, Philadelphia, p. 84.

Meguid, A. A. and Chee, E. B. (1983). *J. Mech. Work. Technol.*, 8: 129.

Miner, M. A. (1945). *J. Appl. Mech.*, 12 (Sept.).

Miner, R. V., Gabb, T. P., Gayda, J. and Hemker, K. J. (1986a). *Metall. Trans. A*, 17: 507.

Miner, R. V., Voigt, R. C., Gayda, J. and Gabb, T. P. (1986b). *Metall. Trans. A*, 17(3): 491.

Mitchell, M. R. (1977). *J. Eng. Mater. Technol.*, 99(4): 329.

Morris, W., James, M. and Buck, O. (1981). *Metall. Trans. A*, 12(1): 57.

Mott, N. F. (1958). *Acta Metall.*, 6: 195.

Neate, G. J. (1985). *High Temp. Technol.*, 3(4): 195.

Neuber, H. (1958). *Kerbspannungslehre*, Springer-Verlag, Berlin.

Neuber, H. (1961). *J. Appl. Mech.*, 28E(4): 544.

Nisitani, H. and Takso, K. (1981). *Eng. Fract. Mech.*, 15(3–4): 445.

Orowan, E. (1934). *Zeitschrift für Kristollogrephie* 89: 327.

Orowan, E. (1948). *Rep. Prog. Phys.*, 12: 185.

Palmgren, A. (1924). *Z. Ver. Dtsch. Ing.*, 68.

Pangborn, R. N., Weissman, S. and Kramer, I. R. (1981). *Metall. Trans.*, 12A(1): 109.

Paris, P. and Erdogan, F. (1963). *J. Basic Eng.*, 85(12): 52.

Pogodin-Aleksiejev, G. I. (1969). *Wytrzymalosc dynamiczna i kruchosc metali*, WNT, Warsaw.

Reuchet, J. and Remy, L. (1983a). *Mater. Sci. Eng.*, 58: 19.

Reuchet, J. and Remy, L. (1983b). *Mater. Sci. Eng.*, 58: 33.

Rice, J. R. (1967). Brown University ARPA SD-86, Report E39.

Rice, J. R. and Rosengren, G. F. (1968). *J. Mech. Phys. Solids, 16*: 1.

Saal, H. (1972). *J. Basic Eng., 94*(3): 243.

Saario, T., Wallin, K. and Törrönen, K. (1984). *J. Eng. Mater. Technol., 106*(4): 173.

Scott, J. A. (1969). The influence of processing and metallurgical factors on fatigue, in *Metal Fatigue: Theory and Design* (A. Madayag, ed.), Wiley, New York, p. 66.

Siebel, E. (1955). *Handbuch der Werkstoffprüfung*, vol. II, Springer-Verlag, Berlin.

Sih, G. C. and MacDonald, B. (1974). *Eng. Fract. Mech., 6*: 361.

Smith, E. (1968). *Int. J. Fract. Mech., 4*(2): 131.

Sors, L. (1971). *Fatigue Design of Machine Components*, Akademiai Kiado, Budapest, Pergamon Press, London (joint edition).

Stroh, A. N. (1954). *Proc. R. Soc. A, 223*: 404.

Taira, S., Ohtani, R. and Komatsu, T. (1979). *J. Eng. Mater. Technol. Trans. ASME H, 101*: 162.

Tanaka, K., Hoshide, T. and Sakai, N. (1984). *Eng. Fract. Mech., 19*(5): 805.

Thum and Buchmann (1932). *Dauerfestigkeit und Konstruktion.*

Tobler, R. L. and Reed, R. P. (1977). *J. Eng. Mater. Technol., 99*(4): 306.

Tobler, R. L., Reed, R. P. and Schramm, R. E. (1978). *J. Eng. Mater. Technol., 100.*

Tomkins, B. (1968). *Philos. Mag., 155*: 1041

Tomkins, B. (1971). *Philos. Mag., 183*: 687.

Ueda, T. and Matsuo, T. (1966). *Trans. Iron Steel Inst. Jpn., 6*: 39.

Vishnevsky, C. and Wallace, J. F. (1968). *AFS Cast Met. Res. J.*, June: 95.

Waku, Y., Masumoto, T. and Ogura, T. (1983). *Trans. Jpn. Inst. Met., 24*(12): 849.

Wareing, J., Tomkins, B. and Sumner, G. (1973). ASTM STP 520, ASTM, Philadelphia, p. 123.

Weertman, J. (1966). *Int. J. Fract. Mech., 2*(2): 460.

Wellinger, K. and Sautter, S. (1973). *Arch. Eisenhuettenw., 44*(1): 47.

Weronski, A. (1984). *Hartowanie Natryskowe*, Politechnika Lubelska, Lublin, Poland, p. 110.

Wilson, W. K. and Chrepko, J. (1983). *Int. J. Fract., 22*: 303.

Wolkov, W. A. and Berzov, W. F. (1983). *Zavod. Lab. 49*: 53.

Wyatt, O. H. and Dew-Hughes, D. (1978). *Wprowadzenie do inzynierii materialowej*, WNT, Warsaw (translated from English), p. 192.

Yamaguchi, K. and Kanazawa, K. (1980). *Metall. Trans. A, 11*: 1691.

Yamaguchi, K. and Kanazawa, K. (1983). *Trans. Nat. Res. Inst. Met., 25*(3): 23.

Yoder, G. R., Cooley, L. A. and Crooker, T. W. (1977). *J. Eng. Mater. Technol., 99*(4): 313.

Yokobori, T. (1954). *Rep. Inst. Sci. Technol. Univ. Tokyo, 8*(1): 5.

Yu, K. O., Domingue, J. A., Maurer, G. E. and Flanders, H. D. (1986). *J. Met., 38*(1).

2

Creep

2.1 INTRODUCTION

The term *creep* denotes slowly proceeding deformation of solid matter under a maintained load. The first systematic studies providing some quantitative information on the nature of creep were those of Andrade (1910). During World War I, research on creep became more urgent. Impetus was given by the rapid increase in steam admission temperatures in power plants (to about 670 K in the 1920s), approaching the creep range of low-alloy steels. Early researchers were concerned with finding the limiting stress below which creep would not occur, but using more accurate experimental techniques, this idea was subsequently shown to be false.

The temperature at which creep becomes important for the designer is about 0.4 of the melting temperature of the material considered, but numerous exceptions bring this rule into question. For example, creep is observed in titanium at lower temperatures than in iron-based alloys, despite the higher melting point of the former. Stress levels under which creep is observable are always much lower than the strength of the material. A creep curve, which is the graphical presentation of the dependence of the strain on time under constant stress and temperature, is shown in Fig. 2.1. The strain ε_0 is developed immediately upon loading; the period of time between ε_0 and ε_1 is called *primary creep*. Between ε_1 and ε_2 the creep rate remains almost constant; this portion of the curve is termed *secondary* (or *steady-state*) *creep*:

$$\dot{\varepsilon}_s = \frac{\varepsilon_2 - \varepsilon_1}{t_2 - t_1}$$

The creep rate increases beyond ε_2, and this period is called *tertiary creep*. It is convenient to obtain experimental data under a constant tensile load. As creep proceeds, the true stress increases continually, giving rise to a pronounced change in the creep rate. The tertiary creep, where necking is appreciable, is certainly also

54

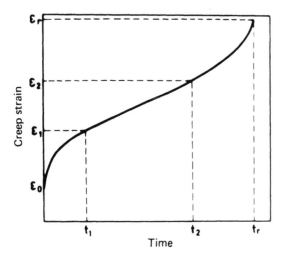

Figure 2.1 Schematic representation of a creep–rupture curve.

affected by microstructural damage accumulated in the material during preceding periods.

Moreover, it should be noted that results of creep tests are extremely sensitive to the state and chemistry of material and may differ appreciably for the same grade. Similarly, size effects appear to a greater degree compared with room-temperature testing. The curve drawn in Fig. 2.1 refers to an idealized case, at high temperatures, and especially in a hostile environment the period of secondary creep may be missing. Another example comes from Glen (1963); normalized 0.1% C, 0.4% Si, 0.5% Mn, 0.5% Mo steel when tested at 823 K demonstrated the very sudden increase in a creep rate above a certain value of total elongation. Silicon nitride was formed during testing initially reducing the creep rate, and later it spheroidized, thus increasing the creep rate. The net result was a sharp peak in the stress–strain curve.

Creep tests are usually aimed to estimate either the value of creep limit [i.e., the stress necessary to obtain the predetermined strain (usually, 0.05%, 0.1%, or 0.2%) at the termination of the isothermal test lasting for a specified period of time (say 500, 1000, 5000 h)] or the creep strength. The latter is the value of stress, which causes rupture of a specimen at a given temperature after a specified time.

Considering the effect of the test temperature, three distinct, special cases are anticipated:

1. At low temperatures, where recovery does not significantly alter material properties and material gradually hardens with strain, the strain is proportional to the logarithm of time, which in this instance is called the *logarithmic creep*.
2. At moderate temperatures, where recovery and strain hardening compete with each other, and after some time, a constant creep rate is established that is

much more stress and temperature dependent than in the former case. This is
called *recovery creep*.
3. At high temperatures, where diffusion is of prime significance, the creep rate
is strongly temperature dependent and in practice varies linearly with stress.
This case is termed *diffusional creep*.

2.2 CREEP–TIME RELATIONS

Variables affecting creep were already identified. The relation we are looking for
is of the form

$$f(\varepsilon,\sigma,t,T) = 0$$

where

ε = strain

σ = stress

t = time

T = temperature

It was experimentally evidenced that the data fit reasonably well with various
relations, depending on the material and test conditions. For T/T_m ratios (also called
homologous temperature) of 0.05 to 0.3, the time-dependent creep strain is often
given by

$$\varepsilon = \alpha \ln t + c$$

where α and c are independent of time. The relation agreed well with results
obtained for a number of face-centered cubic, body-centered cubic, and close-
packed hexagonal metals and alloys (Garofalo, 1955). The relation was, however,
satisfactory for a narrow range of temperatures and small creep strains. Thus the
relation above describes primary rather than secondary creep. The relation predicts
an unreasonably high initial rate of deformation. This objection can be eliminated
by letting $c = \alpha \ln v$, where v is a constant. Finally, without introducing much error,
$\varepsilon = \alpha \ln(1 + vt)$.
   In the range 0.2 to $0.7T_m$, two relations describe experimental data:

$$\varepsilon = \varepsilon_0 + \beta t^m$$

where ε_0 is the initial strain ($t = 0$) and β and m are constants independent of time.
If stresses are high enough for the secondary creep to be observed in a laboratory,
the linear term is added and the relation becomes

$$\varepsilon = \varepsilon_0 + \beta t^m + kt$$

where k represents the steady-state creep rate. Both of these relations satisfy experimental results, deviations being seen at low strains, particularly at high stresses. The last equation when letting $m = \frac{1}{3}$ reduces to the relation originally obtained by Andrade (1910).

The equation avoiding the objection concerned the initial moment $(t = 0)$ was proposed by McVetty (1934) and Garofalo (1963) for ferritic and stainless steels within the range 0.4 to $0.6T_m$:

$$\varepsilon = \varepsilon_0 + \varepsilon_t(1 - e^{-rt}) + \dot{\varepsilon}_s^t$$

where

ε_0 = strain observed just on loading

ε_t = limiting (in the sense of the asymptotic value) transient creep strain

r = ratio of transient creep rate to transient creep strain

$\dot{\varepsilon}_s$ = secondary creep rate

In general, increasing temperature makes structural constituents of a material more active, opposing the strain hardening of the material. Consideration of this effect in terms of activation energy clearly leads to an exponential equation of the form

$$\dot{\varepsilon} \approx \exp\left[\frac{-\Delta H(T,S)}{RT}\right]$$

where ΔH is the activation energy dependent on temperature T and material structure S and R is the universal gas constant. If creep is controlled by independent mechanisms with different activation energies (Sherby et al., 1957), we can write

$$\dot{\varepsilon} = \varphi(\sigma,S) \sum_i w_i \exp\left(\frac{-\Delta Hi}{RT}\right)$$

where w_i is the weight factor for the ith process.

A rough estimation shows that below about $0.5T_m$ creep is controlled by the nondiffusional mechanisms, including cross-slip, intersecting of dislocations, and lattice friction. At high temperatures, as numerous experiments show, the creep rate is diffusion controlled. In pure metals the activation energy is close to that for self-diffusion (Dorn, 1956). In alloys, creep may depend on the diffusion of an impurity or an additive.

2.2.1 Stress Dependence

The stress dependence for relatively high temperatures is

$$\dot{\varepsilon}_i = A_i\sigma^n$$

where $\dot{\varepsilon}_i$ is the creep rate in the initial primary stage and A_i and n are constants independent of stress. Similarly, for the secondary creep rate (Norton, 1929; Bailey, 1929)

$$\dot{\varepsilon}_s = A\sigma^n$$

or otherwise (Garofalo, 1963)

$$\dot{\varepsilon}_s = A''(\sinh \alpha\sigma)^n$$

where A, A'', α, and n are coefficients independent of stress.

Odqvist (1953) included an additional term to take into account the effect of primary creep:

$$\dot{\varepsilon} = \frac{d}{dt}\left(\frac{\sigma}{\sigma_0}\right)^{n_0} + \left(\frac{\sigma}{\sigma_c}\right)^{n}$$

where σ_0, σ_c, m_0, and n are constants.

2.2.2 Theoretical Considerations

In the thermal activation theory of slip (Becker, 1925, 1926) it was supposed that slip would occur when the applied stress (τ) was raised by thermal fluctuation to the value of the molecular shear cohesion within the small volume V. The activation energy is defined as

$$A = \frac{(\tau_0 - \tau)^2 V}{2G}$$

where τ_0 is the molecular shear cohesion and G is the shear modulus. Then assuming that slip is a dominant mechanism, its contribution to the creep rate is

$$\dot{\varepsilon} \approx \Delta e^{-A/kT}$$

where Δ is the elementary slip advance and K is the Boltzmann constant.

The alternative approach (Orowan, 1946–47; Mott and Nabarro, 1948) assumes the "soft spots" of a relatively low activation energy to arise only during the early phase of deformation. Further deceleration of the creep rate is attributed to the exhaustion of soft spots. Thus for deformation to proceed, more energy is needed, and thermal fluctuations rarely bridge the gap between the molecular shear and the applied stress.

In the dislocational approach, the steady creep rate is controlled by:

Dislocation multiplication
Dislocation glide
Stopping of dislocations by obstacles (e.g., Guinier–Preston zones, pinned dislocations, precipitates)

Recovery and climb, causing the rearrangement of dislocations and the annihilation of dislocations of opposite sign

Weertman (1955) considered creep in pure metals, in which slip can occur on more than the single slip system. The model involved dislocation multiplication by the Frank–Read source and their holdup by obstacles (i.e., Lomer immobile dislocations). The resultant pileup of stresses temporarily blocks the Frank–Read source, as the dislocation can escape from the occupied position through the climb. The rate of dislocation climb is naturally controlled by the vacancy diffusion. The rate of creep obtained in this way is

$$\dot{\varepsilon}_s = \text{const} \, \frac{\sigma^\alpha}{kT} \exp\left(-\frac{Q}{kT}\right)$$

where Q is the activation energy for self-diffusion. The great virtue of this approach is the correct stress exponent, $\alpha = 3$.

Nabarro (1967) assumed a specific, regular array of edge dislocations having two sets of Burgers vectors, whose spacing is a function of applied stress. The calculated steady creep rate assumed to be governed by the climb motion of these dislocations was also dependent on the third power of stress. Lindroos and Miekkoja (1968) proposed a model in which the creep rate was determined by the annihilation of two groups of dislocations having opposite signs and climbing up and down along the subboundary.

It was experimentally proven that in some solid-solution-strengthened alloys, the stress exponent in the steady creep rate–stress relation,

$$\dot{\varepsilon}_s = A\sigma^n \exp\left(-\frac{\Delta H}{RT}\right)$$

where ΔH is an activation energy and A is a material-structure-dependent quantity, is equal to 3, while in others it remains about 5. The former type of alloy is classified as class I and the latter as class II. In some of them, however, the transition from class II to class I behavior was observed with increasing stress. This can be rationalized assuming that creep consists of two successive processes: viscous glide motion of dislocations, obeying the third-power law of stress, and the climb motion of dislocations, obeying the fifth-power law of stress. The creep-rate-controlling process is obviously the slower of the two. High stacking fault energy, high concentration of solute, and large-size misfit of solute atoms favor class I behavior. The value of the exponent can thus be used to justify which mechanism probably governs creep.

Davies et al. (1973) suggested introducing the friction stress σ_0 into the preceding relation:

$$\dot{\varepsilon}_s = A(\sigma - \sigma_0)^p \exp\left(-\frac{Q_c^*}{RT}\right)$$

In particle-hardened materials, Q_c^*, which bears a close relation to the activation

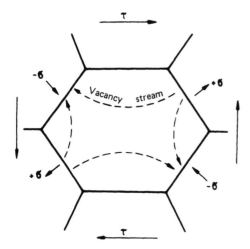

Figure 2.2   Diffusion processes in a grain subjected to a shear stress.

energy for self-diffusion, is equal to the activation energy for creep of the single-phase material under equivalent testing conditions, indicating that the matrix determines creep performance over a wide range of stresses and temperatures.

In polycrystalline material, the concentration of foreign atoms, vacancies, and dislocations is appreciably higher at grain boundaries. The material adjoining them has therefore quite different properties from those of the interior of the grain. Grain boundaries are known to provide paths of easy diffusion and as a place at which high-temperature deformation is concentrated. Thus, when considering high temperatures, grain boundaries should be taken into account. The relevant model of a purely diffusional creep is illustrated in Fig. 2.2. The shear stress causes tensile ($+\sigma$) and compressive ($-\sigma$) stresses to occur and impart on the grain boundaries. Tensile stresses facilitate the formation of vacancies, whereas the compressive stresses heal them. In consequence, matter is transported from zones in compression to zones in tension. The creep rate is here governed by diffusion and two paths are possible: either along grain boundaries (Coble, 1963) or through the volume of the grain (Nabarro, 1948; Herring, 1950).

In a series of experiments, Harper and Dorn (1957) observed the Newtonian viscous flow (the stress exponent of 1) in pure aluminum at temperatures slightly below the melting point and found that creep rates were almost three orders of magnitude higher than those predicted by models formerly proposed. Yavari et al. (1982) summarized results obtained on the Al–5% Mg solid–solution alloy tested in the Harper–Dorn creep and arrived at the conclusion that the rate-controlling mechanism was of a dislocation type, more precisely, a climb of jogged dislocations in saturated conditions (i.e., the distance between jogs being lower than one characteristic length for vacancy diffusion).

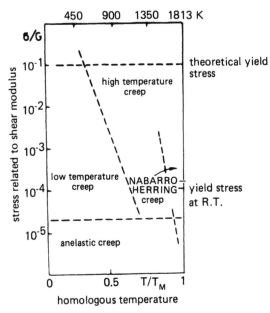

Figure 2.3 Diagram showing stress and temperature conditions under which different creep mechanisms occur: pure iron. (From Hornbogen, 1973.)

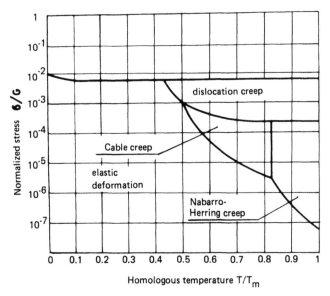

Figure 2.4 Diagram showing stress and temperature conditions under which different creep mechanisms occur: silver. (From Ashby, 1972.)

Figures 2.3 and 2.4 demonstrate the conditions at which different creep mechanisms are operative.

2.3 CREEP RUPTURE

The tertiary creep shown in Fig. 2.1 as terminating the life of a component does not necessarily have to be present, as in tests carried out in compression or torsion. If present, it involves a phase with an increasing creep rate, due to the formation and growth of inhomogeneities in the material. In the early stage of creep, when strain hardening is operative, the deformation is essentially homogeneous. Later, when recovery dominates, the deformation becomes more and more localized.

Garofalo (1965) stressed that modes of deformation observed in tensile tests with a low rate of stress increase resemble those in creep. One can observe the following on surfaces of specimens that have experienced creep:

Slip lines whose spacing increased with increasing temperature and with decreasing stress, the waviness of which evidences the cross-slip.

Twinning, lattice bending, and rotation.

Specific dislocation arrangements, including tangles, cells, and subgrain boundaries; isolated dislocations formed in the early stages of creep (in the iron-based austenitic alloy tested at 980 K, subgrounds were found at the onset of secondary creep).

Grain boundary sliding, migration of grain boundaries, formation of folds near triple points (i.e., where the three grains meet). Folding in grains is associated with grain-boundary sliding and is unique to polycrystalline material. Estimates of the contribution of grain boundary sliding to the total creep strain ranged from very small fractions to very large ones.

Cavities.

It was found in early experiments that creep rates increased considerably (even over 100-fold) when the specimen surface was electrochemically etched to remove several micromicrometers per minute. The creep rate can be markedly decreased by the appropriate surface alloying treatments. It appears, therefore, that the near-surface sources multiply dislocations far more effectively than those in the interior. Kramer et al. (1984) proved experimentally that for aluminum in high-temperature creep, the surface dislocation density was greater than the bulk density by one order of magnitude, and increased with increasing stress, contrary to the density in the interior, where it remained constant. Moreover, the surface layer was found to control this dislocation density below it.

Chen and Argon (1981) investigated creep cavitation in 304 stainless steel at $0.5T_m$. The material was given heat treatment to precipitate the well-defined, stable grain boundary carbides. Specimens allowed to creep at the required stress for the desired period of time were withdrawn from the rig, cooled, and then broken at a cryogenic temperature to reveal the size and profile of intergranular creep cavities. Both the number and size of cavities were found to be larger at grain corners.

Cavities were found as early as in the primary creep stage. Wedge microcracks growing by linkage of grain boundary cavities were also found. The concentration of cavities increased with creep time and with increasing stress. Cavity diameters followed Poisson's distribution: The peak of the distribution gradually shifted toward larger diameters as creep proceeded. Creep cracks emanating from the surface were formed by grain boundary sliding, the process being intensified by cavitation. Cavity shapes were of two distinguishable types: equiaxed, when the cavity grew by diffusion, and elongated, due to high grain-boundary sliding rates.

In general, the experimental data proved that cavities nucleate at inhomogeneities in the grain boundary, such as folds; at precipitates; at intersections of grain boundaries with subgrain boundaries or at twin boundaries and their intersections with grain boundaries; and at grain boundary triple points. Models of grain boundary cavity nucleation are shown in Fig. 2.5: (a) at grain boundary folds, (b) at the intersection of grain boundary with the slip band, (c) shearing of lamellar precipitate, (d) at noncoherent particles, (e) shearing of the second-phase precipitates, and (f) at the intersection of grain and subgrain boundaries. In austenitic steels and creep-resisting alloys containing intermetallic phases the pattern shown in (e) is frequently observed. Ferritic steels under moderate stress and temperature conditions follow (d), and in more severe tests, (f). Grain boundary cavities are, however, also found in high-purity metals. Grain boundary precipitates are preferred sites for the cavity to form, as the stress concentration on it assists nucleation and facilitates attainment of a stable size.

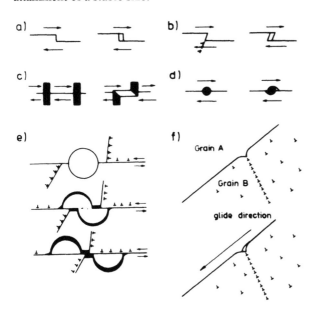

Figure 2.5 Models of cavity nucleation at grain boundaries. Adapted from Presland and Hutchinson, 1963; Regener, 1975; Rukweid, 1972.

Internally oxidized alloys are well suited to an investigation of a cavity nucleation. Varying the composition of the alloy and oxidizing conditions, a wide variation in the average size and spacing of oxide particles can be produced (Finie and Weertman, 1984). Results support the thesis of the role of the dislocation pileup in cavity nucleation. In the nickel-based superalloy Astroloy, with its closely spaced $M_{23}C_6$ carbides, large differences in elastic moduli between matrix and particle and in their deformabilities produce large concentrations of stress at the matrix–carbide interface. These stresses fall off over a distance that is comparable with the particle size, and therefore voids would be less likely to occur at the site of small particles (Shiozawa and Weertman, 1982). Moreover, stresses developed at the interface drive vacancies into the cavities or heal them, depending on their orientation relative to stress. Diffusion is considered as one possible mechanism of cavity growth. Segregation of the solute and impurity atoms to the grain boundaries affects the grain boundary diffusivities. Cavity growth can be constrained by the neighboring grains being unable to accommodate the strain introduced by cavity growth. This accommodation is influenced by both grain size and the morphology of the boundary.

Small cavities almost always grow by diffusion. When they attain greater sizes, the properties of the surrounding material determine their growth rate. The former mechanism was treated theoretically by Raj and Ashby (1971), the latter by Cocks and Ashby (1980). For a variety of situations the coupled mechanism is valid. Finally, this damage is partly reversible. Wortman (1985) proved that turbine blades made of Nimonic 108 can be rejuvenated by pressing at high temperature followed by aging. The life of rejuvenated specimens that had previously reached the tertiary stage of creep was at least 75% that of virgin specimens.

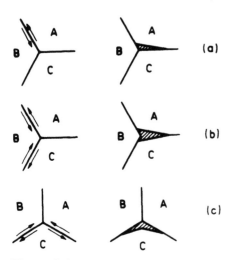

Figure 2.6 Model of triple-point crack nucleation. Arrows indicate direction of grain boundary siding. (From Chang and Grant, 1956.)

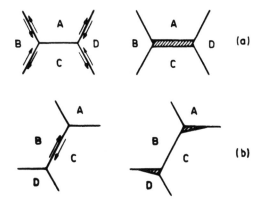

Figure 2.7 Model of intercrystalline crack nucleation. Arrows indicate direction of sliding.

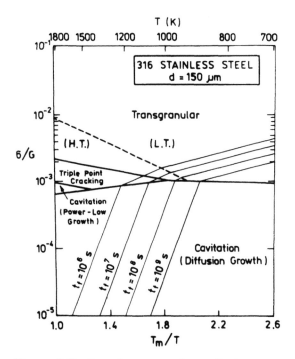

Figure 2.8 Creep facture map of normalized stress versus the reciprocal of homologous temperature for type 316 stainless steel having a grain size of 150 μm. (Reprinted with permission from ASM International.)

Models of cracking characteristic for creep are shown in Fig. 2.6. Triple-point cracks (Fig. 2.7) nucleate in general at low strains during creep under a relatively large stress. Under tensile stress grain boundaries slide, which increases the stress at a triple point, which may be relieved either by the formation of an intragranular fold or by the nucleation of the crack. Further growth of cracks is also governed by the process of grain boundary sliding. Fracture occurs when the reduced volume of sound material is not capable of carrying the load. Wedge-shaped cavities can also originate at intersections of the grain boundary with the low-angle boundary.

Miller and Langdon (1979) plotted maps of creep fracture processes for type 316 stainless steel. The map (Fig. 2.8) is drawn with coordinates of normalized stress and homologous temperature. The expressions for time to failure were derived by Ashby and Raj (1975) for transgranular fracture, Watanabe (1978) for triple-point cracking (tpc), Taplin et al. (1978) for diffusion growth of cavities, and Taplin et al. (1977) for the power-law growth of cavities. The thick lines in Fig. 2.8 represent theoretical boundaries between fields within which a specific mechanism leads to fracture. The transgranular field consists of two parts showing two conditions: high-temperature (H.T.) flow controlled by lattice diffusion and low-temperature (L.T.) flow controlled by diffusion along the dislocation cores. This division obviously does not continue into the field of cavitation failure, which is controlled by the grain boundary diffusion. For comparison, some experimental data are included. Mechanisms are thought here to operate independently, which is often an oversimplification.

2.4 CREEP CRACK GROWTH

The extrapolation of creep data is necessary because the service lives of numerous components are well beyond times found in laboratory tests and map approach 10^5 h. Moreover, th occurrence of a crack is not necessarily equivalent to a need for urgent replacement of a cracked element provided that we are able to estimate when it should be done. For that, we need a parameter than would precisely describe crack growth in as universal a state, independent of material, as is possible. A number of approaches have been adopted and it is now agreed that their efficiency depends on the material being considered and on the conditions involved. The potent parameters are:

1. The stress intensity factor, the crack growth law as described by Paris's law, the approach being justifiable when the cumulative effects are insignificant and the plasticity ahead of the crack tip is negligible.
2. The net section stress, when there is no bending component of the stress (i.e., applied load divided by the area of the uncracked ligament. The implicit assumption is that there is no stress concentration ahead of the crack tip; in other words, the stresses had relaxed. Unlike the above, the correlating equation is of the power form: $da/dt = A\sigma_{net}^n$, where A and n are material constants and da/dt denotes crack advance per time unit.

3. The third set of parameters introduced below [$C*$ (also named $J*$), J_{mod}, $\dot{J}$, creep J, etc.] is applicable under conditions of widespread plasticity, but however attractive, is not directly measurable.

In the case of substantial plasticity, the stress and strain fields near the crack tip in a power-law hardening material are characterized by (Rice and Rosengren, 1968; Hutchinson, 1968)

$$\sigma_{ij} \propto \left(\frac{J}{r}\right)^{1/N+1}$$

$$\varepsilon_{ij} \propto \left(\frac{J}{r}\right)^{N/N+1}$$

where r is the distance from the crack tip. The third equation making the set is the power law of material hardening (i.e., $\varepsilon = B_N \sigma^N$). The J integral can then be written in the form

$$J = -\frac{1}{B}\frac{du}{da}$$

where

u = potential energy

B = thickness of the material

a = crack length

The integral can be interpreted as the change in the potential energy associated with the crack increment. Trying to extend it to creep problems, from the Norton–Bailey creep law,

$$\dot{\varepsilon} = B_n \sigma^n$$

and replacing strains and displacements by the corresponding rates, we can redefine

$$\sigma_{ij} \propto \left(\frac{C*}{r}\right)^{1/(n+1)} \qquad \dot{\varepsilon}_{ij} \propto \left(\frac{C*}{r}\right)^{n/(n+1)} \qquad C* = -\frac{1}{B}\frac{du*}{da}$$

where $u*$ is the functional in the stress–strain rate field identical in form to the potential energy functional u. For a compact tension specimen (CTS; Kubo, 1975),

$$C* = \frac{2n}{n+1}\frac{P\dot{\Delta}}{Bb}$$

where

P = load

b = remaining ligament

B = specimen thickness

$\dot{\Delta}$ = displacement rate at the point of load application

The time-dependent crack growth can be divided into two successive stages (Riedel, 1984):

Stage I, immediately upon an application of the load and on attainment of test temperature; the crack advance is well described by the K values.

Stage II, in which the controlling parameter is C^* and the specimen is intensively creeping.

Both stages are separated by the characteristic time, which for plane strain conditions equals

$$t_1 = \frac{K_I^2(1 - v^2)/E}{(n + 1)C^*}$$

where

E = modulus of elasticity

n = stress exponent of the steady-state creep rate law (i.e., $d\varepsilon/dt = B_n\sigma^n$)

v = Poisson's ratio

Fortunately, this time is typically minutes or hours, the important exception being creep-resistant nickel-based superalloys, where a t_1 value can be as long as several

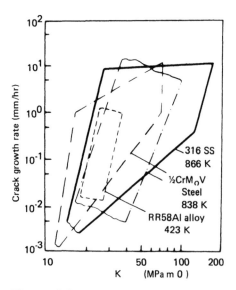

Figure 2.9 Relationship between creep crack growth rate and stress intensity factor. (Reprinted with permission from Elsevier Applied Sci. Publishers.)

Table 2.1 Material
Composition (wt %)

Cr	0.42	P	0.012
Mo	0.51	S	0.017
V	0.25	Sn	0.01
C	0.11	Ti	0.002
Mn	0.51	W	0.002
Ni	0.13	Al	0.001
Si	0.18		

Source: Pilkington et al. (1981).

years, and therefore experimental data should be expressed in terms of the K factor. It is, however, agreed that further work is needed to describe adequately environmental effects in creep and to achieve better accuracy in the low-K-factor range. The power form equations with the C^* integral fit the experimental data reasonably well if steady-state creep is well established. The C^* integral seems independent of temperature.

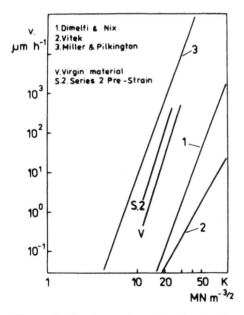

Figure 2.10 Comparison of predicted and experimental crack growth rates as a function of stress intensity factor. (From Pilkington et al., 1981.)

The ½% Cr–½% Mo–¼% V steel was tested by Neate (1977) and Webster and Nikbin (1981). The first researchers studied material that was given optional heat treatment in two conditions: brittle, where K-factor approach was suitable, and ductile, where no such correlation was found. Other researchers used material of intermediate ductility, and good correlation of creep crack growth rate with the C^* integral was found. Such behavior can be rationalized in terms of negligible plasticity at the crack tip in brittle material, whereas in ductile it was substantial.

Radhakrishnan (1985) compiled data obtained by Neate (1977), Nikbin et al. (1977), and Sadananda and Shahinian (1981). The materials studied were AISI type 316 stainless steel, ½Cr-Mo-V steel, and RR58Al alloy. Poor correlation is seen (Fig. 2.9). Singh (1981) investigated AISI type 316L steel and found the crack opening displacement applicable.

Finally, Pilkington et al. (1981) examined the applicability of the theoretical models. Table 2.1 contains the chemical composition of the alloy tested. Specimens were prestrained before the creep test, and after the test were broken at liquid-nitrogen temperature to reveal cavity radii and spacings. The first of the models was that proposed by Dimelfi and Nix (1977), who assumed the cavity to grow by the plastic flow at the wedge grain boundary crack.

The model suggested by Vitek (1978) involved the removal of atoms from a crack tip by stress-induced grain boundary diffusion. The modified model of Dimelfi and

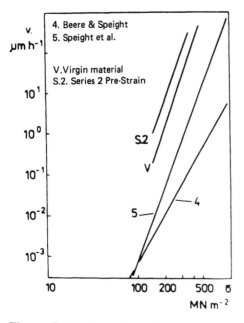

Figure 2.11 Comparison of predicted and experimental crack growth rates as a function of net section stress. (From Pilkington et al., 1981.)

Nix (1977) and Miller and Pilkington (1980) admitted plastic strain concentration at the crack tip.

Beere and Speight (1978) considered the combined effects of grain boundary diffusion and surface diffusion. In a subsequent paper, Speight et al. (1978) assumed that successive processes of surface and grain boundary diffusion occurred between two elastic grains with no plastic deformation of neighboring grains. The models presented are approximate (compare Figs. 2.10 and 2.11).

2.5 NOTCH EFFECT

It is known that notched members usually display shorter lives, and on occasion longer lives, than unnotched members of the same overall dimensions. Figure 2.12 illustrates the latter (Ng et al., 1981). Two casts of ½Cr-Mo-V steel heat treated to simulate the heat-affected zone material of the weld were creep tested at 838 K. The results are plotted against a notch actuity ratio R/r. It is seen that intermediate sharpness notches strengthened the material. Cast 4F was certainly more susceptible than cast 10G to cracking, despite the same unnotched creep properties.

Prediction of the notched member life makes use of creep data usually obtained on standard-shaped specimens tested in pure tension; for that there is a need to find a parameter correlating them. In seeking such a parameter the following are considered (Ellison and Wu, 1983):

Net section stress, σ_n

A type of reference stress, here $\sigma_n K/L$, where L is the constraint factor, equal to the ratio of the strength of the notched bar to that of the unnotched bar for the same

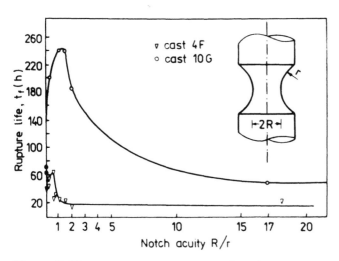

Figure 2.12 Rupture life versus notch acuity ratio.

cross-sectional area as the minimum section of the notched bar, and K_σ is Neuber's plastic stress concentration

The stress that controls creep rupture (i.e., maximum principal stress or von Mises equivalent stress)

Stress concentration factors

Ellison and Wu (1983) tested a 1Cr-Mo-V alloy steel at 823 K. The specimens were crept in torsion and in bending. Types of used specimens are as follows:

1. Circumferentially notched round bar (CNRB) with a cut wide notch or thin slit; the letters N and S refer to the specimen type.
2. Single-edge notch bending specimen (SENB), either plane strain (pl. ε or plane stress (pl. σ), the latter being relatively thinner
3. Compact tension specimen (CTS)

The degrees of constraint are in the following order:

CTS > SENB(pl. ε) > CNRB > SENB(pl. σ)

Two competing processes occur: stress redistribution and damage accumulation. The stress distribution is a function of degree of constraint. Notch weakening (see Fig. 2.13) occurs at high stress concentration because the inadequate material creep ductility does not enable the high elastic–plastic stresses at the notch tip to relax before the onset of cracking. In this instance the crack initiation was found to be very quick. For a milder constraint, notch strengthening is due to the triaxial stress

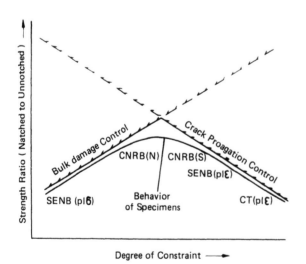

Figure 2.13 Diagram showing the effect of constraint on the creep life of notched bars. (Reprinted with permission from Elsevier Applied Sci. Publishers.)

in the notch region, reducing the equivalent stress $\bar{\sigma}$ at the notch tip to a value below that in an unnotched specimen subjected to the same nominal stress. The creep deformation of the notched specimen is hence lower. At longer exposures, however, this effect gradually disappears, due to stress redistribution. The fractographic examinations revealed (1) major crack propagation in heavily constrained specimens and (2) global damage—numerous cracks in specimens that displayed notch strengthening. The reference stress method was found reliable for the CNRB and SENB(pl. σ), that is, for milder constraints. For high constraint, CTS(pl. ε) and SENB(pl. ε), where damage was localized, the crack propagation parameter was appropriate to correlate the data.

2.6 COMBINED CREEP–FATIGUE CONDITIONS

Ignoring momentary crack retardation due to the overload, the major parameters influencing fatigue crack growth rates are frequency, stress ratio, and temperature. At high frequencies of stress cycling, typically of several hertz, the fatigue crack growth is mostly of transgranular nature. At lower frequencies, below some critical value, time-dependent effects appear; these are mostly creep and corrosion—crack propagation is intergranular. For materials with poor corrosion resistance, the oxidation damage is strongly frequency dependent. The crack growth rate-controlling mechanisms here are (Gell and Leverant, 1973) either surface diffusion of oxidant to the crack tip or diffusion ahead of the crack tip through the lattice or along grain boundaries. It appears that oxidation damage is significant in some range of frequencies. Solomon (1973) demonstrated that the lower and upper bounds of this region can be separated by even more than four orders of magnitude for materials with poor oxidation performance.

The crack growth rate under creep conditions can be written (Ohmura et al., 1973)

$$\frac{da}{dN} = \frac{da}{dt}\frac{1}{\nu}$$

where ν is the frequency. This is true for very low frequencies and for very long hold times.

In the model of Carden (1973) and Döner (1976) the contributions of fatigue, oxidation damage, and creep to crack propagation are summed up linearly:

$$\frac{da}{dN} = C_0(0)f^m(K,R,n) + C_1(0)f^m(K,R,n)\left(\frac{1}{\nu}\right)^k$$

$$\times \exp\left(-\frac{Q_{ox}}{RT}\right) + C_2(0)f^m(F,R,n)\frac{1}{\nu}\exp\left(-\frac{Q_{cp}}{RT}\right)$$

where C_0, C_1, C_2, m, n, and k are empirical material constants. The functional form of f may be a hyperbolic sine function (Annis et al., 1976) or a conventional power law:

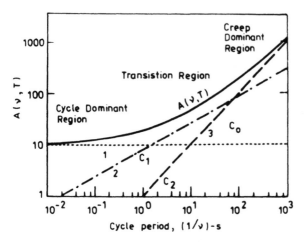

Figure 2.14   Schematic representation of $\log[A(v, T)]$ versus $\log(1/v)$ at constant temperature, illustrating various time-dependent contributions to the proportionality constant $A(v, T)$. (From Döner, 1976, after Fu, 1980.)

$$f^m(K,R,n) = \Delta K^m(1 - R)^{m+2n}$$

where R is the stress ratio. The F parameter characterizes the crack tip under the particular creep conditions. Letting $F = K$ for simplicity, we derive

$$\frac{da}{dN} = A(v,T)f^m(K,R,n)$$

The $A(v,T)$ factor is represented schematically in Fig. 2.14.

The creep–fatigue interaction effects are as follows (Fu, 1980):

1. For given T and ΔK the crack growth rate usually increases monotonically with temperature.
2. For given T and ΔK it increases with the decreasing frequency of stress cycling; above some critical frequency, da/dN becomes frequency independent.
3. Under creep conditions crack growth rates appear as the lower limit of the fatigue creep crack growth rates.
4. At a given temperature and frequency, da/dN increases with increasing stress ratio R.

2.7   CREEP LIFE ASSESSMENT

We are concerned with two problems here: the extrapolation of creep data and the cumulative damage in creep (i.e., the situation when either stress or temperature or both simultaneously are varying). The commonly used extrapolation methods can be divided into the following groups:

1. Graphic methods where the experimentally derived creep curve is elongated to longer times.
2. Graphically numeric methods where the family of creep curves is transformed into one master curve. This new curve is drawn in the coordinates log σ versus P, P being the transformation parameter: for example (Larson and Miller, 1952), $P = T(C + \log t)$, where T is the temperature, C a constant, and t is the time to rupture. The constant C varies from one material to another, and for a given one, its value at lower stresses differs appreciably from that for higher stresses. All the experimental curves are thus equivalent regardless of test conditions. The master curve can be represented by a polynomial and fitted to the experimental points. The time to rupture or admissible stress can thereby be found analytically. The Larson–Miller parameter is capable of detecting the service-induced damage. Data for such materials are outside the scatterband for virgin material.
3. Numeric methods based on the application of regression techniques to describe the set of experimental points and to extrapolate results into longer lives.

Granscher and Wiegand (1972) compared the above-mentioned methods and found the second to be the most feasible.

The other philosophy is accelerating creep tests by increasing either stress or temperature above the value of interest. The second method considered more appropriate is reviewed below in detail. Melton (1983) investigated 10 heats of various ferritic steels to assess the applicability of this isostress method. The test temperatures ranged from 723 to 873 K and the corresponding lives from a few hours to several hundred. It appeared from results that:

1. There is a linear relationship between the logarithm of the time to rupture and the temperature of the test, the correlation coefficient being typically of 0.98 or better.
2. The accuracy of the isostress extrapolation was better for the service-exposed material, as the microstructural changes that could affect the creeplife had taken place before test.
3. The method was proved capable of producing reasonable expectancies even for 100-fold longer periods of time than test periods. In most cases the error was less than ±2 in lifetime.

Finally, empirical relationships between the steady state creep rate and the time to rupture can be used to predict creep life. For wrought alloys the most commonly used relation is (Monkman and Grant, 1956)

$$t_r \dot{\varepsilon}_s^m = C$$

where

$\dot{\varepsilon}_s$ = secondary creep rate

t_r = time to rupture

m, C = material constants

Obviously, if constants m and C are known, the time to rupture can be assessed simply by measuring the creep rate. The Monkman–Grant relationship for IN-738 is illustrated in Fig. 2.15. The data were collected from Mazzei (1978), Stevens and Flewitt (1981), Tipler (1981), Tipler et al. (1978), Beck and Burke (1981), Castillo et al. (1987), and International Nickel INCO Alloy IN-738: Technical Data (International Nickel Company, Inc.).

Koul et al. (1984) proposed to isolate the tertiary creep life and strain from the overall creep curve. Primary plus secondary creep lives $(t_p + t_s)$, normalized by the corresponding primary and secondary creep strains $(\varepsilon_p + \varepsilon_s)$, revealed a straight-line relationship:

$$\frac{t_p + t_s}{\varepsilon_p + \varepsilon_s} \dot{\varepsilon}^m = K$$

A survey of creep data showed that the total creep deformation ε_c (i.e., the difference between deformation at rupture and instantaneous deformation) exhibits scatter similar to that for creep rates. Therefore, it can be expected that the Monkman–Grant relation normalized by the total creep strain would appropriately describe the dependence between the secondary creep rate and time to fracture. Dobes and Milicka (1976) wrote such a relationship in a more general form:

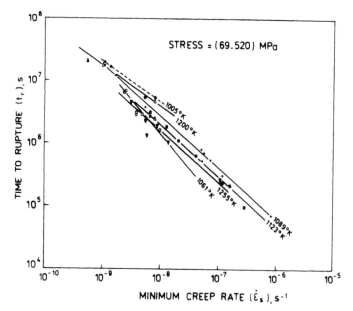

Figure 2.15 Dependence of the time to rupture on the minimum creep rate for IN-738 LC. (From Castillo et al., 1987.)

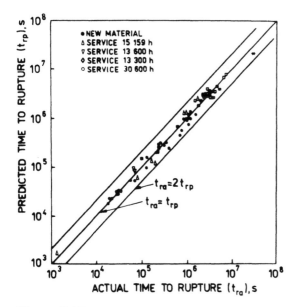

Figure 2.16 Comparison between the measured creep life and those calculated using a modified relation. (From Castillo et al., 1987.)

$$\log\frac{t_r}{\varepsilon_c} + m' \log \dot{\varepsilon}_s = C'$$

where m' and C' are constants, and proved it to be superior over the original Monkman–Grant relation with regard to the scatter of data for a range of laboratory alloys.

Castillo et al. (1987) observed for IN-738 LC a strong correlation between the tertiary creep life and the rupture life that enabled them to write the Monkman–Grant relation in modified form:

$$(-bt_r^{n_1} + t_r)\dot{\varepsilon}_s^{m_1} = K_1$$

where b, K_1, n_1, and m_1 are material constants. Figure 2.16 presents the correlation between the measured creep lives and the predicted ones, based on the modified Monkman–Grant relation. The new approach produced consistently better predictions (compared with Fig. 2.15).

Robinson (1938) proposed the life-fraction rule to calculate the damage caused by creep periods at different loads and/or temperatures:

$$\sum_i \frac{t_i}{t_{ri}} = 1$$

where t_i is the time spent at σ_i stress and T_i temperature and t_{ri} is the time to rupture under i-th conditions. Goldhoff and Woodford (1972) verified this rule experimentally and found that it is adequate for Cr-Mo-V steel when small variations in stress and temperature occur. Some other rules are listed below:

$$\sum_i \frac{\varepsilon_i}{\varepsilon_{ri}} = 1 \quad \text{strain-fraction rule (Lieberman, 1962)}$$

$$\sum_i \left(\frac{t_i}{t_{ri}}\right)^{1/2} \left(\frac{\varepsilon_i}{\varepsilon_{ri}}\right)^{1/2} = 1 \quad \text{(Voorhees and Freeman, 1959)}$$

$$K\sum_i \frac{t_i}{t_{ri}} + (1-K) \sum_i \frac{\varepsilon_i}{\varepsilon_{ri}} = 1 \quad \text{(Abo-El-Ata and Finnie, 1971)}$$

where ε_i and ε_{ri} are strains defined as above and K is the constant intended to account for the creep cracking mode of the material. Time and strain fractions are generally considered not equivalent. All the methods described above are used frequently to correlate data obtained in thermal fatigue tests.

REFERENCES

Abo-El-Ata, M. M. and Finnie, I. (1971). *A Study of Creep Damage Rules*, ASME Paper 71-WA/MET-1.

Andrade da, E. N. (1910). *Proc. R. Soc. 84a*: 1.

Annis, C. G., Wallace, R. M. and Sims, D. L. (1976). *An Interpolative Model for Elevated Temperature Fatigue Crack Propagation*, AFML-TR-76-176.

Ashby, M. F. (1972). *Acta Metall., 20*: 887.

Ashby, M. F. and Raj, R. (1975). *The Mechanics and Physics of Fracture*, The Metals Society, London, p. 148.

Bailey, R. W. (1929). *Trans. 3rd World Power Conference*, Tokyo, p. 1089.

Beck, C. G. and Burke, M. A. (1981). Research Memo 81-124-STABL-M2, Westinghouse Research and Development Center.

Becker, R. (1925). *Phys. Z., 26*: 919.

Becker, R. (1926). *Z. Tech. Phys., 7*: 547.

Beere, W. and Speight, M. V. (1978). *Met. Sci., 12*: 593.

Carden, A. E. (1973). Parametric analysis of fatigue crack growth, *Proc. International Conference on Creep and Fatigue at Elevated Temperature Applications*, Paper C230/73.

Castillo, R., Koul, A. K. and Toscano, E. H. (1987). *J. Eng. Gas Turbines Power, 109*: 99.

Chang, H. C. and Grant, N. J. (1956). *Trans. AIME, 206*: 544.

Chen, I. W. and Argon, A. S. (1981). *Acta Metall., 29*: 1321.

Coble, R. L. J. (1963). *J. Appl. Phys., 34*: 1679.

Cocks, A. C. F. and Ashby, M. F. (1980). *Metall. Sci. 14*, no. 8–9:395.

Davies, P. W., Nelmes, G., Williams, K. R. and Wilshire, B. (1973). *Met. Sci. J., 7*: 87.

Dimelfi, R. J. and Nix, W. D. (1977). *Int. J. Fract., 13*: 341.

Dobes, F. and Milicka, K. (1976). *Metall. Sci. 10*, no. 1: 382.

Döner, M. (1976). *J. Eng. Power 98a*, no. 4: 473.

Dorn, J. E. (1956). Creep and fracture of metals at high temperatures, *NPL Symposium*, Her Majesty's Stationery Office, London, p. 8.

Ellison, E. G. and Wu, D. (1983). *Res. Mech.*, 7: 37.

Finie, M. E. and Weertman, J. R. (1984). *Proc. 11th Canadian Fracture Conference*, Ottawa (Krausz, A. S., ed.), Martinus Nijhoff, Hingham, Mass., p. 93.

Fu, L. S. (1980). *Eng. Fract. Mech., 13*: 307.

Garofalo, F. (1963). *Trans. Metall. Soc. AIME, 227*: 351.

Garofalo, F. (1965). *Fundamentals of Creep and Creep Rupture in Metals*, Macmillan, New York, p. 11.

Gell, M. and Leverant, G. R. (1973). STP 520, ASTM, Philadelphia, p. 37.

Glen, J. (1963). *J. Basic Eng.*, 85: 595.

Goldhoff, R. M. and Woodford, D. A. (1972). The evaluation of Creep Damage in a CrMoV Steel, STP 515, ASTM, Philadelphia, p. 89.

Granacher, J. and Wiegand, H. (1972). *Arch. Eisenhuettenw.*, 42: 699.

Harper, J. and Dorn, J. E. (1957). *Acta Metall.*, 5: 654.

Herring, C. J. (1950). *Appl. Phys.*, 21: 437.

Hornbogen, E. (1973). Mechanisms to control creep resistance, in *Proc. Symposium on High-Temperature Materials*, Baden (Sahm, P. R. and Speidel, M. R., eds.), Elsevier, Amsterdam.

Hutchinson, J. W. (1968). *J. Mech. Phys. Solids, 16*: 13.

Koul, A. K., Castillo, R. and Willett, K. P. (1984). *Mater. Sci. Eng.*, 66: 213.

Kramer, I. R., Feng, C. R. and Arsenault, R. J. (1984). *Metall. Trans. A, 15*: 1571.

Kubo, S. (1975). Ph.D. thesis, Osaka University, Japan.

Larson, F. R. and Miller, A. (1952). *Trans. ASME, 74*: 765.

Lieberman, V. (1962). *Metalloved. Term. Obrab. Met., 4*: 6.

Lindroos, V. K. and Miekkoja, H. M. (1968). *Philos. Mag., 17*: 119.

Mazzei, P. J. (1978). Technical Reports PAL 78-76 and PAL 78-77, Westinghouse Canada Inc.

McVetty, P. G. (1934). *Mech. Eng., 56*: 149.

Melton, K. N. (1983). *Mater. Sci. Eng., 59*: 143.

Miller, D. A. and Langdon, T. G. (1979). *Metall. Trans. A, 10*: 1635.

Miller, D. A. and Pilkington, R. (1980). *Metall. Trans. A, 11*: 595.

Monkman, F. C. and Grant, N. J. (1956). *Proc. ASTM, 56*: 593.

Mott, W. F. and Nabarro, F. R. N. (1948). *Report on a Conference on the Strength of Solids*, University of Bristol, Physical Society of London, London, p. 1.

Nabarro, F. R. N. (1948). *Report on a Conference on the Strength of Solids*, University of Bristol, Physical Society of London, London, p. 75.

Nabarro, F. R. N. (1967). *Philos. Mag., 16*: 231.

Neate, G. J. (1977). *Eng. Fract. Mech., 9*: 297.

Ng, S. E., Webster, G. A. and Dyson, B. F. (1981). Notch weakening and strengthening in CRE 1/2 Cr 1/2 Mo 1/4 V steel, Proc. International Conference on Fracture and Fatigue, Cannes.

Nikbin, K. M., Webster, G. A. and Turner, C. E. (1977). *Proc. 4th International Conference on Fracture*, University of Waterloo, Vol. 2, p. 627.

Norton, F. H. (1929). *Creep of Steel at High Temperatures*, McGraw-Hill, New York.

Odqvist, F. K. G. (1953). *Proc. IUTAM Congress*, Istanbul, Vol. 1, p. 99.

Ohmura, T., Pelloux, R. M. and Grant, N. J. (1973). *Eng. Fract. Mech., 5*: 909.

Orowan, E. (1946–47). *J. West Scotl. Iron Steel Inst., 54*: 45.

Pilkington, R., Miller, D. A. and Worswick, D. (1981). Creep crack growth and cavitation damage, in *Proc. 3rd Symposium on Creep in Structures*, Leicester (Ponter, A. R. D. and Hayhurst, D. R., eds.), Springer-Verlag, New York.

Presland, A. and Hutchinson, R. J. (1963). *J. Inst. Met.*, *92*: 264.

Radhakrishnan, V. M. (1985). *Res. Mech.*, *13*: 23.

Raj, R. and Ashby, M. F. (1971). *Metall. Trans.*, *2*: 1113.

Regener, D. (1975). Untersuchungen zum Zeitstandverhalten eines niedriglegierten CrMoV Stahls, Ph.D. thesis, Technical University of Magdeburg.

Rice, J. R. and Rosengren, G. F. (1968). *J. Mech. Phys. Solids*, *16*: 1.

Riedel, H. (1984). Mechanics and micromechanics of creep crack growth, in *Subcritical Crack Growth Due to Fatigue, Stress Corrosion and Creep* (Larson, L. H., ed.), Elsevier, New York.

Robinson, E. L. (1938). *Trans. ASME*, *60*: 253.

Ruckweid, A. (1972). *Metall. Trans.*, *3*: 3003.

Sadananda, K. and Shahinian, P. (1981). *Proc. 2nd International Symposium on Elastic–Plastic Fracture Mechanics*, ASTM, Philadelphia.

Sherby, O. D., Lytton, J. L. and Dorn, J. E. (1957). *Acta Metall.*, *5*: 219.

Shiozawa, K. and Weertman, J. R. (1982). *Scripta Metall.*, *16*: 735.

Singh, G. (1980). Application of non-linear elastic fracture mechanics to aged Type 316 stainless steel, in *Proc. 3rd Symposium on Creep in Structures*, Leicester (Ponter, A. R. S. and Hayhurst, D. R., eds.), Springer-Verlag, New York, p. 592.

Solomon, H. D. (1973). *Metall. Trans.*, *4*: 341.

Speight, M. V., Beere, W. and Roberts, G. (1978). *Mater. Sci. Eng.*, *36*: 155.

Stevens, R. A. and Flewitt, P. E. J. (1981). *Acta Metall.*, *29*: 867.

Taplin, D. M. R., Collins, A. L. W., Gandhi, C. and Ashby, M. F. (1977). In *Fracture Mechanics and Technology* (Sih, G. C. and Chow, C. L., eds.), Vol. 1, p. 127.

Taplin, D. M. R., Sidey, D. and Gandhi, C. (1978). *Trans. Indian Inst. Met.*, *31*: 163.

Tipler, H. R. (1981). Maintenance in Service of High Temperature Parts, AGARD, CP–317, paper no. 12–1.

Tipler, H. R., Lindblom, Y. and Davidson, J. H. (1978). *Proc. Conference on High Temperature Alloys for Gas Turbines*, Leige, Applied Science Publishers, Barding, Essex, England, p. 359.

Vitek, V. (1978). *Acta Metall.*, *26*: 1345.

Vorhees, H. R. and Freeman, J. W. (1959). *Notch Sensitivity of Aircraft Structural and Engine Alloys*, Part II, *Further Studies with A-286 Alloy*. Wright Air Development Center Technical Report.

Watanabe, T. (1978). *Proc. 23rd International Symposium on the Strength and Fracture of Materials*, The Society of Materials Science, Tokyo, p. 1.

Webster, G. and Nikbin, K. M. (1981). History of loading effects on creep crack growth in $\frac{1}{2}$ Cr $\frac{1}{2}$ Mo $\frac{1}{4}$ V steel, in *Proc. 3rd Symposium on Creep in Structures*, Leicester (Ponter, A. R. S. and Hayhurst, D. R., eds.), Springer-Verlag, New York. p. 576.

Weertman, J. (1955). *J. Appl. Phys.*, *26*: 1213.

Wortman, J. (1985). *Mater. Sci. Technol.*, *1*: 644.

Yavari, P., Miller, D. A. and Langdon, T. G. (1982). *Acta Metall.*, *30*: 871.

3

Creep-Resisting Materials

3.1 GENERAL CHARACTERISTICS

The problem of creep resistance is considered in the temperature range 470 to above 2200 K; the lower limit is fixed by plastics, the upper by tungsten filaments. The most frequently met temperatures in industrial applications are below about 1500 K. Considering material properties that determine its applicability, the melting point is undoubtedly crucial. This property favors ceramics and refractory metals. Unfortunately, such materials are brittle and become ductile only at temperatures slightly below the melting point. The next criterion is the dependence of normalized yield stress (σ_y /E) on homologous temperature (T/T_M). This dependence is shown in Fig. 3.1. As seen, the best properties possess solids with a covalent bond. In pure metals, the creep activation energy strictly correlates with the self-diffusion activation energy (see Fig. 3.2). It is seen that metals fall into five groups. The structure effect should be noted; austenitic steels behave better. Figure 3.3 shows the maximum working temperatures of creep-resisting alloys for the case where the only significant factor is creep.

Easier plastic deformability of pure metals at high temperatures has two prime causes:

1. At moderate temperatures, the one possible dislocation glide is the conservative; at high temperatures, cross-slip, dislocation climb, and grain boundary sliding become possible.
2. At moderate temperatures, deformation is accompanied by strain hardening; at high temperatures, recovery and recrystallization exclude it.

The processes of recovery and recrystallization are governed by the formation and absorption of vacancies and thus by the coefficient of self-diffusion.

Structural materials must meet a number of requirements specified in the stage of production and during their service. For example, requirements written for a gas turbine blade include (Hernas and Maciejny, 1989):

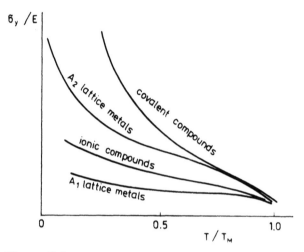

Figure 3.1   Temperature dependence of normalized yield stress for various materials.

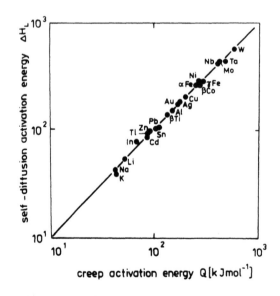

Figure 3.2   Correlation between creep activation energy and activation energy of self-diffusion. (From Sherby and Weertman, 1979.)

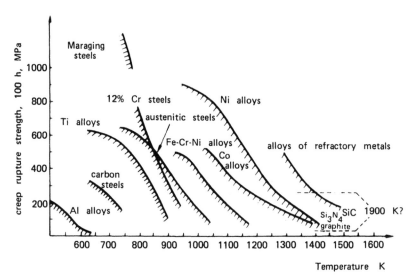

Figure 3.3 Maximum service temperatures of various creep-resisting materials.

Adequate strength and creep resistance
Resistance to hot corrosion
Resistance to thermal and mechanical fatigue
Ductility and resistance to service-induced embrittlement
Structural stability
Formability, weldability, and other properties making fabrication easy

The mechanisms from which creep strength is derived are given in Table 3.1. The low-temperature behavior is included for comparison. The contributions of these mechanisms are shown in Fig. 3.4.

The alloying additives that strengthen the solid solution increase creep resistance through (Hernas and Maciejny, 1989):

Increasing melting point and temperatures of effective recovery and recrystallization
Increasing the energy of lattice bonding
Reducing stacking fault energy
Increasing lattice friction and causing chemical pinning of dislocations
Retardation of dislocation glide and climb
Increasing stability of solid solution
Increasing valence

Table 3.1 Basic Hardening Mechanism Affecting Low-Temperature Deformation and High-Temperature Creep Life

	Solute atoms; solid solution hardening	Dislocations work hardening	Grain boundaries; grain boundary hardening	Particles; precipitation, dispersion hardening	Anisotropy	
					Crystal structure	Micro-structure
Low-temperature deformation	+	++	++	++	+	+
Creep	+	$\pm^{a}$	–	++	+	+

Source: Hornbogen (1973).

++, increases strongly; +, increases; –, decreases. [a] ± + if dislocations are pinned by stable particles; – if recrystallization takes place during creep.

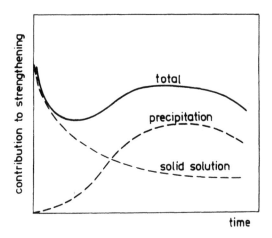

Figure 3.4 Contribution of various strengthening mechanisms as a function of time, 16 M steel, creep tested at 923 K. (From Baird et al., 1972.)

The strengthening action of second-phase particles relies on:

Locking of dislocation glide
Retarding grain boundary slip
Hindering recovery by pinning dislocation networks

The minimum secondary creep rate corresponds to a certain specific interparticle distance and particle size. The efficiency of this mechanism is seen by the exponent in the Norton–Bailey equation:

For pure metals and solid solutions, it is within 3 to 5.
For precipitation hardened solutions, it is 5 to 15.
For dispersion hardened solutions, it is even 40.

Stability of precipitation hardening is related to:

Crystalline structure of phases and their interfaces
Stereological parameters (i.e., volume fraction of particles, particle size and shape, interparticle spacing)
Structural stability, understood as the lack of phase transformations of second-phase particles; low solubility and diffusivity of atoms constituting the particle and the matrix

Considering the interaction between dislocations and particles at elevated temperatures, two cases are distinguished (see Fig. 3.5). In the first, hard, incoherent particles force dislocations to bow around them and below some critical stress, the value of which is inversely proportional to the interparticle spacing, straining takes

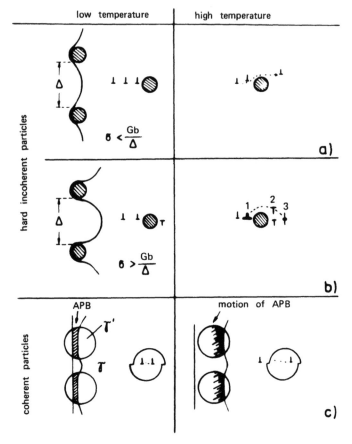

Figure 3.5 Interaction of dislocations and particles at elevated temperatures: (a) bypassing by climb; (b) annihilation of an Orowan loop by climb; (c) rearrangement of antiphase boundary in an ordered particle inhibits dislocation pair formation. (From Hornbogen, 1973.)

place by climb of dislocations over particles. Above this critical stress Orowan loops are formed. For coherent particles (e.g., in nickel-based superalloys), dislocation cutting through particles produces an antiphase boundary that can move out of its original plane at higher temperature; the next dislocation requires the application of external stress to destroy the restored order. The creep strength increases in such alloys up to approximately 1070 K.

Grain boundaries are the weakest places in the microstructure at high-temperature creep, contrary to low-temperature straining, where they add some strength. The rate of grain boundary sliding is a function of both the diffusion coefficient for the grain boundary (D_{GB}) and that for self-diffusion (D_{SD}) as the process requires the interaction of lattice and grain boundary dislocations. Lower values of the D_{GB}/D_{SD} ratio are advantageous. Mechanisms by which grain boundary sliding is controlled include segregation of solute atoms to grain boundaries and precipitation of particles there. At relatively high temperatures

$$\dot{\varepsilon}_s = f\left(\frac{b}{d}\right)^p$$

where

$\dot{\varepsilon}_s$ = secondary creep rate

b = Burgers vector

d = grain diameter

$p = \begin{cases} 2 \text{ for Nabarro–Herring creep} \\ 3 \text{ for Coble creep} \end{cases}$

It is acknowledged that columnar-grain, single-crystal, eutectic alloys (in situ composites) are promising, and this has been proven by a vast number of tests. The most promising effect was found when the γ' phase was fine and spheroidal.

Powder metallurgy offers the advantages of very high purity, homogeneity of chemical composition, and isotropic properties. It also enables us to produce near-shape elements.

At low temperatures, a large number of metallic materials are used in a state far different from thermodynamic equilibrium. With increasing service temperatures, the rate of diffusion processes, formation of crystalline nuclei, and growth of the new, more stable phases increases. These spontaneous processes, usually detrimental to the material, are also affected by external load and temperature variations. The processes mentioned can be generally divided into:

1. Annealing of internal stresses
2. Grain coarsening
3. Polymorphous transformations
4. Coagulation of finely dispersed phases
5. Decomposition of oversaturated solutions
6. Exchange of constituents between particular phases
7. Chemical reactions between structural constituents and formation of new phases
8. Changes in a substructure

Creep-resistant materials are discussed in more detail in later sections of this chapter.

3.2 CORROSION

Working environments at elevated temperatures can be divided into three basic groups: gases, molten electrolytes, and molten metals. A gaseous environment can be either oxidizing or reducing. In an oxidizing environment, the layer of the corrosion products gradually thickens at the expense of the parent material. In the presence of a second oxidant such as sulfur or carbon, the attack can be enhanced as the second oxidant destroys the ability of the alloy to maintain the continuous layer of the oxide formed by the reaction between metal and the first, prime oxidant. The following mechanisms are proposed to account for this effect in ferrous alloys (Stringer, 1983):

1. The second oxidant or a compound including is deposited within the protective scale, causing its spalling. In CO/CO_2 mixtures when gas is transported toward the surface through the pores in the scale, the possible reactions are $3Fe + 4CO_2 \rightarrow Fe_3O_4 + 4CO$, which increases the content of CO in the mixture, and consequently, $2CO \rightarrow CO_2 + C$ becomes important. The volume change caused by the deposited carbon can disrupt the scale.
2. The second oxidant can introduce short-circuit paths within the protective oxide, allowing rapid transport of the base metal. For example, Fe–20% Cr develops Cr_2O_3 scale in oxygen, and in the presence of sulfur, the outward diffusion of iron through the scale may occur. The iron oxide nodules can disrupt the underlying layer of Cr_2O_3 scale.
3. The second oxidant migrates through the oxide and reacts preferentially with the element forming the protective scale, which can preclude continued development of the protective scale. A product of the reaction with the second oxidant can locate at the grain boundary, weakening it.

Corrosion in molten electrolytes follows a typical electrolytic pattern where in some areas on the material's surface, cathodic or anodic processes occur. The controlling mechanisms are ionic conduction of electrolyte and the migration of depolarizors.

In molten metals, constituents of the solid metallic phase can dissolve in the liquid phase. Possible reactions are the formation of intermetallic compounds on the metal surface, and corrosion caused by constituents dissolved in the molten metal, which is undoubtedly the case with ingot mold, for example.

In general, environment-induced effects are:

1. Dissolution of environmental components in the metal, with the consequent modification of its chemical composition, which can induce phase transformation in the extreme case.
2. Chemical reaction on the metal surface or below it, with the formation of corrosion products on the surface or in the interior without observable losses in material.

The latter is termed *internal corrosion*. The consequence of this process is a depletion of the surface layer in the elements that had reacted, the formation of new phases, corrosion products in the metal, and the appearance of cavities below the surface. An example is the corrosion of petrochemical installations, where the exposure of 25Cr-20Ni steel to the carburizing environment causes the carbide precipitates to form: M_7C_3 in the outer zone, M_7C_3–$M_{23}C_6$ carbides in the intermediate zone, and $M_{23}C_6$ in the inner zone (Harrison et al., 1979).

Internal corrosion can precede the formation of a scale on the metal surface or appear alone in favorable conditions. It is worth noting that internal oxides constitute intermediate layers between the metal and the scale, providing a better bond in thermal fatigue conditions. The depth of the internal oxidation can be of 100 μm, which is a considerable fraction of the thickness of the hollow blades used in gas turbines.

As a rule, the corrosion aggressiveness increases with the temperature rise. An exception is the temperature at which a structural transformation occurs in the scale. In gas turbines running on a contaminated fuel there is no simple relation between a corrosion rate and temperature. Moreover, a threshold temperature is observed below which virtually no corrosion is noted. Such a temperature dependence is determined by the two characteristic temperatures of deposits formed on the blades: their melting point, where corrosion starts, and their dew point, where it ends.

The corrosion products provide some protection against further corrosion advance. For this reason, apart from static tests where the scale grows continuously with almost no spalling, the tests developed to study the scale adhesion are of prime concern. The risk of the accelerated tests is that what is measured is not a corrosion rate but a result of corrosion during the incubation period. According to the Piling and Bedworth (1923) theory, the condition of the dense scale formation is a molar ratio of the corrosion products to the metal that exceeds 1. The theory is valid for the inward diffusion of the oxidizing agent, whereas in most metals, the outward diffusion of metal atoms is a major factor. The possible sources of internal stresses in the scale are:

The formation of new compounds having larger molar volumes than the metal
Structural redevelopment of the scale
The coagulation of vacancies

3.3 CORROSION KINETICS

If the corrosion product is volatile, it does not provide adequate protection of the surface, and the mass decrement of the specimen being corroded is constant per time unit. Similarly, if the scale is discontinuous, the corrosive medium is admitted to the surface and the mass gain is linear. When the growth of the scale is governed by diffusion as it is at high temperatures, the mass gain law takes the parabolic form

$$(\Delta m)^2 = at + b$$

where

Δm = mass gain

a, b = constants

t = time

It is also a case of internal corrosion.

At low and intermediate temperatures, in some instances scale growth follows the logarithmic equation

$$\Delta m = a \log(bt - c)$$

or the inversely logarithmic equation

$$\frac{1}{\Delta m} = a' \log(b't - c')$$

–where a, b, c, a', b', and c' are constants. Simplifying it somewhat, with increasing temperature the scale grows continually faster—logarithmically at first, then according to the power law, and finally linearly at the highest temperatures.

The corrosion rate in flowing combustion bases depends on their residence time at the metal surface; for the gas turbine blade it is highest at its leading and trailing edges. Corrosion can penetrate unevenly into the material. The probable causes are:

1. Solubility of aggressive agent contained in the environment in the material being corroded and its migration along grain boundaries
2. Penetration of the liquid substances and the formation of the low-melting-point eutectics, which is most probable at grain boundaries

3.4 EFFECT OF CORROSION ON CREEP PERFORMANCE

The usually observed effect of corrosion is a reduction in component life. There are, however, numerous exceptions contradicting this pattern. The mechanisms evoked to explain crack growth retardation in creep–fatigue experiments are (Yuen et al., 1984):

Crack closure (i.e., the crack is closed by corrosion products throughout some part of the stress cycle)
Corrosion fracture surface roughness; also causes the crack to close
Crack branching; when numerous cracks grow slowly
Crack tip blunting, reducing stress severity

Creep strengthening is often associated with the formation of a strongly adhering, integral oxide layer on the metal surface, and the effect is attributed to the ability of a surface scale to inhibit dislocation generation at the surface and their egress. Internal corrosion affects material properties in a number of ways. It can cause:

Precipitation of new phases that have poor dispersion with respect to creep performance
Alteration of chemical and phase composition of parent metal
Internal stresses built by corrosion products
Diffusion of metal atoms to a scale leaving vacancies behind which can either be captured by traps (such as grand boundaries, structural defects) or can coagulate and form cavities

Considering the latter, the range of vacancy diffusion is small and therefore its effects on diffusion creep are negligible. In turn, cavities obviously reduce the life of the component.

Moreover, pollution atoms affect internal friction, form gas bubbles as a product of, for example, oxygen–carbon reaction, and also cause grain boundary pinning. Internal corrosion is undoubtedly enhanced by creep, during which corrosion scale loses its integrity, facilitating inward and outward diffusion. In this case, dislocation multiplying in creep provides easy diffusion paths. The effect of internal corrosion is illustrated by the following two examples. Solberg and Thon (1984) tested two nickel alloys in creeping corrosion, the test environment being combustion gas. The first alloy, which had a small quantity of alloying additives, was strengthened by inward diffusion of pollution atoms. The second, highly alloyed, had a considerably reduced life that was an effect of large oxide inclusions. Guttmann and Bürgel (1983) tested two austenitic steels in a carburizing environment. The effects of the carbonaceous atmosphere included increased creep ductility and a reduced creep rate with increasing test duration, although short- and medium-term exposures did not show consistent effects.

3.5 NONISOTHERMAL CONDITIONS

During cooling from the oxidizing temperature the phase transformations can take place either in metal or in corrosion products. In addition, the difference in thermal expansion coefficients causes stresses to appear at the interface of the parent metal–oxide. If the processes of cooling and heating are repeated, cycling of these stresses will produce oxide cracking and detachment with consequent acceleration of corrosion. In alloys that are dependent on the formation of a protective scale involving one of the alloy elements, repeated spalling and regrowth depletes the surface of the alloy in the oxide-forming element and the protective layer is not restored.

Internal corrosion was found to promote scale adhesion in nonisothermal conditions. In cobalt alloys with simple chemical formulas, this characteristic is lacking; the scale completely detaches during a few temperature variations.

The enhancing effect of lanthanides was first established in heater alloys. Lanthanides segregate to grain boundaries and when oxidized, bond the scale mechanically with the metal. This is called the *keying effect* and relies on the whisker crystals of the oxide phase reaching deeply into the metal. The introduction of lanthanides is troublesome, as the full profit is gained when they are introduced to the metallic phase. The other most frequently cited mechanisms accounting for enhanced scale adhesion are:

Prevention of vacancy coalescence at the scale–metal interface by providing alternative coalescence sites
Enhancement of scale plasticity
Modification of the scale growth processes

Table 3.2 Mass Loss per Unit of Surface Area (g/m^2) of Dispersion-Strengthened NiCr Heat-Resistant Alloys in Cyclic Corrosion Tests[a]

Alloy composition	Time (h)			
	24	48	96	504
20% Cr, balance Ni	−20.7	−22.0	−41.3	−324.0
NiCr	+1.2	+1.4	+0.3	−4.5
20% Cr, 2.4% ThO_2, balance Ni	+1.6	+2.3	+4.7	+2.7
20% Cr, 1.13% Y_2O_3, balance Ni	+0.9	+0.9	+2.3	−0.5
20% Cr, 1.48% Al_2O, balance Ni	−4.2	−4.7	−5.3	−9.7
20% Cr, 1.27% Li_2O, balance Ni	−8.6	−9.0	−17.8	−439.0

Source: Mrowec and Werber, (1982).
[a]Dry air, T_{max} =1370 K; cycle duration, 24 h.

Formation of graded oxide layer, which lessens differences in the mechanical and thermal properties of the metal and the scale
Enhanced bonding by the preferential segregation of lanthanides at the metal–scale interface

The theory put forth by Funkenbusch et al. (1985) is that the role of such elements (originally, yttrium) can be to interact with the indigenous sulfur, reducing the extent of its segregation and concentration at the critical sites with respect to scale adhesion.

Apart from that, profitable effects of finely dispersed oxide phase were found (see Table 3.2). The most encouraging results were observed for $Y_2O_3 \cdot Li_2O_3$; Li_2O_3 has virtually no influence, contrary to the isothermal oxidizing tests. The operative mechanism here is scale growth based on the inward diffusion of oxygen instead of outward diffusion in the alloy of nonaltered composition.

3.6 LIGHT ALLOYS

Titanium and aluminum alloys have attractively high ratios of strength to specific density and relatively good corrosion resistances. The heat-resistant aluminum alloys are applied in the temperature range 470 to 630 K. Their creep data are collected in Fig. 3.6.

Titanium alloys are developed either as a single α or β phase, or as a duplex α + β structure. Hexagonal-close-packed (hcp) α titanium has better creep resistance and inferior yield strength than those of the body-centered-cubic (bcc) β form. The α phase is stabilized by aluminum, whereas molybdenum and vanadium stabilize the β phase. These elements are the major ones; other additives as Sn, Zr, and Si. The highly alloyed α matrix [aluminum equivalent (Al + ⅓Sn + ⅙Zr + 4Si) % higher than 8%] forms a brittle intermetallic α_2 phase. Titanium alloys do not

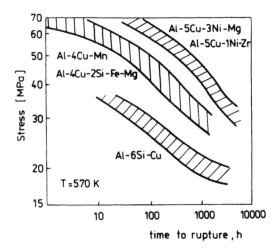

Figure 3.6 Creep rupture diagram of cast aluminum alloys, test temperature of 570 K. (From Weller, 1967.)

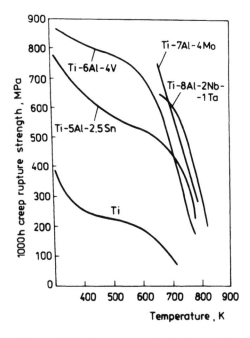

Figure 3.7 Temperature dependence of creep rupture strength for Ti-based alloys. (From Sedlacek, 1986.)

offer high creep strength and corrosion resistance above 870 K, the temperature dependence of the former being shown in Fig. 3.7. There are two intensively studied, promising approaches: the first using a rapid solidification technology that enables us to incorporate elements having negligible solubility in titanium, where the conventional ingot metallurgy fails (Whang, 1986); and the second making use of protective coatings (e.g., composite $TiAl_3 + TiN$ coatings), the first of which serves as an antioxidation film, with a second coating providing adequate strength.

3.7 FERROUS ALLOYS

Pure iron is not resistant to corrosion at temperatures above 840 K. At these temperatures, the wistite phase (FeO) becomes stable and the scale consists of FeO, Fe_3O_4, and Fe_2O_3. The latter is unstable above 1470 K, so the scale is again two-phase. The concentration of lattice defects is much higher in wistite than in other phases, so that its presence raises the corrosion rate. The addition of a few percent nickel increases the temperature of the wistite occurrence to some 1040 K; a similar chromium-induced effect is also known.

Chromium steel in which the chromium varies between 3 and 30% has increased the corrosion resistance relative to carbon steels. The minimum oxidation rate corresponds to a Cr content of 25 to 30%. At lower contents the long-term mass gain data point to structural transformation in the oxide layer with resultant scale perforation. The oxide layer is restored in places where it got detached when the chromium content was high enough to balance this process. The aluminum concentration in the scale is much higher than that indicated by the chemical formula of the alloy. Manganese undergoes preferential attack, thereby reducing the corrosion resistance of the alloy.

In general, scales adhere better to austenitic than to ferritic steels, due to nickel-enhancing scale adherence. This effect is rationalized in terms of internal oxidation. The working temperatures of carbon steels are below 720 K; of low- and medium-alloy steels are 820 to 850 K, and of high-chromium martensitic or bainitic, 870 K and exceptionally, 920 K.

Manganese intensifies the action of other elements, forming nitrides and enhancing the fine-grained structure. The strengthening effect of carbon and nitrogen dissolved in ferrite relies on the segregation to dislocations, on hindering dislocation climb and cross-slip by reducing the stacking fault, energy, and on the increase in friction forces.

The elements intensively increasing creep resistance are chromium, molybdenum, vanadium, niobium, and boron. The creep-resisting carbon steels are used in the normalized state and their structures consist of pearlite and ferrite. The alloy steels are applied in the heat-treated condition and the variety of microstructures can be here formed. Heat treatment should provide the optimum morphology of phases and partition of elements.

Relative to ferritic steels, the austenitic steels display higher corrosion resistance, higher thermal expansion, and lower heat conduction. The austenitic structure also provides a lower value for the stacking fault energy and lower values of diffusion coefficients. The elements present in steels can be divided into the following groups:

Those constituting solid solution: Fe, Ni, Cr
Those that are interstitial: C and N
Those that are soluble in austenite: Mo, W, V, Co
Those stabilizing carbon and nitrogen: Ti, Nb, and Zr
Those refining secondary phases and modifying the precipitation processes at grain boundaries: B and Zr
Those providing strengthening with intermetallic phases, mostly γ' : $Ni_3(Ti,Al)$

The chromium content is usually kept below that in acid-resistant steels, and nickel is usually present at a level of 13 to 16%. The purpose of it is to avoid ferrite in the initial structure, which transforms into $Cr_{23}C_6$ and σ phase above 870 K. The σ phase undoubtedly decreases plasticity; however, its effect on creep performance is moderate. The σ-phase precipitation is hindered by a high nickel content. The C and N contents are suited to meet the required plasticity requirements. The N content is kept below 0.15%. The vanadium-alloyed steels have working temperatures below 920 K. Boron causes carbide refinement and forces transcrystalline cracking. Boron atoms having higher radii than interstitials segregate to the disordered regions (i.e., grain boundaries making their glide more difficult), hindering diffusion processes and the formation of $M_{23}C_6$ and σ phase there. Cobalt increases the recrystallization temperature, reduces the stacking fault energy, and intensifies precipitation processes.

During aging of Cr-Ni steels the precipitating phases are $M_{23}C_6$, MC, M_6C carbides, intermetallic σ, χ, and Laves phases. The $M_{23}C_6$ carbides precipitate first to grain boundaries, then to twin boundaries, and finally in the grain interior. The plastic strain facilitates its nucleation but simultaneously reduces its rate of growth. The term σ *phase* includes intermetallic phases, either binary or multicomponent, formed by a combination of Nb, V, Cr, Mo, or V with Mn, Fe, or Co. The χ phase is the intermetallic compound whose formula can be identified with $M_{18}C$. The term *Laves phases* denotes a group of intermetallic compounds of formula AB_2. The main constituents are molybdenum and iron, and the minor constituents are chromium and tungsten. At higher carbon concentrations in austenite, precipitation of these phases is hindered.

In the high-nickel steels, the solubility of carbon is appreciably reduced, and moreover, contrary to ferritic steels, the carbide-forming action of Ti is limited. At high enough Ti + Al contents, the intermetallic $\gamma' = Ni_3(Al,Ti)$ phase is formed. For high Ti contents and high Ti/Al ratios, the room-temperature hardness assumes large values, such as that of the yield strength. Unfortunately, the creep performance is in such cases strongly affected by high-temperature exposure and decreases rapidly with increasing time.

Lai and Wickens (1979) reported long-term (up to 80,000 h) aging data for type 316 stainless steel and made an attempt to correlate the microstructures observed with the creep data. They observed that:

1. The first phase to appear in aged specimens was $M_{23}C_6$ carbide; later, σ phase was formed.
2. In specimens that experienced the highest temperatures and the longest exposures, the χ phase was also present.
3. The precipitation process was accelerated by the stress applied.

The corresponding failure modes were:

Wedge cracks in specimens least affected by the exposure

Voids, some formed at the grain boundary precipitates and some intragranularly, presumably at oxide or sulfide inclusions

In the most exposed specimens, cracking at the σ/austenite interfaces and transgranular cracking

3.8 SUPERALLOYS

The predecessor of the modern nickel-based superalloys was Monel K500, developed in the United States. Nickel is the major constituent of this family of alloys, having an advantageous A1 crystalline structure and a moderately high melting point of 1723 K. In contrast to ferrous alloys, nickel develops only one type of oxide above 670 K. The chromium content as an alloying element is usually maintained above 10% to provide adequate corrosion resistance. At high temperatures where chromium oxide volatilizes, the corrosion protection relies on aluminum, also usually added to nickel-based superalloys. Elements that constitute Ni-based superalloys are shown in Fig. 3.8, where the height of the blocks provides an idea of the amount of elements usually present. Elements added deliberately in small, controlled amounts are distinguished by crosshatching; deleterious elements are shown horizontally shadowed. The γ′ phase, on which the creep resistance is based, is especially important. The right γ′ phase is an intermetallic compound $Ni_3(Al,Ti)$. The derivative phases have the general formula A_3B, where the A location take metals of higher electronegativity (i.e., Ni, Co) and the B location, metals of lower electronegativity (i.e., Al, Ti, V, Nb, Ta); metals having an intermediate affinity (Fe, Cr, Mo, W) can take either A or B. The corrosion performance of various Nimonic alloys is compared in Table 3.3.

Today's applied superalloys differ appreciably from their prototypes and typically contain 10 to 18 alloying elements. These alloys can be classified on the basis of the major elements composing their matrix as follows: Fe-Ni-Cr (Table 3.4), Ni-Fe-Cr or Ni-Cr (Table 3.5), and Ni-Co-Cr and Co-Ni-Cr (Table 3.6). The effect of alloying elements onto strengthening mechanisms in superalloys is shown in Fig. 3.8. Their temperature dependence is depicted in Fig. 3.9.

Iron can partly replace nickel in the γ matrix, which has the advantage of low cost. At a higher content, however, it reduces the adherence of oxides to the alloy and increases liability to the unwanted, brittle intermetallic σ-phase formation. The Co-additive stabilizes the gamma solid solution and increases the solubility of other elements, mostly Al and Ti; it also reduces the stacking fault energy. Chromium forms the stable $M_{23}C_6$ carbides providing creep resistance. This carbide precipitates mostly to grain boundaries and structural defects. Boron increases creep resistance, affecting grain boundary morphology similarly to Zr and Hf. The rare earth elements increase the resistance to hot corrosion. Yttrium is similarly added, to improve the scale adherence. Herchenroeder et al. (1983) gave an example of the Ni-based Cabot M214 with 0.01% yttrium, which when present at higher contents caused incipient melting at grain boundaries. Molybdenum and tungsten strengthen the γ matrix and reduce the stacking fault energy in the cobalt-based superalloys.

It is thought that undesirable impurities reduce the value of the surface energy, thereby reducing the critical size of the cavity over which it is stable. This increases the degree of cavitation and induces it at a comparatively earlier stage of creep life. Nowadays, all the high-strength materials are produced by vacuum melting and casting, raw materials being carefully selected. The small amount of lead, 2 ppm, which is a common value, reduces the creep life of IN-939 by a factor of 2, and bismuth, which is even more surface active, by a factor of 5, when present in the same amounts (Osgerby and Gibbons, 1986).

The cobalt-based alloys offer higher (relative to nickel-based alloys) resistance to corrosion in the combustion gases containing aggressive ashes and sulfur

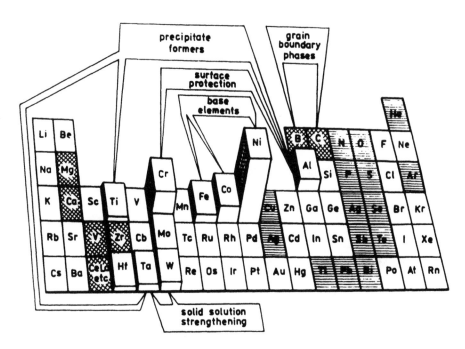

Figure 3.8 Elements that may be present in Ni-based superalloys and their effects. (From Meetham, 1982, after Gibbons, 1985.)

Table 3.3 Mass Loss of Nimonic Alloys After 100 h (mg/cm^2)

Alloy composition	Temperature (K)				
	1070	1170	1230	1370	1470
Nimonic 75					
0.08–0.15% C, 0.2–0.6% Ti, 18–21% Cr, 1.0% Si, 1.0% Mn, 5% Fe	0.55	1.18	4.00	6.66	8.92
Nimonic 80A					
0.1% C, 1.8–2.7% Ti, 18–21% Cr, 0.5–1.8% Al, 1.0% Si, 1.0% Mn, 5.0% Fe	0.64	2.62	3.96	5.96	—
Nimonic 90					
0.1% C, 1.8–3.0% Ti, 18–21% Cr, 0.8–2.0% Al, 1.5% Si, 1.0% Mn, 5.0% Fe, 15–21% Co	0.46	2.52	5.50	10.40	11.23
Nimonic 95					
0.15% C, 2.3–3.5% Ti, 18–21% Cr, 1.4–2.5% Al, 1.0% Si, 1.0% Mn, 5.0% Fe, 15–21% Co	0.50	2.95	5.03	7.38	12.41
Nimonic 100					
0.3% C, 1.0–2.0% Ti, 10–12% Cr, 4.0–6.0% Al, 0.5% Si, 2.0% Fe, 18–22% Co, 4.5–5.5% Mo	0.08	0.27	0.87	1.61	11.61

Source: Wood and Hodgkiess (1966).

Table 3.4 Chemical Composition of Fe-Based Alloys

Alloy	Composition (%)								
DIN	ASTM	C	Cr	Ni	Co	Mo	W	Nb	Other
X1O-Ni-Cr-Mo25 16	Timken 16-25-6	0.12	16	25		6			0.15 N
X5Ni-Cr-Ti26 15	A-286	0.06	15	25.5		1.3			2.1 Ti
X12Cr-Co-Ni21 20	N-155 Multimet	0.12	21	20	20	3	2.5	1	0.15 N
X40Co-Cr-Ni20 20	S-590	0.40	20	20	20	4	4	4	

Source: *Metal Progress*, Mid-June 1977, Data Book; Dienst (1978).

Table 3.5 Chemical Compositions of Iron/Nickel-Based Alloys

Alloy		Composition (Wt %)									
DIN	ASTM	C	Ni	Cr	Co	Fe	Al	Ti	Mo	Nb	Other
Ni-Cr-15Fe	Inconel 600	0.1	75	15		8					
Ni-Cr-20Ti	Nimonic 75	0.12	75	20		<5		0.4			
Ni-Cr-22Fe-Mo	Hastelloy X	0.1	47	22	1.5	18.5			9		0.6 W
Ni-Cr-22Mo-9Nb	Inconel 625	0.06	60	22		<5	0.2		9	4	
Ni-Cr-15Fe-Ti	Inconel X-750	0.06	73	15		7	0.7	2.5		1	
Ni-Cr-20Ti-Al	Nimonic 80 A	0.07	72	20		<5	1.4	2.4			
Ni-Cr-20Co-18Ti	Nimonic 90	0.07	54	20	18	<5	1.4	2.4			
Ni-Fe-Cr-16Mo-Al-Ti	Nimonic PE 16	0.06	44	16		33	1.2	1.2	3.2		
Ni-Fe-Cr-12Mo-Ti	Incoloy 901	0.05	42.5	12.5		35	0.2	2.8	6		0.015 B
Ni-Cr-19Fe-Nb-Mo	Inconel 718	0.06	53	19		18	0.6	0.9	3	5.3	
Ni-Cr-20Co-Mo-Ti	Waspaloy	0.06	55	20	14	<2	1.3	3	4.3		0.1 Zr, 0.007B
Ni-Cr-19Mo-Ti	René 41	0.1	52	19	11	<5	1.5	3.2	10		0.007 B
Ni-Cr-18Co-Mo-Al-Ti	Udimet 500	0.1	52	18	17	<4	2.9	2.9	4		
Ni-Co-20Cr-15Mo-Al-Ti	Nimonic 105	0.1	53	15	20	<1	4.6	1.2	5		
Ni-Cr-15Co-Al-Ti-Mo	Nimonic 115	0.15	56	15	15	<1	5	4	3.7		
Ni-Co-17Cr-15Mo-Al-Ti	Udimet 700	0.1	52	15	17	<4	4.2	3.3	5		0.03 B
G-Ni-Cr-13Mo-Al	Inconel 713 C	0.15	71	13	1	<2	6.1	0.8	4.5	2.3	0.1 Zr, 0.01 B
G-Ni-Co-15Cr-10Mo-Al-Ti	IN-100	0.18	60	10	15	<1	5.5	5	3		1.0V, 0.06 Zr, 0.015 B
G-Ni-W-12Co-Cr-9Al	MAR-M 200	0.15	59	9	10	<1	5	2		1	12.5 W, 0.06 Zr, 0.015 B
G-Ni-Co-10Cr-8Mo-Al	B-1900	0.1	63	8	10	<1	6	1	6		4 Ta, 0.10 Zr, 0.015 B

Source: Metal Progress, Mid-June 1977, Data Book; Dienst (1978).

Table 3.6 Chemical Compositions of Co-Based Alloys

Alloy		Composition (wt %)						
DIN	ASTM	C	Co	Ni	Cr	Fe	W	Other
Co-Cr-20W-15Ni	L-605, HS-25	0.1	50	10	20	3	15	
Co-Cr-20Ni-20W	S-816	0.4	43	20	20	5	4	4 Mo, 4 Nb
G-Co-Cr-27Mo	Vitallium HS-21	0.25	62	2.5	27	2		5.5 Mo
G-Co-Cr-25Ni-10W	X-40, HS-31	0.5	55	10	25	2	7.5	
G-Co-Cr-20W-128	Haynes 151	0.5	62	1	20		12.8	0.18 Ti, 0.05 B
G-Co-Cr-22W-10Ta	MAR-M302	0.85	55	1.5	22	1	10	9.0 Ta, 0.25 Zr, 0.01 B
G-Co-Cr-22Ni-10W	MAR-M509	0.6	55	10	22		7	3.5 Ta, 0.5 Zr, 0.2 Ti

Source: Metal Progress, Mid-June 1977, Data Book; Dienst (1978).

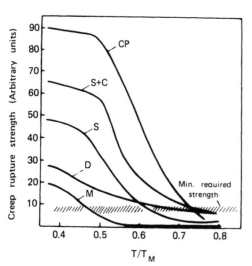

Figure 3.9 Effect of temperature on strengthening mechanisms in superalloys. M, base metal strength; D, dispersion strengthening; S, solid solution strengthening; S + C, solid solution and carbide precipitates; CP, coherent phase. (From Dienst, 1978.)

Table 3.7 Phases Occurring in Multiphase Ni-Based Superalloys

Phase	Structure and composition
γ'	Cubic face centered, Cu_3Au type $(Ni,Co,Fe,Cr,...)_3$ (Al,Ti)
γ'' or $\gamma*$	Tetragonal body-centered Ni_3Nb
η	Hexagonal close-packed Ni_3Ti
δ	Orthorombic Ni_3Nb
σ	Tetragonal body centered A_xB_y type $(Cr,Mo)(Fe,Ni)$
Laves phase	Hexagonal closepacked, A_2B $(Fe,Cr,Mn,Si)_2$ (Mo,Ti,Nb)
μ	Rhombohedral, A_6B_7 type, $(Mo,W)_6$ $(Ni,Fe,Co)_7$
G	Cubic face centered, $A_6B_{23}Hf_6N_8Al_{15}$ type
MC	Cubic face centered, $(Ti,Mo,Nb,Ta,W)C$
Carbides	
M_7C_3	Trigonal, Cr_7C_3
$M_{23}C_6$	Cubic face centered, $(Cr,Mo,Co,W,Nb)_{23}C_6$
M_3C	$(Ni,Co)_4(Mo,W)_2C$
Borides	
M_3B_2	Tetragonal, $(Mo,Ti,Cr,Ni)_3B_2$
$M_{23}(C,B)_6$	Cubic face centered
Nitrides and	MX cubic face centered
carbonitrides	$TiN(C,N), M_{23}(C,N)_6$
Other phases	$Ni_x(Mo,Cr)_y(C,Si)$, $Ti_nC_2S_2$, ZrS_x

compounds. The binary Co-Cr alloys display a minimum corrosion rate for Cr contents of approximately 30%. Unfortunately, internal oxidation is practically missing, which causes oxide spallation and, in effect, virtually no resistance to thermal cycling. Aluminum added as the third constituent does not improve this behavior. The commercial MAR-M-509 alloy incorporates W and Ta, known as spinel formers, in its chemical composition. The scale consists of the outermost layer of CoO, the intermediate chromium spinel $CoCr_2O_4$ layer, and the innermost, dense layer of Cr_2O_3. Silicon intensifies Laves's phase precipitation, thereby reducing plasticity, and additionally reduces the Mo and W contents in the matrix, lessening its degree of strengthening.

The alloys we are talking of are multiphase alloys. Table 3.7 juxtaposes phases occurring in Ni-based superalloys. The fraction of precipitated γ' phase, its morphology, and the properties of the alloy are a function of chemical composition, mostly Al, Ti, Nb, and Ta (see Fig. 3.10). In the multicomponent alloys the chemical composition is selected so that the γ' and γ matrix parameters were very close to

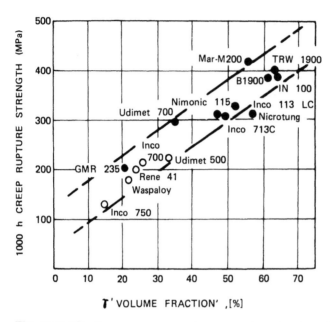

Figure 3.10 Creep rupture strength versus volume fraction of γ' phase. (From Taira and Ohtani, 1986.)

each other. Moreover, the index of γ'-phase stability is the Ti/Al or (Ti + Nb)/Al ratio. If the former exceeds 3 and the latter 5, a brittle phase incoherent with the matrix is formed. The precipitates of this phase are usually in the form of large lamellas harmful to alloy strength. In alloys containing above 4 wt % Nb or 10 wt % Ta, the metastable Ni_3Nb phase appears or the Ni_3Ta phase coherent with the γ matrix, designated by γ'' or γ^*. This phase forms fine lamellas and has a strengthening effect.

In cobalt-based alloys, γ' strengthening is unstable. However, it is hoped that properly selecting γ' elements and matrix elements will enable us to attain a high-volume fraction of γ' and to avoid transformation into the η phase above 1170 K.

The primary MC carbides are usually distributed in an uneven fashion in the gamma matrix. The stability of these carbides increases in the following order: VC, NbC, TaC, TiC. Chromium, molybdenum, and tungsten, especially the latter two, when dissolved in the MC carbide, weaken its interatomic forces, which leads to the transformation of MC carbide into $M_{23}C_6$ and M_6C.

The M_7C_6 carbide is unstable and transforms above 1370 K into $M_{23}C_6$. This carbide is found in low-Cr alloys free of strong carbide formes. The $M_{23}C_6$ carbide is formed in alloys of high Cr content. In alloys containing Mo and W in addition to Cr, the exact formula can be $Cr_{21}(Mo,W)_2C_6$; moreover, it can also incorporate Ni, Co, and Fe.

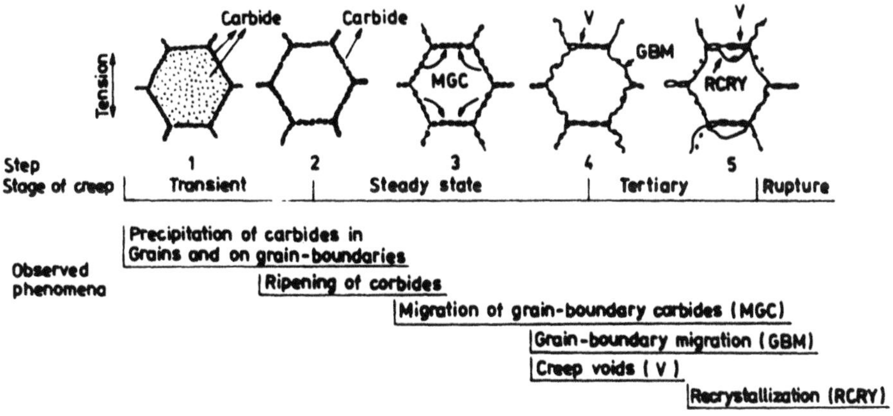

Figure 3.11 Schematic representation of microstructural changes observed in solution-treated Inconel 617 during creep at 1270 K. (From Kihara et al., 1980.)

After long exposures to high temperature, the following facts were established (see also Fig. 3.11):

1. In Ni-based superalloys with relatively large amounts of Cr, Ti, Al, and Ta, $M_{23}C_6$ carbides are favorably formed.
2. Large concentrations of Mo and W facilitated the occurrence of M_6C carbide.
3. Large concentrations of Nb and Ta favored MC carbides.
4. Formation of $M_{23}C_6$ carbides, thus depleting zones adjoining them in chromium and hence σ phase precipitating there.

The σ phases have the formula $(Cr,Mo)_x(Ni,Co)_y$, where x and y are between 1 and 7. The μ phase is its variation; the atomic radii of constituents are dissimilar.

Figure 3.12 shows a tendency toward alloy development. The single-crystal superalloys exhibit large directional coarsening in high-temperature exposure. The γ' particles rapidly coalesce and form continuous lamellae perpendicular to the stress applied. The initial structure is important. In MAR-M200 superalloy the profitable structure contains strongly aligned cuboidal γ' particles, which form after coarsening almost perfect lamellar $\gamma - \gamma'$ array. The straining mechanism here is shearing the $\gamma - \gamma'$ interface, since the continuous structure prevented bypassing mechanisms from occurring (Nathal, 1987).

3.9 REFRACTORY METALS

The discussion is here confined to Nb, Mo, Ta, and W. Among strengthening mechanisms the use is practically made of two: precipitation and dispersion strengthening. The optimum content of oxides is within 0.5 to 2 mol %. The most stable carbide is HfC, due to its low diffusivity. Carbide strengthening can increase

Table 3.8 Tensile Strength and Creep Rupture Strength of Refractory Alloys

Base metal	Alloy	Tensile strength (MPa)				Creep rupture strength (MPa)	
		RT	1270 K	1470 K	1670 K	100 h 1270 K	1000 h 1470 K
Nb	Nb	300	120	50	30	70/50	25/–
	Nb-1Zr	400	270	170	50	130/–	35/–
	Cb-752	600	380	250	110		90/25
	D-43	600				180/160	100/60
	FS-85	700	450	300	130	–/150	110/60
	SU-16	700				–/190	130/80
	F-48	800	460	300	130	350/–	130/80
	B-88	1000					350/–
Mo	Mo(0.02C)	700	270	150	100	140/–	70/–
	Mo-0.5Ti	750	460	280	130	330/–	120/–
	TZM	900	600	500	250	470/400	230/100
	Nb-TZM					550/480	350/250
	TZC	800	500	400	310	–/420	270/210
Ta	Ta	400	200	100	70	50/–	25/–
	T-111	800	530	360	250	–/200	–/90
W	W	800	400	320	250	230/–	130/–
	W-2ThO$_2$			340	250		

Source: Dienst (1978).

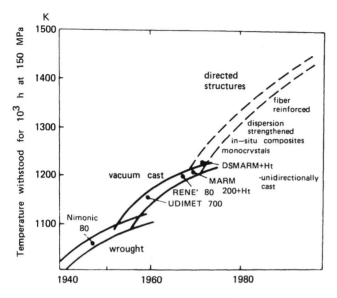

Figure 3.12 Modern creep-resisting materials. (From McLean, 1983.)

creep life by even 10 times at $0.5T_m$. This index rapidly falls down at higher temperatures. Chemical compositions of refractory alloys are presented in Table 3.8, together with the corresponding creep data.

3.9.1 Tantalum and Niobium Alloys

The scale formed on tantalum consists of tantalum pentaoxide. This metal becomes prone to oxidation above 1520 K, where the corrosion rate becomes so high that it is not measurable using conventional methods. The addition of lanthanides and noble metals enhances the corrosion resistance. Niobium forms a number of thermodynamically stable oxides, among which the major are NbO, NbO_2, and Nb_2O_5. Similarly to tantalum, the corrosion rate becomes very high above 1470 K. Elements enhancing corrosion performance here are Cr, Fe, Ti, and Zr. This improvement is produced by the formation of binary oxides and by increasing the plasticity of corrosion products.

Tantalum and its alloys have the disadvantage of high cost and high specific density. Their virtue is good formability at room temperature and good weldability. The demand of room-temperature ductility imposes a limit on the total content of alloying additives, being at most 10 at %. The major elements are tungsten, added to strengthen the solid solution, and hafnium, used to bind the carbon, nitrogen, and oxygen; the latter also serves to produce almost-equal precipitation strengthening and carbide precipitation. For that reason, both methods are often used simultaneously.

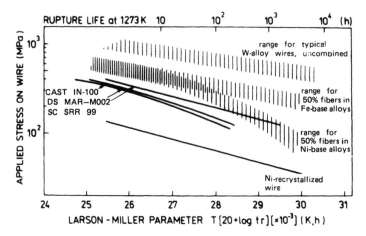

Figure 3.13 Larson-Miller plot of creep behavior of tungsten-reinforced superalloys, t_R is time to rupture or 2% secondary creep strain. (From Warren, 1987.)

3.9.2 Tungsten and Molybdenum

The problem encountered here is the formation of volatile oxides—WO_3 and MoO_3—which have melting points of 1740 and 1070 K, respectively. In humid environments, the situation is even more drastic; the complex compounds formed under such conditions are more volatile. Elements added to increase corrosion resistance are chromium in tungsten (the content necessary to maintain a continuous Cr_2O_3 layer is about 50%) and Ti, Tb, and Ta in molybdenum.

Widely used tungsten alloys are oxide-dispersion strengthened, produced by powder metallurgy methods. Tungsten-fiber-reinforced superalloys are thought to permit increases in blade temperatures of up to 150 K over current values. In practice, their application is limited by a recrystallization of the fiber activated by the matrix atoms and a dissolution of fiber metal (although attempts to apply barrier coatings to it are successful). Their tensile creep behavior is summarized in Fig. 3.13.

Molybdenum alloys are usually strengthened by titanium and zirconium carbides. Tungsten solid-solution-strengthened alloys are designed for use at extremely high temperatures.

REFERENCES

Baird, J. D., Jamieson, A., Preston, R and Cochrane, R. C. (1972). *Proc. Conference on Creep Strength in Steel and High-Temperature Alloys*, Sheffield, p. 207.

Dienst, W. (1978). *Hoch-Temperatur Werkstoffe*, Werkstofftechnologisch. Verlag, Karlsruhe, West Germany.

Funkenbusch, A. W., Smeggil, J. G., and Bornstein, N. S. (1985). *Metall. Trans. A, 16A*: 1165.

Gibbons, T. B. (1985). *Mater. Sci. Technol., 1*: 1033.

Guttmann, V. and Bürgel, R. (1983). The effect of a carburizing environment on the creep behaviour of some austenitic steels, in *Corrosion Resistant Materials for Coal Conversion Systems*, Applied Science Publishers, Barking, Essex, England, p. 429.

Harrison, J. M., Norton, J. F., Derricott, R. T. and Marriott, J. B. (1979). *Werkst. Korros., 30*:785.

Herchenroeder, B. B., Lai, G. Y. and Rao, K. V. (1983). *J. Met.*, (Nov.): 16.

Hernas, A. and Maciejny, A. (1989). *Heat-Resisting Alloys*, Wydawnictwo PAN, Wroclaw, Poland.

Hornbogen, E. (1973). Mechanisms to control creep resistance in high-temperature, *Proc. Symposium on High-Temperature Materials*, Baden, Elsevier, Amsterdam, p. 194.

Kihara, S., Newkirk, J. B., Ohtomo, A. and Saiga, V. (1980). *Metall. Trans. A, 11A*: 1030.

Lai, J. K. and Wickens, A. (1979). *Acta Metall.*, 27:217.

McLean, M. (1983). *Directionally Solidified Materials for High Temperature Service*, The Metal Society, London, p. 296.

Meetham, G. W. (1982). *Metall. Mater. Technol., 14*: 387.

Mrowec, S. and Werber, T. (1982). *Modern Heat-Resisting Materials*, WNT, Warsaw, p. 363.

Nathal, M. V. (1987). *Metall. Trans. A, 18A*: 1961.

Osgerby, S. and Gibbons, T. B. (1986). *Proc. Conference on High Temperature Alloys for Gas Turbines and Their Applications*, Liege.

Pilling, N. and Bedworth, R. (1923). *J. Inst. Met., 29*: 529.

Sedlacek, V. (1986). *Non-ferrous Metals and Alloys*, Materials Science Monograph 30, Elsevier, Amsterdam, p. 251.

Sherby, O. D. and Weertman, J. (1979). *Acta Metall.*, 27: 387.

Solberg, J. K. and Thon, H. (1984). *J. Mater. Sci., 19*: 3908.

Stringer, J. (1983). The role of underlying research in corrosion in coal conversion systems, in *Corrosion Resistant Materials for Coal Conversion Systems* (Meadowcroft, D. B. and Manning, M. I., eds.), Applied Science Publishers, Barking, Essex, England, p. 407.

Taira, S. and Ohtani, R. (1986). *Tieorija Wysokotiempieraturnoj prochnosti matierialow*, Moscow (translated from Japanese).

Warren, R. (1987). Fibre reinforced ceramic and metal matrix composites, their potential as high temperature materials, in *17th International Congress on Combustion Engine*, Paper T-29.

Weller, J. (1967). *JFL Mitteilungen, 6*:4.

Whang, S. H. (1986). *J. Mater. Sci., 21*: 224.

Wood G. and Hodgkiess, T. (1966). *J. Electrochem. Soc., 113*: 319.

Yuen, J. L., Roy, P. and Nix, W. D. (1984). *Metall. Trans. A, 15*: 1769.

Thermal Fatigue

4.1 THERMAL STRESSES AND STRAINS

Thermal stresses arising from thermal transients are commonly divided into:

1. *Strain-controlled stresses*, where the local expansion or contraction of material is hindered by the surrounding material;
2. *Displacement-controlled stresses*, where the overall expansion or contraction of the component is limited by another component.

Displacement-controlled stresses can be calculated only in a limited number of cases. Figure 4.1 shows a number of examples illustrating the former type of stresses. Let us assume that the elastic plate has originally uniform temperature T_0 quenched from one face to the T_f temperature. The relevant temperature distribution and internal strain σ_i are shown in Fig. 4.1. When one direction is restrained against expansion, a σ_m compressive membrane stress must be applied:

$$\sigma_m = -\frac{1}{L} \int_{-L/2}^{L/2} E\alpha(T - T_0) \, dx' = E\alpha\Theta_m(T_0 - T_f)$$

where

L = plate thickness

E = Young's modulus

α = coefficient of linear expansion

Θ_m = nondimensional mean temperature

$x' = x - L/2$, x being measured from the face

T = temperature at the x position

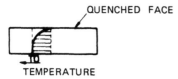

TEMPERATURE

(a) TEMPERATURE DISTRIBUTION

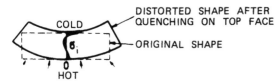

(b) STRESSES – NO RESTRAINT

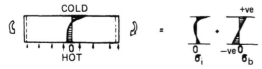

(c) STRESSES – BENDING RESTRAINT

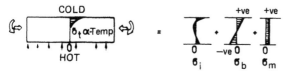

(d) STRESSES – TOTAL RESTRAINT

Figure 4.1 Stress distribution in a plate quenched on one face under various degrees of restraint. (From Clayton, 1983.)

If rotation of any plate section is prevented, then a bending stress of $2\sigma_b x'/L$ must be applied, where

$$\sigma_b = -\frac{6}{L^2} \int_{-L/2}^{+L/2} E\alpha(T - T_0)x' \, dx'$$

If both extension and rotation are prevented, the σ_t stress that must be applied is equal to

$$\sigma_t = -E\alpha(T - T_0) = E\alpha\Theta(T_0 - T_f)$$

where there is restraint in two directions, σ_m, σ_b, and σ_t are multiplied by $1/(1-v)$, v being Poisson's ratio and $\sigma_t = \sigma_i + \sigma_b + \sigma_m$. In a fully restrained stainless steel element, yielding is produced by a temperature change of 40 K; in structural steels it is about 100 K. For this reason the plastic components of strain are usually present. Considering a thin cylinder subject to cyclic external quenching when the intermittent shock quenching is followed by slow rewarming, the distributions of stresses and strain are shown in Fig. 4.2.

In ASME Code Case N47 multiaxial stress–strain conditions are assessed by use of the equivalent strain to relate the behavior of material to uniaxial tests. This equivalent stress, $\bar{\sigma}$, for principal orthogonal stresses $\sigma_1, \sigma_2, \sigma_3$ is defined as follows:

$$\bar{\sigma} = \frac{\sqrt{(\sigma_1 - \sigma_2)^2 + (\sigma_2 - \sigma_3)^2 + (\sigma_3 - \sigma_1)^2}}{\sqrt{2}}$$

Yielding of material occurs when $\bar{\sigma} = \sigma_y$, σ_y being the uniaxial yield stress.

The elastic equivalent strain is given by

$$\bar{\varepsilon}_e = \frac{1}{2(1+v)} \sqrt{(\varepsilon_{e1} - \varepsilon_{e2})^2 + (\varepsilon_{e2} - \varepsilon_{e3})^2 + (\varepsilon_{e3} - \varepsilon_{e1})^2}$$

where ε_{e1}, ε_{e2}, and ε_{e3} are the elastic components of principal strain. The plastic equivalent strain $\bar{\varepsilon}_p$ is given by the assumption of constant volume ($v = 0.5$):

$$\bar{\varepsilon}_p = \frac{\sqrt{2}}{3} \sqrt{(\varepsilon_{p1} - \varepsilon_{p2})^2 + (\varepsilon_{p2} - \varepsilon_{p3})^2 + (\varepsilon_{p3} - \varepsilon_{p1})^2}$$

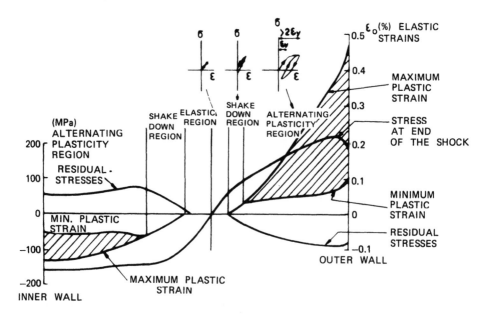

Figure 4.2 Stress distribution and plastic strain distribution in an uncracked tube wall during repeated thermal shocks. (From Clayton, 1983.)

where $\varepsilon_{p1}, \varepsilon_{p2}$, and ε_{p3} are the plastic components of principal strain. The total equivalent strain $\bar{\varepsilon}$ equals $\bar{\varepsilon}_e + \bar{\varepsilon}_p$; ASME Code Case N47 gives a plastic relationship for total strain, as it is not always simple to differentiate elastic and plastic components:

$$\bar{\varepsilon} = \frac{\sqrt{2}}{3} \sqrt{(\varepsilon_1 - \varepsilon_2)^2 + (\varepsilon_2 - \varepsilon_3)^2 + (\varepsilon_3 - \varepsilon_1)^2}$$

where ε_1, ε_2, and ε_3 are the total principal strains. Kostiuk et al. (1974) found that experimental lives in thermal fatigue testing were predicted with fairly good accuracy by the strain-life relationship based on Manson–Coffin relation, in which the total equivalent strain amplitude was used.

It is convenient to examine the effect of constraint using a two-bar structure (see Figs. 4.3 and 4.4). The geometry of this model dictates the degree of constraint of the thermally fatigued element. The entire two-bar structure may undergo the following modes of behavior, depending on the degree of constraint and range of temperatures and stresses in the cycle: (1) elastic, (2) elastic shakedown after first half-cycle, (3) thermal stress ratcheting, (4) reversed plasticity, and (5) simultaneous thermal ratcheting and reversed plasticity. Thermal stress ratcheting can cause incremental collapse of the structure; reversed plasticity may initiate fatigue cracks and develop them. The combination of both of the above is more harmful to material than either of them in isolation. The parallel-bar model is shown in Fig. 4.3. The net force applied to the assembly is zero. Element 1 is heated from T_1 to

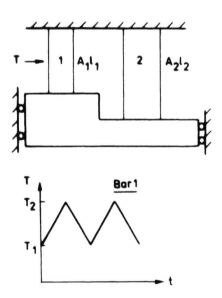

Figure 4.3   Parallel bar model. (From Sehitoglu, 1985.)

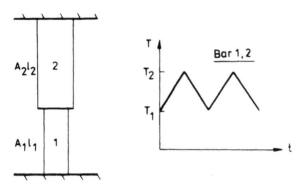

Figure 4.4   Series bar model. (From Sehitoglu, 1985.)

T_2, while the temperature of element 2 is considered constant. Equilibrium and compatibility equations give for bar 1,

$$\varepsilon_e + \varepsilon_p + \varepsilon_c + \varepsilon_{th} = \frac{-S/E_2}{C} = \varepsilon_m + \varepsilon_{th}$$

where the subscripts e, p, c, th, and m denote elastic, plastic, creep, thermal, and mechanical components of strain, $C = A_2 l_1 / A_1 l_2$ is a factor determining the level of constraint (A_1, A_2 and l_1, l_2 denote cross-sectional area and length, respectively).

A more severe form of constraint develops in the second, series bar model (Fig. 4.4). Both bars are heated and cooled, the equilibrium and compatibility equations yield

$$\varepsilon_m + \varepsilon_{th} = -\frac{S/E_2}{C} - \varepsilon_{th}\frac{l_2}{l_1}$$

Bar 2 is assumed to behave elastically and it imposes additional strains on bar 1. This case is termed *overconstraint*, as an additional term exists in the strain equation.

If the element includes a defect or a notch, the material in its vicinity experiences an additional constraint. Neuber's rule is applicable here. The effect of constraint was experimentally verified by Sehitoglu (1985) on 1070 steel. Specimens, either smooth or including a circular notch ($K_t = 2$), were exposed to thermal cycles, in which the minimum temperature of 423 K was held constant throughout the experiment and the maximum temperatures varied from 620 to 970 K. Partial constraint means $C = 1$, total constraint means $C \rightarrow \infty$, and overconstraint means

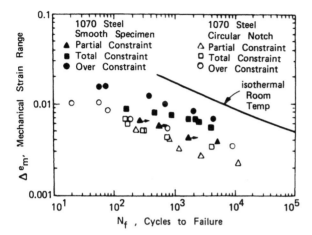

Figure 4.5 Comparison of experimental lives. (From Sehitoglu, 1985.)

$C \to \infty$ and $l_2/l_1 = 1$. Note from Fig. 4.5 that the higher the constraint, the shorter the life.

Finally, let us consider a component cyclically heated and cooled by a stream of flowing gas. The difference between the surface and bulk temperatures is a measure of the thermal gradient experienced by the part of a component near its surface. Assuming the component to be in the form of a turbine blade, the value of this coefficient is theoretically proportional to the reciprocal of the edge thickness. Blunting the leading edge will reduce the magnitude of the thermal gradients. A decrease in the thermal conductivity k of the blade material results in higher thermal gradients. Thermal strains are a direct function of the coefficient of thermal expansion. The factors governing the magnitude of thermal strains are therefore (1) thermal factors: heat transfer coefficient h, thermal conductivity k, thermal expansion coefficient α, and cycle temperature difference $T_2 - T_1$; (2) size and shape of specimen; and (3) nature of cycle. Our discussion on factors affecting component life is based on this juxtaposition.

4.2 EFFECT OF COMPOSITION AND STRUCTURE

The usefulness of comparisons made between various alloys is limited, as the relative merits of materials often vary with test conditions. Nevertheless, such data are often reported; for example, Muscatell et al. (1957) found cobalt-based alloys (both wrought and cast) to be superior to nickel-based alloys. Hunter (1956), in contrast, did not observe any consistent superiority or inferiority. Riddihough

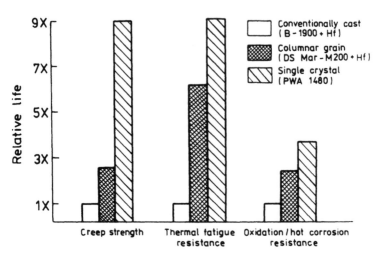

Figure 4.6 Typical performance curves for columnar grain and single-crystal alloys compared to conventionally cast alloys. (From Hoffelner, 1987.)

(1960) showed that cast cobalt-based alloys were inferior to nickel-based alloys in test with $T_{max} = 923$ K, whereas above 973 to 823 K the opposite ranking was true. Figure 4.6, a bar diagram showing a comparison of performances of gas turbine materials, introduces some quantification of this discussion. Similarly, nodular irons were more durable than gray irons in tests.

The effect of elements incorporated into the chemistry of an alloy should be considered in the following aspects:

Material strengthening through, for example, carbide formation, solid solution strengthening with consequent increase in the capacity for withstanding repeated strains and with improved creep performance

Microstructural stability and oxidation resistance

Thermal properties of material: coefficient of thermal expansion (preferred low value), specific heat, and heat conductivity (profitable high values)

Ferritic steels have higher values of thermal conductivities and lower thermal expansion coefficients than austenitic steels. At higher temperatures, say at 1073 K, their thermal conductivities become similar. Successful use of a molybdenum insert in pressure die-casting dies is due partly to their high values of thermal conductivity and therefore lower thermal gradients occurring in service. Bengtsson (1957) stated in his study on die steels that the decrease in thermal conductivity with alloying additions was least for molybdenum- and tungsten-containing alloys. Moreover, in ferrous alloys, increasing silicon, vanadium, manganese, and/or chromium contents is considered profitable. Carbon at high content reduces the life. Sulfur and phosphorus, as well as tin, lead, and antimony, have a negative effect. Zmihorski et al. (1975) studied the group of steels in thermal fatigue (see Table 4.1). Findings are depicted in Fig. 4.7. It is seen that rankings of alloys do not bear a close resemblance to the heat conductivity plot (compare with Fig. 4.8). Among metals only refractory metals have attractively low thermal expansion coefficients. The low-expansion materials have either very specific structure (e.g., uranium, with its anisotropy of thermal expansion coefficient, which causes a gradual increase in the length of a uranium bar with repeated thermal cycles) or magnetic transformation-reducing thermal elongation as it does in the Invar alloy. Unfortunately, both types of materials are not of practical importance under industrial conditions: first because of cost, and second, in a range of working temperatures.

Many structural ceramic materials are susceptible to failure when thermally shocked due to a high Young's modulus and relatively high thermal expansion coefficient, low tensile strength, and thermal conductivity. Zirconium oxide, which is frequently used, exists in three crystalline forms, depending on temperature and pressure. At ambient conditions it has a monoclinic structure and transforms at higher temperatures and/or pressures into a tetragonal phase with an accompanying 3.25% volume change. Such a volume change readily causes a brittle fracture of material. The addition of a stabilizing agent to zirconia cures it of this defect. The commercial grades known as Zircoa 2032, 1027, and 2016 contain 30%, 40%, and 50 to 60% monoclinic phase, respectively. In the first two materials, nearly all of the monoclinic phase is in intragranular precipitates. In the third material, which is superior to the formers, the monoclinic phase is evenly distributed between inter- and intragranular precipitates. The properties of these partially stabilized materials are listed in Table 4.2 and the results of field trials are given in Table 4.3. As seen, Zircoa alloys are consistently better than die steels.

Information on thermal-chock resistance is obtained in a highly popular test based on heating appropriate specimens in the oven at high temperature and rapid immersion into a cooling medium. The critical quenching temperature difference is the quantitative measure of the thermal-stress resistance of a tested material. The maximum thermal stresses develop soon after the specimen is inserted into the medium. This time-to-maximum stress is usually on the order of 0.01 s and can also be called the time-to-fracture, as the tested materials are usually brittle. As far as mechanical properties are concerned, the test conditions are decisive, but the ratio of tensile strength to yield point can be adopted as an approximate index of the potent material capability.

In general, material with a low Young's modulus and a high yield stress is desirable, as the elastic component of the strain is high and the plastic component is low during the thermal cycle. High strength and high ductility are profitable, but high strength is invariably associated with the lower value of the latter. Under given thermal fatigue conditions there is an optimum combination of strength and ductility producing the maximum endurance, particularly when time-induced metallurgical changes occur. In the situation where vast temperature variations are expected, strength is considered of prime importance (Forrest, 1960–61). Franklin and co-workers (1964) found for several wrought nickel-based alloys subjected to various heat treatments: an approximately linear relationship on a double logarithmic plot

Table 4.1 Chemical Compositions of Steels Tested

Steel	Chemical composition (%)								
	C	Mn	Si	Cr	Ni	W	Mo	V	Co
WCL	0.42	0.53	1.03	4.90	0.13	—	1.29	0.47	—
WLK	0.40	0.30	0.49	2.51	—	—	2.74	0.50	3.30
WWS	0.32	0.29	0.90	1.28	0.17	4.60	—	0.35	—
WWN1	0.30	0.38	0.22	2.62	1.30	7.80	—	0.39	—
WCMB	0.36	1.3	0.34	2.22	0.23	—	0.40	—	—
WNL	0.60	0.69	0.35	0.72	1.55	—	0.25	—	—
3H13	0.34	0.5	0.3	13	0.3	—	—	—	—
65	0.63	0.80	0.32	—	0.11	—	—	—	B 0.003

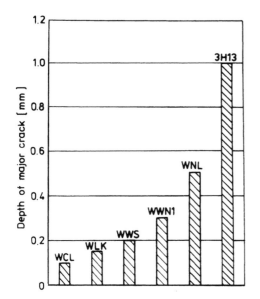

Figure 4.7   Resistance to thermal fatigue of steels hardened to 44HRC.

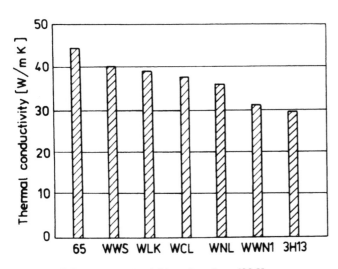

Figure 4.8   Heat conductivities of steels at 620 K.

Table 4.2 Properties of Partially Stabilized Zirconia

	Composition		
	1027	2016	2032
Flexural strength (MPa)	275	170	275
Young's modulus at RT (GPa)	200	170	205
Poisson's ratio	0.32	0.28	0.31
Average thermal expansion coefficient from 300 to 1270 K (10^{-7} K^{-1})			
Heating	75	80	95
Cooling	40	58	78
Volume change [% (v/v)]			
Heating	0.5	1.0	0.06
Cooling	1.1	1.5	0.1

Source: Gulati et al. (1984).

Table 4.3 Field Test Data for PSZ Extrusion Dies (Composition 1027)

	Hoover-Ugine	Reading Industries	McQuay-Perfex
Material	Steel	Copper	Copper
Product	19-mm-diameter rod	67-mm-diameter tubing	44-mm-diameter tubing
Reduction	144:1	100:1	100:1
Extrusion temperature (K)	1533	1118	1198
Extrusion speed (m/s)	15	6	12
Average die life			
PSZ dies	24 extrusions	450–500 extrusions	485 extrusions
Metal dies	0 extrusions	125 extrusions	200 extrusions

Source: Gulati et al. (1984).

Table 4.4 Chemical Compositions

Material	Co	Ni	Fe	Cr	W	Cb+Ta	Mo	Al	Mn
FSX-414	Balance	10.4	0.70	29.3	7.2	—	—	—	0.77
FSX-430	Balance	9.8	<0.10	28.5	7.5	—	—	—	<0.10
MM-509	Balance	10.0	<0.10	23.0	7.0	3.59	<0.10	—	<0.10
René 77									
Heat 1	14.4	Balance	0.15	14.2	—	—	4.2	4.10	<0.10
Heat 2	14.7	Balance	0.10	14.9	—	—	4.24	4.39	<0.10
Heat 3	14.5	Balance	0.15	14.7	—	—	4.24	4.20	<0.10
Heat 4	14.3	Balance	<0.10	14.4	—	—	4.10	4.15	<0.10
IN-738									
Heat 1	8.4	Balance	0.08	16.2	2.4	2.66	1.95	3.55	<0.10
Heat 2	8.95	Balance	0.09	15.8	2.55	2.45	1.90	3.47	<0.10
Heat 3	8.30	Balance	0.03	16.1	2.40	2.70	1.70	3.50	—

Material	Cu	Si	Ti	Zr	C	B	S	P	Y
FSX-414	—	0.82	—	—	0.232	0.008	0.009	0.006	—
FSX-430	—	0.18	<0.10	1.37	0.52	0.008	<0.005	<0.015	0.25
MM0-509	—	<0.10	0.19	0.47	0.63	<0.01	0.004	—	—
René 66									
Heat 1	<0.05	<0.10	3.30	0.02	0.072	0.015	0.003	<0.01	—
Heat 2	<0.05	<0.10	3.38	<0.04	0.077	0.016	0.003	<0.01	—
Heat 3	<0.05	<0.10	3.400	<0.04	0.07	0.015	0.003	<0.01	—
FSX-414	—	0.82	—	1.37	0.232	0.008	0.009	0.006	—
FSX-430	—	0.18	0.10	0.47	0.52	0.008	<0.005	<0.015	0.25
MM-509	—	<0.10	0.19		0.63	<0.01	0.004	—	—
René 77									
Heat 1	<0.05	<0.10	3.48	0.15	0.15	0.007	0.003	—	—
Heat 2	<0.05	<0.10	3.40	0.13	0.15	0.013	0.003	<0.01	—
Heat 3	—	0.05	3.20	0.04	0.18	—	0.003	—	—
Heat 4	<0.10	<0.10	3.30	0.04	0.08	0.015	0.003	<0.01	—
IN-738									
Heat 1	<0.05	<0.10	3.48	0.15	0.15	0.007	0.003	—	—
Heat 2	<0.05	<0.10	3.40	0.13	0.15	0.013	0.003	<0.01	—
Heat 3	—	<0.05	3.20	0.04	0.18	—	0.003	—	—

Source: Woodford and Mowbray, (1974).

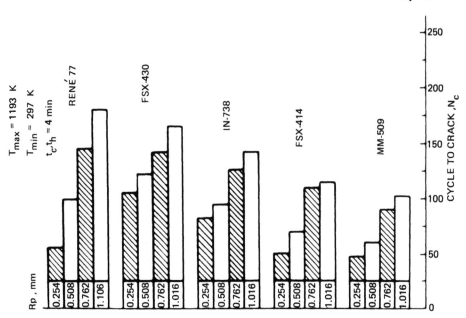

Figure 4.9 Comparison of crack initiation data. (From Woodford and Mowbray, 1974.)

between the number of withstood thermal cycles and an empirical parameter involving yield strength and ductility evaluated in tests conducted at strain rates similar to those involved in a thermal fatigue test. On the other hand, no correlation of the thermal fatigue endurance was found with the creep rupture data.

Woodford and Mowbray (1974) investigated the effect of chemistry and structure on thermal fatigue life. The chemical compositions of alloys studied are presented in Table 4.4, and the results are shown in Figs. 4.9 and 4.10. Experiments were carried on specimens in the form of tapered disks (see Fig. 4.11). Peripheral radii of the specimen between 0.254 and 1.016 mm were used to change the value of the heat transfer coefficients, which were there proportional to the reciprocal peripheral radius. Woodford and Mowbray found that carbides played an important role in thermal fatigue; all cracks proceeded between them. The rankings of alloys in this test were accounted either by the type and morphology of carbides or by environmental effects, which also seems convincing.

In discussing the effect of grain size, the following, well-established facts should be noted: (1) fine grains retard crack initiation in thermal fatigue, and (2) coarse grains, in turn, retard crack propagation. The as-cast grain size distribution is therefore profitable.

The directional solidification produces materials of high resistance to thermal fatigue (see Fig. 4.12). The test specimen was cut from a directionally solidified ingot. Cracks grew in a zigzag fashion along the directions of preferred carbide

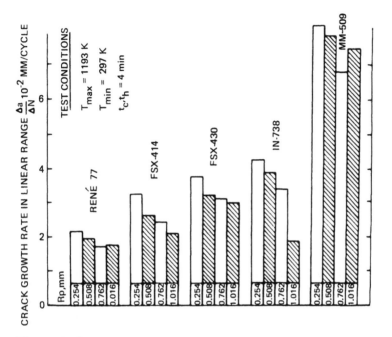

Figure 4.10 Comparison of crack growth rate data. (From Woodford and Mowbray, 1974.)

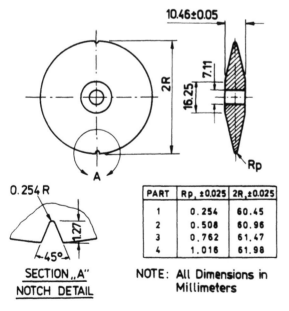

Figure 4.11 Tapered disk thermal shock specimen. (From Woodford and Mowbray, 1974.)

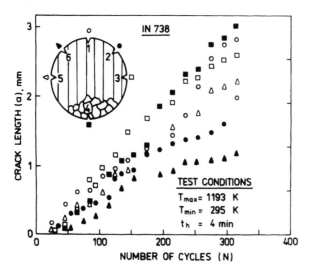

Figure 4.12 Crack propagation at various angles to the growth direction in a specimen of IN-738 machined from a directionally solidified ingot. (From Woodford and Mowbray, 1974.)

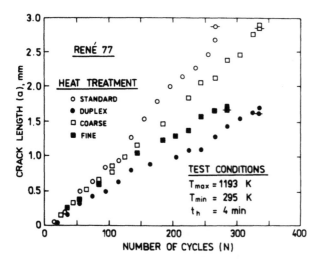

Figure 4.13 Effect of heat treatment on crack growth rate for René 77. (From Woodford and Mowbray, 1974.)

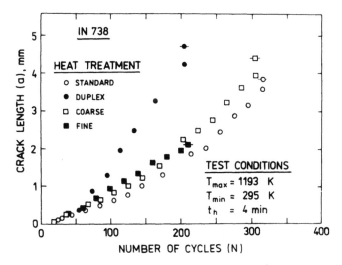

Figure 4.14 Effect of heat treatment on crack growth rate for IN-738. (From Woodford and Mowbray, 1974.)

alignment, which accounted for the results. The premachined notches served as crack starters.

Finally, the effect of heat treatment is illustrated in Fig. 4.13 and 4.14. The heat treatments were selected to produce different γ' sizes. The cycle temperatures were high enough to alter the initial microstructure. The results showed contradictions; moreover, rankings seemed independent of the number of cycles. It was not clear what structural feature has a decisive effect.

4.3 EFFECT OF TEMPERATURE AND HOLDING TIME

A separation of effects produced by these factors is inexpedient, as both are interrelated. In the range of temperatures of interest the degree of temperature-induced changes is determined by the amount of time spent at the highest temperatures of the cycle. In tests where rates of heating and cooling are high (e.g., fluidized beds), time spent at a maximum temperature is obviously or prime importance. The real, industrial temperature histories are usually more complicated, and phases of heating and cooling can occupy a considerable time: Here cycle time becomes an important parameter. Moreover, the entire test procedure terminated at fracture usually takes a long enough time to change a microstructure.

The maximum temperature in the cycle is considered the most important parameter, as it corresponds to considerably lower mechanical properties, mostly yield point, which at the same stress value gives a higher plastic component of deformation. It should be noted, however, that mechanical properties can change in a nonmonotonic way and go through local minima. The increase in temperature also changes the mode of cracking. At high temperatures we can expect recovery, recrystallization, relaxation of stresses by creep, and so on. The minimum life for the vast group of materials corresponds to a cycle time of $t = 3$ to 5 min. This peculiarity is explained by stress relaxation, which is most intensive in the beginning of the hold period and is practically completed in a few minutes. For this, extension of cycle time beyond 3 to 5 min increases the total time withstood in the thermal fatigue test by the material. At lower temperatures in the cycle one can expect work hardening and strain aging. For fast cycles the fatigue process is periodic (i.e., the processes of weakening mentioned are in transient equilibrium with strengthening process). Woodford and Mowbray (1974) demonstrated that rankings of alloys in thermal fatigue tests depend on the test procedure. The cobalt-based alloys displayed longer lives relative to nickel-based alloys in tests with long hold times and high temperatures, while the moderate test conditions preferred the opposite order. This reversal appeared to be caused by a microstructural instability. The detailed examination of René 77 alloy showed that by increasing the temperature above some value (while maintaining hold time constant), crack growth rate is decreased and similarly beyond certain hold times (while holding the second test variable unchanged) also reduces the crack growth rate. The fractographic examination did not reveal any associated changes in the fracture mode which remained principally transgranular. The tested alloy gradually coarsened during the exposure to high temperatures and a continuous γ' film developed at grain boundaries. This followed a reaction $MC + \gamma \rightarrow M_{23}C_6 + \gamma'$. As the solvus temperature is exceeded, the increasing amounts of γ' are taken into the solution. This accounts for a crossover in crack growth rates of tested alloys with increasing temperature. For short crack lengths the ordering was monotonic; after some time, microstructural changes appeared in different extents.

4.4 EXPERIMENTAL ESTIMATION OF THE REMANENT LIFE

It is known that a number of material properties change with an increasing number of thermal cycles. The problem encountered here is that these properties do not often change monotonically. Because of that, the relationships used must be thoroughly scaled and used with great care. Some examples of experimental life estimation are given below.

Zuchowski (1978) found in thermally fatigued H23N18 steel a gradual decrease in yield point, ultimate tensile strength, Young's modulus, and the energy needed to rupture the specimen in the tensile test. The reduction in area and the ultimate elongation found in static tensile tests on fatigued specimens are also sensitive to the history of the material. Unfortunately, both do not vary monotonically.

Tochowicz and co-workers (1976) measured the room-temperature resistances of thermally fatigued heat-resistance alloys. The electrical resistances of most materials remained constant over a substantial part of the life. Their high-temperature values were more sensitive to exposure, slowly increasing with the number of cycles. Poor materials displayed larger changes.

Impact toughness and stress rupture data of the service-exposed material differ considerably from those of the virgin material (Viswanathan and Dolbeck 1987). The scatter in material properties renders impractical the life estimation of the entire turbine blading based on a single turbine blade taken out of service. The technique of taking in situ plastic replicas on all the turbine blades is promising. When examined under a scanning electron microscope the replicas evidence damage to the material, especially cavitation.

Weronski and Szewczyk (1987) found a correlation between the results of nondestructive ultrasound tests and the degree of microstructural damage. The method was proved useful in power-generating plants.

4.5 NOTCH EFFECT

The effect of surface roughness is less appreciable in thermal fatigue than in isothermal fatigue at ambient temperatures, which is illustrated blow (Dulnev and Kotov, 1980). The decrease in value of the average surface roughness from 0.63 μm to 0.16μm increases the life of thermally fatigued specimens by some 40 to 50%. Polishing, which increases lives in isothermal tests three to four times, extends thermal fatigue durabilities only twice (S-816 alloy, exposed to 298 ↔ 1253 K thermal cycles). Casting skin produces typically three- to fivefold lower lives than those for polished specimens (WZL 12U alloy, thermal cycles with 10.7 min holding at T_{max} = 1323 K). The circumferentially notched specimens (the theoretical stress concentration factor of 4.2) revealed lives to be reduced by 50 to 80% relative to smooth specimens (maximum temperature of the thermal cycle of 1123 K).

In Coffin's test, the notch effect appears as the frequency dependence of the life due to the localization of the highest temperatures and stresses at the notch, obviously caused by the smallest cross-sectional area there. It is generally agreed that:

The notch effect is more prominent in the stage of crack nucleation, contrary to crack growth, where it is often negligible.

In most thermal fatigue situations, the surface zone is in a state of alternating plasticity which makes the notch less serious than in isothermal fatigue, where alternating plasticity appears only at the notch room. Neuber's rule is helpful in describing the notch effect.

In his study on the material of the stream turbine valve retired after about 10^5 h of service, Mishchenko et al. (1988) stated that oval inclusions did not act as crack starters. The material in the immediate vicinity of cracks was found depleted in vital elements, especially carbon, due to easier diffusion through the open crack with consequent drop in microhardness. Such a situation causes a pileup of stresses at a crack.

The notch action of oxides depends on their type: Fe_2O_3 and FeO oxides have lower coefficients of thermal expansion compared to the host low-alloy steels, whereas for Fe_3O_4 the opposite is true. In practice, geometry dependence can be more complicated than follows from simple notch considerations. The cooling passages in the blade, for instance, act as stress raisers, but their presence reduces the constraint of surface layers. Hare and Malley (1966) demonstrated that such a blade can be more durable than a solid blade of the same external geometry.

4.6 CRACK PROPAGATION

Among mechanisms of crack initiation the following are considered relevant in thermal fatigue:

The Fujita mechanism
Dislocation pileup with locally exceeded value of theoretical strength
Coagulation of vacancies
Interaction of dislocations with agglomeration of vacancies
Triple-point cracking

It is seen that the models described in Chapters 1 and 2 practically exhaust the subject.

The mode of cracking, and therefore its rate, depend primarily on the actual combination of three independent, important variables describing the thermal cycle: (1) the maximum temperature in the cycle, (2) the strain amplitude, and (3) the hold time. The moderate value of T_{max} and low strain amplitude $\Delta\varepsilon_a$ with no holding produce the fatigue-like transcrystalline pattern of cracking. Increasing $\Delta\varepsilon_a$ and T_{max} and introducing holding at T_{max} change the pattern into the mixed transintercrystalline cracking. The map of failure for some material is shown in Fig. 4.15.

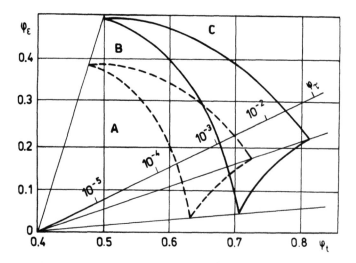

Figure 4.15 Map of failure. (From Dulnev and Kotov, 1980.)

The normalization procedure enables the description of vast groups of materials widely differing in properties. The coordinates then become

normalized strain

$$\varphi_\varepsilon = \frac{\Delta \varepsilon_a}{\varepsilon_e + \varepsilon_p}$$

normalized temperature

$$\varphi_T = \frac{T_{max}}{T_{melt}}$$

normalized holding time

$$\varphi_\tau = \frac{\tau_c}{\tau_r}$$

where

$\Delta \varepsilon_a$ = strain amplitude

$\varepsilon_e + \varepsilon_p$ = rupture strain in tensile test

T_{max} = maximum temperature in the cycle

T_{melt} = melting temperature

τ_c = duration of thermal cycle

τ_r = time to rupture under the creep stress equal to maximum stress in the cycle

Oxidizing of grain boundaries and depletion in alloying additives has a prominent effect on cracking. It was found (Glenny and Taylor, 1959–60) that endurances were significantly longer in tests carried out in argon than in those conducted in air. Preoxidation is also harmful to lives, which was also revealed in tests.

Kuwabara and Nitta (1979) introduced the index of intergranular cracks to describe quantitatively the cracking mode. This index is

$$F_{IG} = \frac{2n_{IG} + n_{mix}}{N}$$

where N is the total number of cracks found on the surface and n_{IG} and n_{mix} are the numbers of intergranular and mixed intergranular–transgranular cracks, respectively. Thus provided that all the cracks run intergranularly, $F_{IG} = 2$, while if all cracks are transgranular, $F_{IG} = 0$. The calculated fractions are plotted in Fig. 4.16 against the total strain range in the test. *In-phase* refers to a test in which the external tensile load was applied at the high temperature and compressive load at the lower one. For a comparison, the isothermal fatigue data were also included. The conclusions that can be drawn are in accord with the statements above. In a similar type of test on Inconel X-750, Marchand and Pelloux (1985) found that under in-phase conditions cracks developed faster. The results were rationalized introducing the closure stress concept.

It should be pointed out that phases of crack initiation and growth are facilitated by different temperatures:

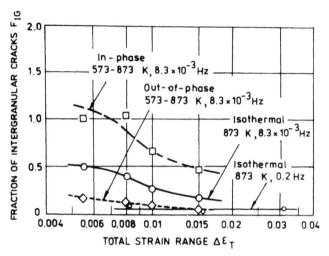

Figure 4.16 Variation of fraction of intergranular cracks with strain range. (Reprinted with permission from Pergamon Press, Inc.)

Numerous cracks form early in life at low temperature
In the high-temperature portion of the scale, a single crack propagates rapidly.

In their study of the chromium cast irons, Feldman et al. (1973) found cracks initiating:

At carbide eutectics due to the difference in thermal expansion coefficients
At primary carbides
In ferrite dendrites

Rezai-Aria and co-workers (1988a,b) found cracks initiating early at preferentially oxidized MC carbides. The same locations were revealed in LCF tests carried out at a constant high temperature. In the material tested, the cobalt-based MAR-M509 alloy, general matrix oxidation is less developed, which is due to different oxidation kinetics of interdendritic areas and of matrix dendrites.

4.7 MICROSTRUCTURAL CHANGES

The long-lasting exposure to elevated, varying temperatures and repeated stresses as it takes place in thermal cycling naturally causes changes in the microstructure. The extent of these changes relative to the exposure to constant temperature and stress depends on the material tested and the test conditions, so that published data and conclusions often conflict with each other. Some examples are given below.

Prowans and Ustasiak (1972) found in thermally fatigued WWN 1 steel (C = 0.3%, Cr = 2.7%, W = 9%, Ni = 1.5%, V = 0.3%) tested at lives of 10^3 to 10^4 cycles a pattern of changes comparable to that in creep. Cailletaud and co-workers (1984) studied IN-100 in two cyclic tests: 1173 ↔ 1323 K and 998 ↔ 1398 K. The material tested was unstable at these temperatures. In the high-temperature portion of the cycle, a part of γ' precipitates was dissoluted, while at lower temperatures, new, very fine precipitates appeared. The curves of cycle stress-strain responses revealed significant hardening during the tensile part of the first type of cycle and moderate for the second type. The curves drawn at maximal temperatures of the cycles showed a normal behavior comparable with that recorded in isothermal fatigue at these temperatures. Moreover, this strengthening had no effect on specimen life.

Rezai-Aria et al. (1988a,b) came to similar conclusions in tests on cobalt-based MAR-M509 alloy. Test conditions involved flame heating to 1473 K and cooling with an air blast to 473 K. The electron microscopy observations showed that it resulted in the precipitation of fine carbides, mainly of $M_{23}C_6$ type. The extent and type of the process were closely simulated by proper heat treatment to study this effect in isolation. Tests on such heat-treated specimens showed that stress-strain responses resembled those of virgin materials. The precipitation did not alter the life of the material tested.

In certain stainless steels there is the tendency for carbides and α phase to precipitate at intergranular sites, thus forming a path of easy crack growth (Glenny, 1961). The formation of large, discrete particles is considered less harmful.

The diffusion of the strongly oxidizable elements such as aluminum and titanium leave in their wake regions denuded of γ'. It is especially evident in the immediate vicinity of the crack. The phase transformations found at the crack are caused by the depletion of a particular phase stabilizer.

Yoda et al. (1978) investigated the tungsten fiber/copper composite and found the preferential extension of the matrix along the fibers. Sliding at the fiber–matrix interface was thought to result from the relaxation of internal stresses that arose from the thermal expansion mismatch. In a subsequent paper (Yoda et al., 1979), porosity was reported in this region.

Ginsztler (1986) tested the three grades of steel under the following conditions:

Grade A: nonalloyed ferritic-pearlitic, thermal cycle 293 ↔ 773 K, life to crack initiation of 140 to 180 cycles

Grade B: Cr-Mo-V ferritic-bainitic steel, cycle 293 ↔ 838 K, life of 300 to 400 cycles

Grade C: Cr-Ni austenitic stainless steel, cycle 293 ↔ 873 K, life of 600 to 640 cycles

The observed microstructural changes were in A, cell structures after 120 cycles; B, cell structures and heavy plastic deformation after 100 cycles; and C, high dislocation density, formation of subgrains after 600 cycles. The cycle duration was 33 s in each test.

Kamachi et al. (1984) studied two steels: one-phase SUS 316 and duplex SUS 329, which consisted of a mixed bcc and fcc structure. Specimens resistively heated were very thin (2 mm), which enabled them to attain in 1 s temperatures well above 1000 K. It was demonstrated that SUS 329 was inferior to SUS 316, which was due to the interphasal stresses. These stresses were about 200 MPa in the fcc phase and 140 MPa in the bcc phase at room temperature. With increasing temperature the magnitude of these stresses passed zero at 600 K and changed signs thereafter. The x-ray method strain measurements proved interphasal stresses to relax partly after a small number of cycles. Remarkable grain growth was found in SUS 329 steel, while in SUS 316 steel, only moderate changes were revealed. The alternating interphasal stresses account for this.

4.8 CRACK GROWTH DESCRIPTION

The methods used to correlate crack growth rate with other quantities described here can be divided into those originally developed to describe isothermal fatigue or creep data, and those based on the oxidation–fatigue interaction. Contrary to isothermal fatigue in most situations, cracks grow under conditions of considerable plastic deformation and creep. For this reason, in seeking the correlating parameter, attention is paid to the crack opening displacement COD, cyclic J integral, and $C*$ integral rather than to the cyclic stress intensity factor. Both types of methods are illustrated below.

Marsh (1981) carried out the thermal shock study on reactor-grade AISI type 316 stainless steel. Specimens of rectangular cross section were heated by passing current, while the central zone of narrower faces was quenched by water spray. Figure 4.17 shows a series of experimentally determined isochronous temperature

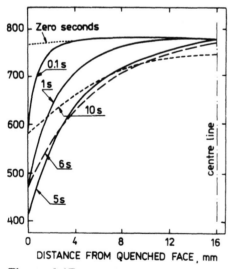

Figure 4.17 Isochronous temperature profiles during a 4-s thermal shock. (Reprinted with permission from Pergamon Press, Inc.)

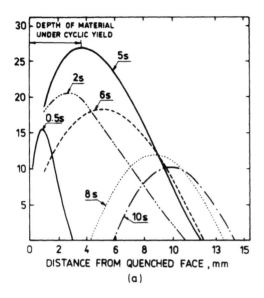

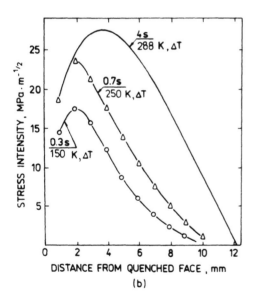

Figure 4.18 (a) Calculated stress intensity profiles during a 4-s quench from 770 K. (b) Peak stress intensity curves for quench times of 4, 0.7, and 0.3 from 770 K bulk temperature with effective temperature amplitude experienced by the surface of 288 K, 250 K, and 150 K, respectively. (Reprinted with permission from Pergamon Press, Inc.)

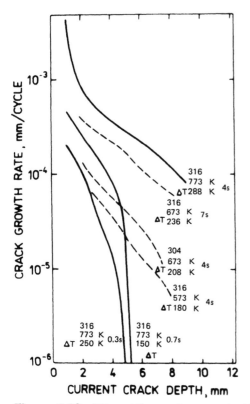

Figure 4.19 Effect of distance from quenched face on crack growth rate. Note different bulk temperatures and effective temperature amplitudes. Results for AISI type 304 stainless steel added for comparison. (Reprinted with permission from Pergamon Press, Inc.)

profiles. Figure 4.18a and b show a set of calculated stress intensity curves. It is seen that the peak stress intensity shifts with passing time into the material, simultaneously reducing its value. The variation in crack growth rate with crack length is shown in Fig. 4.19. Note different quenching times in the thermal cycle (0.3 to 7 s) and that the crack became arrested at some depth. Obviously, the latter holds true if creep effect is negligible. In zones of alternating plasticity just below the surface, crack growth was not stress-intensity dependent, contrary to deeper layers of material.

Okazuki and Koizumi (1987) in the study on type 304 stainless steel, 1Cr–1Mo ferritic cast steel, and 12Cr-Mo-V-W steel found a cyclic J. integral to correlate crack growth data. Tests consisted of isothermal fatigue and thermomechanical fatigue where specimen temperature and strain were varied simultaneously. The readings taken from the strain detector attached to the specimen surface were compensated for thermal free expansion. The crack growth rate followed the equation

$$\frac{da}{dN} = C(\Delta J)^m$$

where

$$C = (4.31 \text{ to } 15.9) \times 10^{-5}$$
$$m = 1.19 \text{ to } 1.52$$

depending on the material and test conditions. In tests where thermal cycles are superimposed on a constant stress, a gradual transition is observed from fatigue-like to creep-like growth with increasing stress magnitude (Shimizu et al., 1983).

Jordan and Meyers (1986) tested Hastelloy X and found the correlation of the crack growth rate with the strain intensity factor $\Delta K_\varepsilon = \Delta\varepsilon\sqrt{\pi a} \cdot f$ (where $\Delta\varepsilon$ is the strain range, a the crack length and f a geometric correction factor) and with the J integral as well. None of them was superior; the COD was similarly judged.

Skelton (1978) investigated 0.5Cr-Mo-V steel and found a crack growth rate to obey the law

$$\frac{da}{dN} = Ba^n$$

where n ranged from 1.3 to 2.3, depending on the test details.

Rezai-Aria and co-workers (1988b) evaluated the oxidation-fatigue model developed by Reuchet and Remy (1983). The crack growth rate was assumed to be a sum of fatigue and oxidation components:

$$\frac{da}{dN} = \left(\frac{da}{dN}\right)_f + \left(\frac{da}{dN}\right)_{ox}$$

The first, fatigue term was estimated using Tomkins's (1968) model, which is also applicable under thermal fatigue conditions (i.e., $da/dN = Ba$). The second, oxidation component consisted of two terms describing the contribution due to oxidation of the matrix and the carbides, respectively:

$$\left(\frac{da}{dN}\right)_{ox} = (1 - f_c)\alpha_M(1 + K_M\Delta\varepsilon_{IN})\Delta t^{1/2} + f_c\alpha_c g(\Delta\varepsilon_{IN})\Delta t^{1/4}$$

where

α_M, α_c = oxidation constants under zero stress of matrix and carbides, respectively

f_c = effective volume fraction of carbides on the crack path

$$g(\Delta_{IN}) = \begin{cases} \Delta\varepsilon_{IN}/\Delta\varepsilon_0 \text{ when } \Delta\varepsilon_{IN} > \Delta\varepsilon_0 \\ 1 \text{ when } \Delta\varepsilon_{IN} \leq \Delta\varepsilon_0 \end{cases}$$

$g(\Delta\varepsilon_{IN})$ = a function of strain

$K_M, \Delta\varepsilon_0$ = constants

$\Delta\varepsilon_{IN}$ = strain range

Δt = cycle period

Furthermore, the fracture of the oxides was assumed to take place at each tensile stroke. Oxidation constants used here have the meaning of average values obtained by the integration of the oxidation equation over the entire temperature cycle. The model gives a fairly good description of the thermally fatigued MAR-M509 alloy. The lives predicted were within the factor of 2 for most cases.

REFERENCES

Bengtsson, K. I. (1957). *Met. Treat.*, *24*: 227.

Cailletaud, G., Culié, J. P. and Kaczmarek, H. (1984). Thermal fatigue of a thermally unstable alloy, in *Proc. 4th International Conference*, Stockholm, 1983 (Carlsson, J. and Ohlson, N. G., eds.), Pergamon Press, Elmsford, N.Y., p. 255.

Clayton, A. M. (1983). *Progr. Nucl. Energy*, *12*: 57.

Dulnev, R. A. and Kotov, P. I. (1980). *Termisheskaya ustalost metalov*, Machinostrojenie, Moscow.

Feldman, B. A., Neugebauer, G. O. and Kushnareva, (1973). *Izv.Vysch. Uchebn.Zaved.Tshern. Metall.*, (2): 134.

Forrest, P. G. (1960–61). *J. Inst. Met.*, *89*: 432.

Franklin, A., Heslop, J. and Smith, R. (1964). *J. Inst. Met.*, *92*: 535.

Ginsztler, J. (1986). *Int. J. Pres. Ves. Piping*, *26*: 181.

Glenny, E. (1961). *Met. Rev.*, *6*(24): 387.

Glenny, E. and Taylor, T. A. (1959–60). *J. Inst. Met.*, *88* 449.

Gulati, S. T., Hansson, J. N. and Helfinstine, J. D. (1984). *Met. Progr.*, *125* (2): 21.

Hare, A. and Malley, H. (1966). *SAE Paper 660053*.

Hoffelner, W. (1987). Materials for future gas turbines, in *Proc. 17th International Congress on Combustion Engines*, Cimac '87, Warsaw, Paper T2.

Hunter, T. A. (1956). STP 174, ASTM, Philadelphia.

Jordan, E. H. and Meyers, G. J. (1986). *Eng. Fract. Mech.*, *23*: 345.

Kamachi, K., Tani, N., Ishida, T. Kawano, M. and Kuhohori, T. (1984). Thermal fatigue test by direct passage method of large electric current on stainless steels, *Proc. 4th International Conference*, Stockholm, 1983 (Carlsson, J. and Ohlson, N. G., eds.), Pergamon Press, Elmsford, N.Y., p. 175.

Kostiuk, A. G., Truchnij, A. D. and Mitschulin, W. N. (1974). *Machinovedenie*, (5): 62.

Kuwabara, K. and Nitta, A. (1979). *Fatigue Eng. Mater. Struct.*, *2*: 293.

Lamain, L. G. (1981). A fatigue analysis of a tube under thermal loading, in *Proc. 6th Structural Mechanics in Reactor Technology Conference*, Paris, Paper G9/4.

Marchand, N. and Pelloux, R. (1985). Thermal-mechanical fatigue crack growth in Inconel X-750, in *Proc. 11th Canadian Fracture Conference*, Ottawa (Krausz, A. S., ed.), Martinus Nijhoff, Hingham, Mass., p. 167.

Marsh, D. J. (1981). *Fatigue Eng. Mater. Struct.*, *4*(2): 179.

Mishchenko, L. D., Pisarenko, I. N., Tarabanova, V. P. and Dyachenko, S. S. (1988). *Probl. Prochn.*, 222(2):114.

Muscatell, F. L., Reynolds, E. E., Dyrkacz, W. W. and Dalheim, J. H. (1957). *Proc. ASTM*, *57*: 947.

Okazaki, M. and Koizumi, T. (1987). *J. Eng. Mater. Technol.*, *109*(2):114.

Prowans, S. and Ustasiak, M. (1972). *Hutnik, 39*: 479.

Reuchet, J. and Remy, L. (1983). *Metall. Trans. A, 14*: 141.

Rezai-Aria, F., François, M. and Remy, L. (1988a). *Fatigue Fract. Eng. Mater. Struct., 11*: 227.

Rezai-Aria, F., François, M. and Remy, L. (1988b). *Fatigue Fract. Eng. Mater. Struct., 11*: 291.

Riddihough, M. (1960). *Metallurgia, 62*: 53.

Sehitoglu, H. (1985). *J. Eng. Mater. Technol., 107*: 221.

Shimizu, M., Brown, M. W. and Miller, K. J. (1984). Fatigue crack propagation in stainless steel subjected to repeated thermal shock, in *Proc. 4th International Conference*, Stockholm, 1983 (Carlsson, J. and Ohlson, N. G., eds.), Pergamon Press, Elmsford, N.Y., p. 207.

Skelton, R. P. (1978). *Mater. Sci. Eng., 35*: 287.

Tochowicz, S., Kamiński, Z. and Szydlo, A. (1976). *Hutnik, 43*: 87.

Tomkins, B. (1968). *Philos. Mag., 18*: 1041.

Viswanathan, R. and Dolbec, A. C. (1987). *J. Eng. Gas Turbines Power, 109*: 115.

Weronski, A. and Szewczyk, S. (1987). Effect of working conditions on life of low-alloy, creep-resistant steels, Report, Technical University of Lublin.

Woodford, D. A. and Mowbray, D. F. (1974). *Mater. Sci. Eng., 16*: 5.

Yoda, S., Kurihara, N., Wakashima, K. and Umekawa, S. (1978). *Metall. Trans A, 9*: 1229.

Yoda, S., Takahashi, R., Wakashima, K. and Umekawa, S. (1979). *Metall. Trans A, 11*: 1796.

Zmihorski, E., Kowalski, W. and Zolciak, T. (1975). *Metalozn. Obrobka Cieplna, 13*: 36.

Zuchowski, R. (1978). *Proc. 8th Seminar on Metals Fatigue*, Czestochowa, Poland, p. 19.

5

Experimental Methods

5.1 INTRODUCTION

Thermal fatigue is by definition a process of nucleation and subsequent, gradual development of damage in elements exposed to cyclically or periodically repeated temperature changes. The resultant failure of the element should be widely understood as a state of partial or complete dysfunction, including rupture as an extreme case, or for instance the appearance of thermal fatigue cracking on the surface that can easily be reclaimed, or finally, as the excessive creep rate due to metallurgical changes in the fatigued material. Thermal fatigue can be accompanied by (1) stresses introduced by machining or during former service, (2) aggressive environment (e.g., corrosive or erosive media, neutron flux exposure), and (3) external load: steady or varying.

According to Spera (1976), thermal fatigue situations can be differentiated between those with an external constraint and those with an internal constraint. The former occurs when either external load is applied to the element or thermal expansion of the entire element is constrained. Practically, we deal more frequently with the latter group of situations, as the former is painstakingly avoided by designers where possible. The internal constraint can be (1) a temperature gradient existing in the element or (2) structural anisotropy with a consequent difference in thermal expansion coefficients of particular phases and therefore alternating stress at the interface.

It can be anticipated in advance that testing methods used to study thermal fatigue are more or less a logical continuation of those used in studying creep or mechanical fatigue. Obviously, creep and mechanical fatigue, being the two limiting cases, do not occur here in pure form. In contrast, thermal fatigue generally involves temperature and stress varying simultaneously with time. Moreover, considering thin elements, we can conclude that temperature gradients in heated or cooled elements are rather negligible, whereas in thick elements they must be taken into account. Finally, to characterize the thermal cycle, two variables must be specified for a thick element: (1) temperature gradients in the body and over the surface and (2) state of stress. In practice, an extensive effort is needed to establish them. Temperature gradients can, of course, be calculated if surface temperatures and properties of material are known or better, measured accurately. Temperature profiles are usually taken by means of pyrometers, which provide data relevant to the surface with an error caused by progressive damage to the surface, with consequent changes in optical properties or with thermocouples that enable us to get an insight in the body of the element. The disadvantages of the latter are a lag in temperature records due to thermal capacity of the junction, changes caused by introducing foreign material, and finally, which is common to both methods, the fact that abrupt changes need fast recording, preferably microcomputer-aided.

The state of stress, in turn, can be found experimentally as deduced from strains measured on the surface. There are two useful methods: (1) observing a deflection of the mesh of fine noble-metal wires welded to the surface (Dulnev and Kotov, 1980) or observing an interferometric pattern formed by notches cut on the surface (Odqvist and Ohlson, 1968); and (2) calculating, which is of ever-increasing importance due to the progress in microcomputing and the availability of useful codes (Mackerle, 1986).

Thermal fatigue tests should make it possible:

To get an insight into the nature of the thermal fatigue process (basic-type tests)
To study particular aspect of design
To compare the performance of materials in simulated service conditions to expect consequences of replacement of one material by another

Life-expectancy techniques, which must also be taken into account when considering types of tests, require knowledge of:

Strain-cycle number dependence (Manson–Coffin equation)
Steady creep rate
Hysteresis in the strain–stress coordinate system
Total elongation in the static test
Fatigue endurance values in common mechanical tests with strain amplitudes or stress amplitudes equal to maximal values in the thermal cycle and a knowledge of time to rupture in a creep test carried out at the highest temperature and stress values
Crack advance per thermal cycle
A change in physical properties associated with exhausting life

The last term comprises changes:

In the static strength of material (e.g., Zuchowski, 1978).
In the microstructure (e.g., morphology of carbides, their distribution, presence of grain-boundary cavities, migration of grain boundaries, dislocation structures, and finally, micro- and macrocracks)
In the electrical resistance (e.g., Tochowicz et al., 1976)
In magnetic properties [e.g., coercive force (Karpinski, 1979)]
In the number of acoustic emission counts (Zuchowski, 1986)

The variety of experimental techniques compels us to focus attention on some of them, as will be done further. The experimental problem concerning all of them is the periodicity, where phases of momentary stability of properties are followed by abrupt changes. The extent of their occurrence renders mechanical fatigue.

Finally, we should determine which materials are considered with respect to thermal fatigue resistance. The answer is that the choice of materials is often imposed by reasons other than thermal fatigue resistance. In the Sunprobe Helios-A, the surface materials were aluminized Teflon, fused silica, and silicon, the temperatures varying between 170 and 520 K (Winkler, 1983). The element resistant to thermal fatigue can be of composite design, comprising a load-carrying core and surface layer of different material possessing adequate resistance to the influence of the surrounding environment. The matters of interest in this case are primarily adhesion of this surface coating to the core, and core protection against an excessive temperature rise (thermal barrier coating) or against an aggressive environment. We discuss first the methods used to estimate the performance of protective coatings.

5.2 ASSESSMENT OF COATING PERFORMANCE

As methods used to assess coating performance are often sophisticated, we restrict our discussion to those useful in thermal fatigue situations. The adhesion of coatings, which is of crucial importance, is conventionally determined by the tensile adhesion test (TAT): ASTM C633-69. In this method, one end of the 25-mm-diameter rod is sprayed with coating material and then epoxied to another rod, thus forming a tensile specimen. Provided that the epoxy has a tensile strength greater than that of the adhesion, the latter can be measured. The results obtained exhibit wide scatter.

Coating failures can be classified as follows: adhesive, where fracture surface runs along the coating–substrate interface; and cohesive, where cracking runs through the coating. Zones of different failure types often neighbor each other. Shankar et al. (1983b) connected the AE (acoustic emission) monitoring device with the TAT specimen. The AE data were found to correlate with the mechanical behavior of coatings, cohesive failure involving more deformation and cracking being associated with more intense acoustic emission. Saganuma et al. (1985) used this method to determine the effect of repeated thermal shocks on the adhesion of coating. The thick layer of alumina was solid-state bonded to AISI type 405 steel, and the coating was interlayered with Nb or Mo to alleviate the difference in thermal expansion coefficients. TAT data showed some deterioration of coating adhesion strength consequent upon repeated thermal cycles.

Another group of methods are body-force methods used to assess adhesion. They are based on acceleration or deceleration forces directed into the coating. In the example shown in Fig. 5.1, a coated specimen playing the role of a rotor was levitated magnetically and rotated in the evacuated chamber until the coating was detached. The force acting on the film could thus be calculated.

The scratch test (sometimes referred to as the stylus method) is commonly

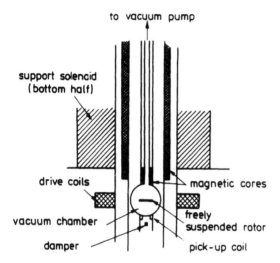

to vacuum pump

support solenoid
(bottom half)

drive coils

magnetic cores

vacuum chamber

freely
suspended rotor

damper

pick-up coil

Figure 5.1 Ultracentrifugal method of adhesion assessment. (From Krongelb, 1978, after Valli et al., 1986.)

accepted. An indenter is pulled across the coating surface and pushed toward it with an increasing load. The load corresponding to the onset of spalling is called a critical load and indicates the adhesion strength. The method, introduced by Heavens (1950), was subsequently used by Benjamin and Weaver (1959a,b) to study adhesion of thin metal films to glass, on which they were deposited. A schematic diagram of the scratch tester currently used is shown in Fig. 5.2. The onset of coating detachment can be detected either by a friction force transducer or an acoustic emission detector, both attached to a stylus, which is usually made of diamond with a tip radius of 200 μm and a cone angle of 120°. Hintermann (1984) used this method to assess the adhesive strength of thin (a few micrometers) chemically vapor deposited coatings. Subrahmanyan et al. (1986) used a modification of this method as a comparative test of coatings. The diamond stylus was pushed with constant force toward the traveling specimen, the mass loss of coating per unit distance being the criterion of the test.

Sumomogi and co-workers (1981) presented two other techniques to evaluate adhesion of sputter-deposited coatings. The first relied on applying an indentation on the coated surface with a Vickers microindenter; the pattern of film peeling was adopted as the measure of film adhesion. In the second method, a compressive force was applied parallel to the film surface until the film peeled off, which provides a measure of adhesion.

Another, quite different group of methods utilizes laser impingement onto the coating or substrate. A block diagram of the testing system used by Almond et al. (1985) is shown in Fig. 5.3. A modulated beam of light emitted by an argon laser is used to generate thermal waves in the sample, which is scanned by a stepping

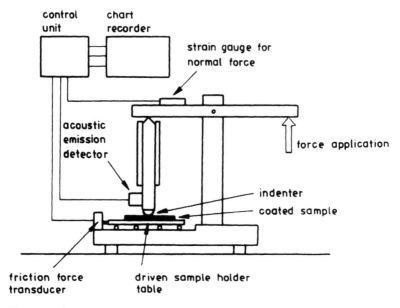

Figure 5.2 Scratch tester.

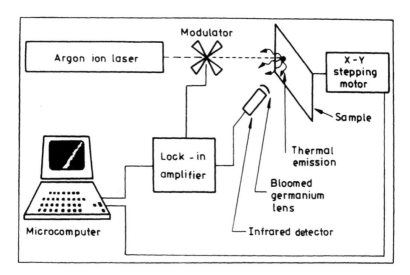

Figure 5.3 Block diagram of the testing system. (Reprinted with permission from Butterworth-Heinemann Ltd.)

motor. The infrared light emitted by the specimen is monitored by a detector. The amplitude of the signal and its phase with respect to modulation are monitored by a microcomputer. The system proved the feasibility of adhesion assessment. Almond et al. (1981) used common ultrasonic attenuation measurements for the same purpose. Hohne (1963a,b) reported that the back-wall echo disappeared completely in regions of poor adhesion.

Acoustic emission techniques were used by Ashary et al. (1983) to supplement the weight-loss method used in corrosion studies. The apparatus is outlined in Fig. 5.4. The waveguide, made of a 1-mm diameter platinum rod, had one end spot-welded to the specimen and the other attached to a stainless cone in contact with the transducer. AE data correlated with weight loss were measured after cyclic thermal exposure. Application of the intermediate layer was found useful, as was the forming of a pattern of fine cracks alleviating stress in the coating—this effect being found in some instances (Shankar et al., 1983a).

Prater and Moss (1983), investigating various steels and superalloys coated with ceramics, used traditional thermal fatigue testing, heat specimens by means of a gas burner to 1270 K and cooling with air or water to 290 K. The criterion was the degree of spalling. In a similar study, Sivakumar (1985) adopted x-ray analysis to reveal phase changes. The tested, partially stabilized zirconia undergoes phase transformation from the original cubic to monoclinic, which alleviates the effect of tensile stress due to the difference in specific volumes of these phases. The excellent

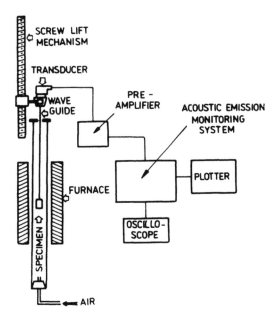

Figure 5.4 Outline of acoustic emission monitoring system used in high-temperature oxidation studies. (From Ashary et al., 1983.)

thermal properties of this material depend on the volume of pores existing in it. The temperature of the element, protected by a thermal barrier on its surface, is lower the higher the porosity (Swain et al., 1986).

Stewart et al. (1986) presented an evaluation of integral multicomponent thermal protection materials with the prospect of applying them to a spacecraft that reenters Earth's atmosphere using aerodynamic forces to decelerate and maneuver. Design requirements include strong mechanical properties at the inner surface and high-temperature capability (the latter estimated in terms of surface recession—linear shrinkage). The test facility provided a high-energy airstream; the temperature developed at the heat surface of the specimen was measured by a standard pyrometer with temperature profiles by means of thermocouples embedded to different depths in the specimen being tested. Aside from the performance of the coating itself, the in-depth temperature profiles developed in a simulated flight must not exceed the softening point of the bond and the admissible temperature of the protected structure.

5.3   ASSESSMENT OF MONOLITHIC ELEMENTS

Testing techniques used to assess monolithic elements are rarely unique to thermal fatigue, the majority of them being variants of those used in studying problems associated with fatigue, creep, and corrosion. For this reason we first discuss some aspects of "common" fatigue tests.

First, specimens used in such testing are uniform to a great degree, contrary to thermal fatigue testing, where specimen shapes and sizes are rather optional. Second, the manner of imposing controlled loads and producing controlled strains is basically the same in both common and thermal fatigue tests. Third, common fatigue tests provide information on crack propagation rate and fatigue life at a constant temperature which is of interest in considering thermal fatigue situations.

It was proved empirically that deep cracks growing in linear elastic stress conditions satisfy the equation

$$\frac{da}{dN} = C\,\Delta K^m$$

where

da/dN = cyclic growth rate

C = a constant

ΔK = stress intensity range

For shallow cracks in a high-strain-fatigue (HSF) regime,

$$\frac{da}{dN} = Ba^n$$

where B depends on the surface plastic strain range (Skelton and Miles, 1984).

Tests can be devised either to model reverse plasticity in surface layers of the element subjected to repeated thermal cycles or to model the growth of deeper cracks trying to render stress conditions at the advancing edge of the crack. Fatigue crack propagation rates are found on specimens of various designs. Standards that slowly emerge are the specimens used in static tests and depicted in Fig. 5.5. The stress intensity factor must be corrected to account for specimen geometry effect, and the specimen width (Fig. 5.5a) is chosen with respect to mechanical properties of the tested material according to the ASTM Designation E399-81 Standard Test Method. The goal of this choice is to keep inaccuracies connected with the departure from the plane stress conditions and the presence of plasticity ahead of the crack tip within reasonable limits. The crack propagation rate is read from extensometers attached to the knife plates screwed to specimen edges or to knife edges of the notch machined in the specimen. A specimen of this type was used, for example, by Tobler and Reed (1977) in fatigue testing of aluminum alloy. Figure 5.5b shows the specimen used in three-point bending. A load is applied on he site opposite the notch, and the specimen is pointwise supported at a point $2w$ distant from the notch. Both specimens are loaded in mode I.

The specimen recommended for crack opening displacement study in three-point bending is shown in Fig. 5.6a. The testing method is covered by British Standard BS 5767:1979, Methods for Crack Opening Displacement (COD) Testing. The specimen depicted in Fig. 5.6b is used in supplementary testing.

The J integral is determined in tests according to the ASTM Designation E813-81, Standard Test for J_{Ic}, a Measure of Fracture Toughness. Specimens used for that purpose are shown in Fig. 5.7. The J integral is proportional to the area below the curve plotted in the coordinates stress–crack advance or stress–load deflection.

Note that the two last cases relate to plasticity ahead of the crack tip that must be taken into consideration. The tests described are static, but despite this, the values of stress intensity factor, crack opening displacement (COD), and J integral corresponding to the onset of crack propagation (denoted by the subscript Ic) are of interest to us, as are those corresponding to the crack arrest; both are generally used in fatigue calculations.

Obviously, the measurement of crack extension at elevated temperatures needs some alteration in the experimental technique. For instance, Neate (1985) used compact specimens (Fig. 5.5a) to determine a crack growth rate at 873 K and a capacitance transducer converting COD into an electrical signal. The device (Gooch and Grimes, 1974) consisted of two vanes, fabricated from a thin platinum sheet, interleaved with each other and attached to the front face of the sample and insulated from it electrically by a wafer of recrystallized alumina. The capacitance of this capacitor is directly proportional to the area common to the vanes and can readily be measured. The alternative method used by Neate (1985), which is also capable of producing high-temperature data, was a dc electropotential technique where COD is deduced from the resistance measured on the current leads attached to the opposite top and bottom surfaces of the specimen. Typically, the current supplied to the specimen is 25 to 50 A, the recorded voltage changes being even as small as

a)

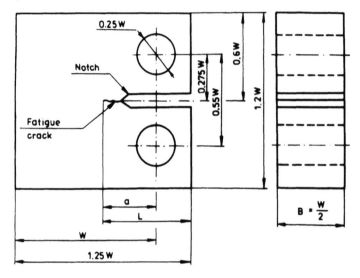

b)

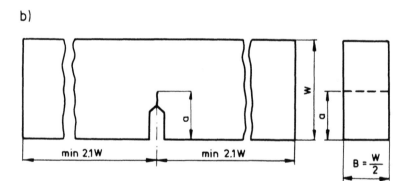

Figure 5.5 Specimens used to assess fracture toughness: (a) compact geometry specimen; (b) specimen used in three-point bending.

a)

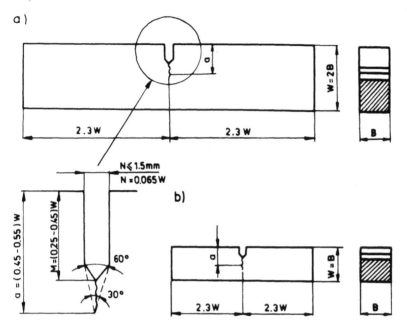

Figure 5.6 Specimens used in crack opening testing: (a) recommended design of specimen; (b) specimens used in supplementary testing.

microvolts. According to Gibson and Druce (1984), the following facts altering the readings should be considered while using this method with compact specimens:

1. The fatigue crack closure and plasticity effects should be known.
2. The quantity deduced from the experimental results is in fact the surface of the crack. The crack extension is calculated assuming a certain crack front shape (e.g., semicircular).
3. The loading pins should be insulated from the specimen and the current leads preferably welded to the top and bottom surfaces of the specimen, which provides higher accuracy than welding to the front face.

Yokobori and Sakata (1979) used a high-temperature microscope to observe crack length continuously during testing, without interruption, in contrast with common optical methods using a traveling microscope. The details of the test facility are shown in Fig. 5.8. The flat, double-edged notched specimen (1) is observed through the piping window (6) using a microscope (7). The temperature is recorded by the thermocouple (3). The specimen is heated by the heater (2), and a load is applied by culverts (4). The lower part of the apparatus is cooled using water admitted through the piping (8). The objective lens is protected from overheating by a moveable shield (5).

When considering the principle of specimen loading is should be pointed out that reverse bending is generally simpler and cheaper; moreover, it is possible to run a

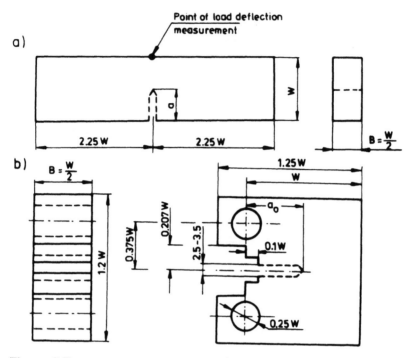

Figure 5.7   Specimens to measure J integral: (a) in three-point bending; (b) in tension.

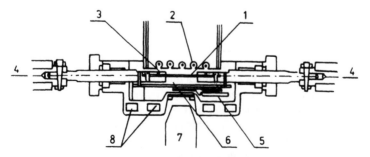

Figure 5.8   Details of the apparatus. (From Yokobori and Sakata, 1979.)

single test out of thousands of cycles. The fatigue endurances obtained compare very favorably with those in a push-pull mode. An example of such a machine designated for higher temperatures was provided by Yavari and Langdon (1982). Another ingenious solution is the "thermoactuator" developed at the U.S. National Bureau of Standards. It is a self-loading clevis, the arms of which are made of materials with appropriate thermal expansion coefficients. Due to the difference in these coefficients relative to that of a specimen, the specimen, rigidly closed with the clevis, is loaded by a known amount at a given temperature (Fields and Smith, 1980).

Methods used to study creep are also in continuing development. The essence of a creep test is the exposure to a given, constant temperature either for a chosen time (typically 10^2, 10^3, 10^4, 10^5 or 2×10^5 h) or for a given load. In the first case we are interested in the stress, causing permanent deformation of the specimen at the termination of the test, which is chosen as a test criterion, or causing its rupture. In the second case, of interest is the time to rupture or the amount of time necessary to cause the given, permanent deformation. Machines to study creep usually make it possible to mount several specimens, causing different stresses in particular specimens while keeping them all at the same temperature. Most data are obtained under tensile loading, although this restriction can easily be circumvented. For example, Findley (1982) described a machine used to perform testing in compression. A tubular specimen was loaded with a dead weight and heated with a quartz lamp located coaxially in the specimen; axial strain was transferred to a differential transformer by a pair of Invar rods.

Provost and co-workers (1984) tried to answer the question of prospects of short-term creep testing. The method used was low-strain-rate tensile testing at rates of 10^{-5} to 10^{-8} mm/mm s. Although attempts to extrapolate results failed, the method an be used for quality control of materials intended for high-temperature service, especially welds, for which creep data are missing almost completely. It was also proved feasible to control the remnent life of service-aged materials.

Endo and Sakon (1984) constructed a machine capable of performing studies on creep–fatigue interaction. A schematic diagram of this machine is shown in Fig. 5.9. The solid, cylindrical, 8-mm-diameter specimen (1) is encircled by an electrical furnace (2) capable of heating it up to 1070 K, coupled with a load cell (3). A reversible motor (4) with a rotating cam (5) is used to produce tension–compression strains controlled by an extensometer. Elastic supports (8) are used to maintain accurate alignment in the displacement restraint state (holding periods). The loading capacity of the machine is ±3 tons.

The problem that naturally arises when considering elevated temperatures is that attack caused by an aggressive environment is enhanced in such conditions. For example, Bates et al. (1980) tested some alloys to obtain their relative corrosion resistance. The tests included (1) creep rupture testing on tensile specimens exposed to liquid lithium at 1173 K, and (2) liquid lithium corrosion tests, where specimen weight loss and depth of leaching penetration were evaluated.

Saunders and Nicholls (1984) discussed various test procedures devised to estimate materials in simulated engine conditions. Test methods included:

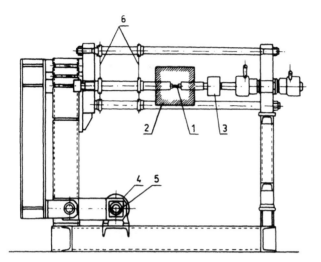

Figure 5.9 Outline of creep-fatigue testing machine.

1. Burner rig tests in which the gaseous environment was produced by the combustion of fuel with the addition of various contaminants
2. Crucible tests where specimens were kept immersed in a molten salt mixture, the composition of which is chosen to render the deposit found on an in-service corroded specimen
3. Furnace tests carried out in controlled atmospheres (e.g., air contaminated with the compound evaporated from the additional source) or surrounding specimens with the synthetic ash designed to simulate the deposit found on blades

The morphology of the corrosion product, which seems to be an essential aspect in simulating real conditions, is not sufficiently well reproduced by all methods; crucible testing, for example, was found to be inadequate. It is generally agreed that test procedures currently employed can distinguish large difference in corrosion resistance. Rankings given by investigators from different laboratories are sometimes quite dissimilar for materials of similar resistance since test parameters are to some degree not within complete control.

Systematic studies on thermal fatigue were pioneered by Coffin. The testing method described by Coffin and Wesley (1954) and Coffin (1954) departs from those described above. The test specimen is hollow and tubular, with a gauge length of 76 mm, an internal diameter of 12.7 mm, and a wall thickness of 0.5 mm. The apparatus was designed to clamp each end of the test specimen rigidly to the end plates so that any longitudinal thermal deformation would be converted directly to elastic or inelastic deformation. The end plates were rigidly connected by heavy columns to prevent their relative motion. The specimen is heated by the passing current, which approaches as much as 1000 A; cooling, in turn, is accomplished by admitting gas flow through the internal orifice. the temperature is read from the thermocouple spot-welded to the specimen. The choice of such a specimen is a compromise between the requirements of even longitudinal thermal distribution, and therefore thin specimen walls, and stiffness, which requires thick walls. The results of the preliminary studies quoted above permitted quantification of the role of prior cold work, which was found beneficial only at cycles of moderate severity, being deleterious in tests involving higher temperatures. Egorov and Pirogov (1972) found that cold work introduced by explosive straining is generally beneficial.

A thermally fatigued, fully constrained specimen usually becomes heavily deformed. Due to the longitudinal thermal gradient, the major flow in compression occurs in the hottest regions of the specimen, increasing its cross-sectional area, and therefore on cooling, the tensile stress developed is lower than before, as there was a stress reversal in the meantime. The repetition of cycles causes bulging in the hot region and necking in the adjacent region.

Albaugh and Venturi (1980) carried out tests on solid cylindrical specimens inductively heated and cooled by a spray of liquid nitrogen. Bulging similar to that noted above was found and assuming that the same volume of material migrates during each cycle, it was feasible to obtain a mathematical equation for the number of cycles to failure expressed by the strain observed in one cycle. However, the model is relevant only to a particular specimen deformation, although one can imagine that such attempts would succeed in solving similar problems. Other investigators introduced methods involving the following factors:

1. Partial constraint. The specimen is attached to the base plate through the intermediary elastic element (Mozarovski, 1963) or loaded with a freely suspended dead weight (Zuchowski, 1978) (see Fig. 5.10).
2. Independent control of strain and temperature (Forrest and Penfold, 1961; Ginsztler, 1986) (see Figs. 5.11 and 5.12).
3. External heating with a burner (Odqvist and Ohlson, 1968) or induction coil (Northcott and Baron, 1956; Baron and Bloomfield, 1961); heating and cooling in a fused salt mixture (Keyes and Krakoviak, 1961).
4. Multiaxial thermal stresses. Taira and Inoue (1971) found a convergence of the uniaxial and multiaxial tests when the data obtained are expressed in terms of equivalent strains.
5. Low temperatures. Stepanov and Kivi (1972) explored the range 77 to 290 K.

Independent stress and temperature control were also considered by Carden (1970). The test method used was based on the Instron machine (Instron Corp., Canton, Massachusetts). Specimens (see Fig. 5.13) were resistively heated, temperatures being controlled by an optical pyrometer. Obviously, the specimen temperature does not follow the command signal exactly. The lag in temperature for a heating rate of 12 K/s was 3 K. However, the longitudinal profile of temperature continues to change for 60 s after reaching the temperature set by a thermocouple junction spot-welded to the specimen. In the present study two extensometers were used: a longitudinal one of 12 mm gauge length and a diametral one. The pro-

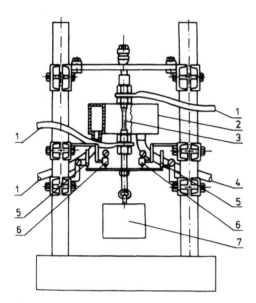

Figure 5.10 Apparatus used by Zuchowski (1978): 1, Current leads; 2, cooling chamber; 3, specimen; 4, inlet of pressurized air; 5, LVDT extensometer; 6, dial gauge; 7, dead weight.

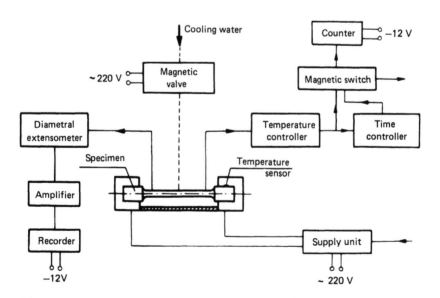

Figure 5.11 Block diagram of the thermal fatigue testing system. (From Ginsztler, 1986.)

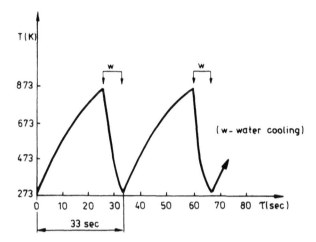

Figure 5.12 Thermal cycles used in studies. (From Ginsztler, 1986.)

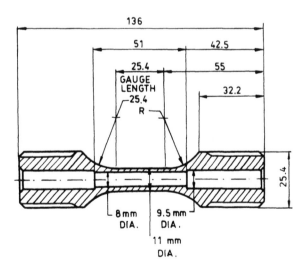

Figure 5.13 Thermal fatigue test specimen dimensions. (Reprinted with permission from American Society for Testing Materials.)

TO LOAD CELL

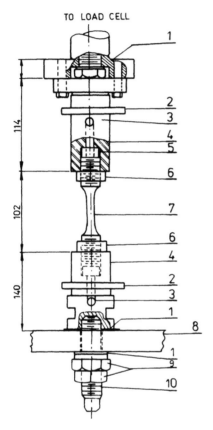

Figure 5.14   Scheme of gripping system (see the text for a description of the numbered elements). (Reprinted with permission from American Society for Testing Materials.)

grammed variables were crosshead extension and specimen temperature. Details of the gripping system are shown in Fig. 5.14. The system is electrically isolated from the rest of the machine, this being accomplished by insulation (1). A threaded-end specimen (7) is attached to the grips (4), and cylindrical copper insert (5) was used to reduce the electrical resistance on the specimen–grip junction and to ensure against their differential thermal expansion. Locknuts (6) were used to tighten the specimen. Electrical current was supplied by leads (2). The cooling air enters through a hole in the lower grip (3) and passes through the specimen to exit through the hole in the upper grip. The test facility was attached to the Instron machine crosshead (8) with a rod (10) and locking nut (9).

The foregoing idea has been altered (Weronski and Hejwowski, 1988) to use a mechanical loading system instead of a servohydraulical system, having the disadvantage of high investment costs. A stepping motor converting electrical pulses into a proportional mechanical response (one step per signal; 100 steps per revolution) is used to drive a ball screw jack and thus load the specimen. The idea was originally proposed by Ceschini (1984). The system being implemented incorporates a microcomputer to produce a load or displacement history as well as to produce a command signal for the specimen heating unit. An orthodox electrohydraulic machine, computer controlled, was used by Kuwabara and Nitta (1979). The specimen was inductively heated by an independently controlled oscillator.

Strain measurement can be performed directly as in the examples above or by the use of remotely fixed extensometers (longitudinal level mounted with the end of the specimen shanks). The latter introduces an extra elastic component of strain from the specimen shanks. The details have been discussed by Skelton (1985); especially noteworthy is the information on controlling fatigue testing using a computer, which enabled the successful use of remote transducers and eliminated elastic follow-up of the test facility.

In turbine blades, radii of leading and trailing edges follow from aerodynamic considerations. When the turbine is started cold or suddenly stopped, temperature gradients arise in it. The tendency of blade edges to expand during heating or to contract during cooling is resisted by the center body, thus producing thermal stresses.

The method used by Glenny and Taylor (1960) and Glenny (1961) involved a fluidized bed that makes use of a bed of fine particles through which a gaseous medium is passed. If the velocity of the gas stream is high enough, the particles will lose contact with each other and become airborne. The thermal cycles consisting of rapid heating and cooling were accomplished by transferring the specimen between two fluidized beds maintained at different temperatures: slow cooling allowing the specimen to cool down in still air; and slow heating, inserting specimens in the furnace. The aspect of the blade that was studied was the radii of the edge curvature and the specimens used were of tapered disk form, the peripheral radii being altered on purpose.

Semenov (1981) used the wedge-shaped specimen to study the same problem. The modified apparatus is shown in Fig. 5.15 The specimens (6), which had keys (3), were mounted into the duct through which the combustion gas stream was admitted (1). Wedge-shaped inserts (2) functioning as vanes were positioned between specimens and protected from slippage by the lock (4). The fixture (5) was attached to the manipulator arm to perform rapid cooling by immersion into liquid gas. The time dependences of temperatures are depicted in Fig. 5.16. The supply of fuel to the combustor is started with a maximum rate at t_0; at t_1, when a stable gas temperature (solid line) has been achieved, the manipulator inserts a container loaded with specimens into the gas stream; at t_2, the auxiliary fuel feeding system is shut, and at t_3, the main fuel feeding system; at t_4, specimens start to cool in a water spray; at t_5, the fuel feeding line of the idle run is shut; at t_6, specimens are immersed in a liquid gas; at t_7, a new cycle is begun.

Avery and co-workers (1967), studying combustor-cooling tubes, found that the modes of failure were buckling caused by steady thermal gradients and thermal fatigue due to temperature changes. The design recommendation was the ratio of

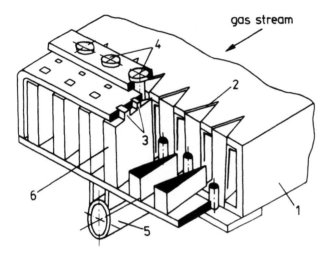

gas stream

Figure 5.15 Container with specimens and vanes. (From Tretiyatchenko et al., 1985.)

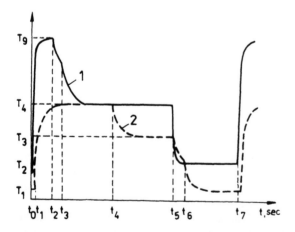

Figure 5.16 Temperatures in the cycle. 1, gas temperature; 2, specimen temperature.

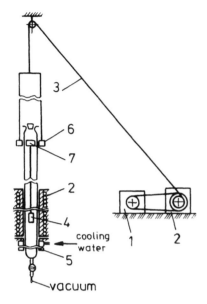

Figure 5.17 Apparatus used by Pokorny and Potuzak (1973).

tube diameter to tube thickness. Thermal fatigue tests were carried out by heating the tubular specimen on one side by a radiant lamp and cooling it on the other. The longitudinal expansion of the tube was fully constrained.

The third group of methods are comparative. Unlike the foregoing methods, here the test conditions only partially indicate the real in-service loadings. An example is given in Fig. 5.17. The specimen (4) is sealed in an evacuated tube a cooled at the lower end to protect the seals (5) from overheating. Being driven by a motor (1) coupled with a gearbox (2), the specimen can perform reversible movements. The lift system consisted of a flexible connector (3) and a magnet (6), transmitting the motion to a steel rod (7). This rig was used to evaluate steam turbine steels.

Close simulation of the working conditions of the mold for centrifugal casting is very expensive. Apart from the rig that would perform casting, we would need a laboratory induction furnace of relatively large capacity. A converse method was proposed by Weronski (1978); see Fig. 5.18. The specimen (1), in the form of a ring, was welded to the cooler (2), through which vigorous water flow was maintained; the inner surface was intermittently heated by an induction coil (3). Thermocouples, spot-welded to the bottom of blind holes 4 to 8 were used to record temperatures over the cycles, and the readings were compared with those derived experimentally on the instrumented mold. Any significant departures were compensated, altering the power of the induction furnace. A series of preliminary

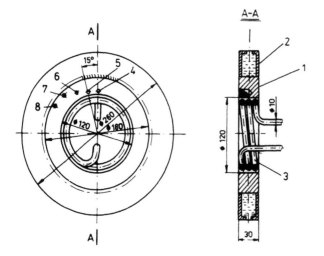

Figure 5.18   Specimen arrangement used by Weronski (1978).

experiments were performed on the rig with a different-shaped coil encircling the
top of a rotating cylindrical specimen while the bottom was cooled in a water bath.

In many industrial applications, the impingement of high-speed particles causes
serious erosion. In their absence, at elevated temperatures, corrosion-resistant
metals would develop protective scales, whereas under real conditions fast particles
perforate the metals, causing detachment (Ives, 1977). The erosion of impellers of
a large radial fan used to exhaust air from the rotary kiln is a significant factor. This
has been partially solved by the use of curved blades. The peripheral velocities of
blades attain a speed of 150 m/s, making erosion a very significant factor. Apart
from erosion, the abrasive action of entrained particles was found. Laboratory tests
intended to extend the service life of impellers included, apart from purely basic
ones, model tests on the two rigs (Weronski and Hejwowski, 1987a,b) to check
recommended changes in the impeller design.

Waschkies et al. (1986) presented the results of thermal shock experiments
performed on an idle reactor pressure vessel within the framework of a nuclear
safety program. The reactor vessel was filled with hot water ($T = 670$ K) and
pressurized at $p = 11$ MPa. The reactor nozzle, which was studied, was sprinkled
with water periodically to reduce its temperature to 290 K, and then allowed to
reheat (by means of the hot water). Acoustic signals generated by the nozzle material
were detected by broadband transducers. A six-channel analysis system was used
to determine the arrival time, rise time, maximum amplitude, duration, and energy
from each AE signal. Two classes of signals were distinguished: those appearing
in the middle part of each cycle and associated with the crack growth, and those
due to crack closure. Moreover, the system was capable of detecting crack location.
Similarly, in their study on steam piping materials and welds, Weisberg and Soldan

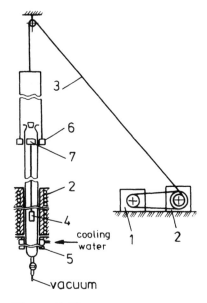

cooling
← water

vacuum

Figure 5.17 Apparatus used by Pokorny and Potuzak (1973).

tube diameter to tube thickness. Thermal fatigue tests were carried out by heating the tubular specimen on one side by a radiant lamp and cooling it on the other. The longitudinal expansion of the tube was fully constrained.

The third group of methods are comparative. Unlike the foregoing methods, here the test conditions only partially indicate the real in-service loadings. An example is given in Fig. 5.17. The specimen (4) is sealed in an evacuated tube a cooled at the lower end to protect the seals (5) from overheating. Being driven by a motor (1) coupled with a gearbox (2), the specimen can perform reversible movements. The lift system consisted of a flexible connector (3) and a magnet (6), transmitting the motion to a steel rod (7). This rig was used to evaluate steam turbine steels.

Close simulation of the working conditions of the mold for centrifugal casting is very expensive. Apart from the rig that would perform casting, we would need a laboratory induction furnace of relatively large capacity. A converse method was proposed by Weronski (1978); see Fig. 5.18. The specimen (1), in the form of a ring, was welded to the cooler (2), through which vigorous water flow was maintained; the inner surface was intermittently heated by an induction coil (3). Thermocouples, spot-welded to the bottom of blind holes 4 to 8 were used to record temperatures over the cycles, and the readings were compared with those derived experimentally on the instrumented mold. Any significant departures were compensated, altering the power of the induction furnace. A series of preliminary

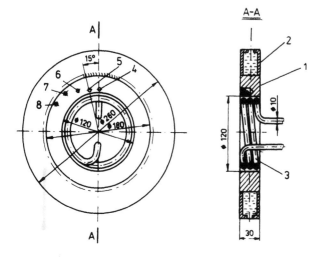

Figure 5.18   Specimen arrangement used by Weronski (1978).

experiments were performed on the rig with a different-shaped coil encircling the top of a rotating cylindrical specimen while the bottom was cooled in a water bath.

In many industrial applications, the impingement of high-speed particles causes serious erosion. In their absence, at elevated temperatures, corrosion-resistant metals would develop protective scales, whereas under real conditions fast particles perforate the metals, causing detachment (Ives, 1977). The erosion of impellers of a large radial fan used to exhaust air from the rotary kiln is a significant factor. This has been partially solved by the use of curved blades. The peripheral velocities of blades attain a speed of 150 m/s, making erosion a very significant factor. Apart from erosion, the abrasive action of entrained particles was found. Laboratory tests intended to extend the service life of impellers included, apart from purely basic ones, model tests on the two rigs (Weronski and Hejwowski, 1987a,b) to check recommended changes in the impeller design.

Waschkies et al. (1986) presented the results of thermal shock experiments performed on an idle reactor pressure vessel within the framework of a nuclear safety program. The reactor vessel was filled with hot water ($T = 670$ K) and pressurized at $p = 11$ MPa. The reactor nozzle, which was studied, was sprinkled with water periodically to reduce its temperature to 290 K, and then allowed to reheat (by means of the hot water). Acoustic signals generated by the nozzle material were detected by broadband transducers. A six-channel analysis system was used to determine the arrival time, rise time, maximum amplitude, duration, and energy from each AE signal. Two classes of signals were distinguished: those appearing in the middle part of each cycle and associated with the crack growth, and those due to crack closure. Moreover, the system was capable of detecting crack location. Similarly, in their study on steam piping materials and welds, Weisberg and Soldan

(1954) connected the test section, consisting of test materials welded together to the main piping of the power station, to form a separate test unit.

Skelton and Miles (1984) investigated the effect of microstructure, circumferential notch, and multiple cracking on crack advance. The test material was supplied as 100-mm-diameter cylinders trepanned from a cast although some were machined internally (Fig. 5.19b) and some had an internal circumferential semicircular groove at their center. The specimens were inductively heated and water quenched from the bore. Crack propagation was detected by an electrical method that was sensitive to large cracks. Therefore, information on crack progress was obtained from fractographic and oxide dating studies. Both were carried out on the termination of the test. The former provided striation spacing measurements; the latter was based on the assumption that the oxide layer found in cracks obeys the parabolic growth rate. Thermally fatigued specimens exhibited circumferential and longitudinal cracks emanating from the bore. Besides major cracking, the mesh of minor, shallow

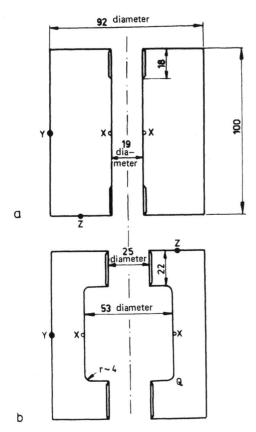

Figure 5.19 Thermal fatigue specimens (dimensions in millimeters). (Reprinted with permission from Butterworth-Heinemann Ltd.)

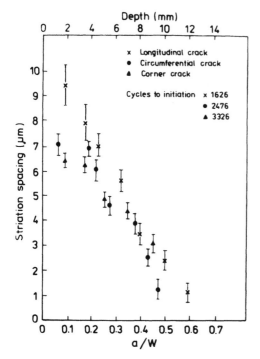

Figure 5.20 Striation measurement data. (Reprinted with permission from Butterworth-Heinemann Ltd.)

cracks was found in many specimens. Point Q in Fig. 5.19b shows the place of corner crack origin, also revealed in specimens. The data obtained from fractographic studies are shown in Fig. 5.20. It can be seen that striation spacing (i.e., crack advance per cycle) decreases with penetration into the section [penetration (a) is expressed as a fraction of wall thickness (W) in this figure]. The estimated numbers of cycles to crack initiation are included. It can be seen that longitudinal cracks propagated considerably faster than others; the tendency of deep cracks to become dormant is also visible. The results were interpreted in terms of stress intensity profiles.

Supplementary push-pull isothermal tests were performed to furnish data on cyclic strain–stress behavior at temperatures involved in the thermal fatigue tests. A wide divergence was found between the rate of crack growth predicted by linear elastic fracture mechanics (LEFM) and that deduced from striation spacings. The crack growth rates based on the stress intensity profiles calculated for each type of specimen, and isothermal push-pull tests carried out at the highest temperature of the thermal cycle, were considerably lower that those found experimentally, which was probably due to corrosion caused by the cooling water.

REFERENCES

Albaugh, H. and Venturi, R. (1980). *Int. J. Fatigue*, (July): 118.

Almond, D. P., Cox, R. L., Moghisi, M. and Reiter, H. (1981). *Thin Solid Films*, *83*: 325.

Almond, D. P., Patel, P. M., Pickup, I. M. and Reiter, H. (1985). *Nondestr. Test. Int.*, *18*(1): 17.

Ashary, A., Meier, G. H. and Pettit, F. S. (1983). *J. Mech. Work. Technol.*, 8(2–3).

Avery, L. R., Carayanis, G. S. and Michky, G. L. (1967). *Exp. Mech.*, *7*(6): 256.

Baron, H. G. and Bloomfield, B. S. (1961). *J. Iron Steel Inst.*, *197*(3): 223.

Bates, D. A., Edwards, G. R. and Ohlson, D. L. (1980). *Mater. Perform.*, *19*(3): 40.

Benjamin, P. and Weaver, C. (1959a). *Proc. R. Soc. A*, *254*: 163.

Benjamin, P. and Weaver, C. (1959b). *Proc. R. Soc. A*, *254*: 177.

Carden, A. E. (1970). Thermal fatigue evaluation, in *Manual on Low Cycle Fatigue Testing*, STP 465, ASTM, Philadelphia, p. 163.

Ceschini, L. J. (1984). *Exp. Tech.*, *8*(2): 34.

Coffin, L. F. Jr. (1954). *Trans. ASME*, *76*(Aug.): 931.

Coffin, L. F., Jr. and Wesley, R. P. (1954). *Trans. ASME*, *76*(Aug.): 923.

Dulnev, R. A. and Kotov, P. T. (1980). *Termicheskaya ustalost metallov*, Maschinostroienie, Moscow, p. 200.

Egorov, W. J. and Pirogov, E. N. (1972). *Probl. Prochn.*, (10): 77.

Endo, T. and Sakon, T. (1984). *Mater. Technol.*, *11*: 489.

Fields, R. J. and Smith, J. H. (1980). *Met. Progr.*, August: 39.

Findley, W. N. (1982). *J. Test. Eval.*, *10*(4): 179.

Forrest, P. G. and Penfold, A. B. (1961). *Engineering*, *192*: 522.

Gibson, G. P. and Druce, S. G. (1984). *Development of the Direct Current-Potential Drop Technique for Determining J-Crack Growth Resistance Curves*, AERE, Harwell, Berkshire, England.

Ginsztler, J. (1986). *Int. J. Pres. Piping*, *26*: 181.

Glenny, E. (1961). *Metall. Rev.*, *6*(24): 387.

Glenny, E. and Taylor, T. A. (1960). *J. Inst. Met.*, *88*: 449.

Gooch, D. J. and Grimes, F. J. (1975). *J. Phys. E Sci. Instrum.*, *8*: 350.

Heavens, O. S. (1950). *J. Phys. Radium*, *11*: 355.

Hintermann, H. E. (1984). *Wear*, *100*: 381.

Hohne, K. (1963a). *Schweisstechnik*, *13*: 55.

Hohne, K. (1963b). *Schweisstechnik*, *13*: 403.

Ives, L. K. (1977). *J. Eng. Mater. Technol.*, *99*(4): 126.

Karpinski, T. (1979). *Przegl. Mech.*, *19*(16): 482.

Keyes, J. J., Jr. and Krakoviak, A. I. (1961). *Nucl. Sci. Eng.*, *2*(1): 462.

Krongelb, S. (1978). *Adhesion Measurement of Thin Films, Thick Films and Bulk Coatings*, STP 640 (K. L. Mittal, ed.), ASTM, Philadelphia, p. 107.

Kuwabara, K. and Nitta, A. (1979). In *Mechanical Behaviour of Materials* (Miller, K. J. and Smith, R. F., eds.), Pergamon Press, New York.

Mackerle, J. (1986). *Comput. Struct.*, *24*(4): 657.

Mozarovski, S. (1963). *Probl. Prochn.*, (6): 243.

Neate, G. J. (1985). *High Temp. Technol.*, *3*(4): 195.

Northcott, L. and Baron, H. G. (1956). *J. Iron Steel Inst.*, 385.

Odqvist, F. K. G. and Ohlson, N. G. (1968). *JUTAM Symposium*, Stockholm, p. 189.

Pokorny, R. and Potuzak, L. (1973). *Neue Huette*, *18*(3): 557.

Prater, J. T. and Moss, R. W. (1983). *Thin Solid Films, 101*: 455.

Provost, W., Steen, M. and Dhooge, A. (1984). *Short Term Determination of the Creep Properties of High Temperature Materials*, MT 160, Research Center of the Belgian Welding Institute, Ghent, Belgium.

Saunders, S. R. J. and Nicholls, J. R. (1984). *Thin Solid Films, 119*: 247.

Semenov, G. P. (1981). *Probl. Prochn., 12*(150): 23.

Shankar, N. R., Berndt, C. C. and Herman, H. (1983a). *Am. Ceram. Soc. Bull., 62*(5): 614.

Shankar, N. R., Berndt, C. C. and Herman, H. (1983b). Characterization of the mechanical properties of plasma-sprayed coatings, in *Advances in Materials Characterization* (Rossington, D. R., Condrate, R. A. and Snyder, R. L., eds.), Plenum Press, New York, p. 473.

Sivakumar, R. (1985). *Mater. Lett., 3*(9–10): 396.

Skelton, R. P. (1985). *High Temp. Technol., 3*(4): 179.

Skelton, R. P. and Miles, L. (1984). *High Temp. Technol., 2*(1): 23.

Spera, D. A. (1976). *What Is Thermal Fatigue: Thermal Fatigue of Materials and Components*, STP 612, ASTM, Philadelphia, p. 3.

Stepanov, G. A. and Kivi, E. A. (1972). *Probl. Prochn.*, 1374.

Stewart, D. A., Leiser, D. B., Kolodziej, P. and Smith, M. (1986). *J. Spacecr. Rockets, 23*(4): 420.

Subrahmanyan, J., Srivastava, M. P. and Sivakumar, R. (1986). *Mater. Sci. Eng., 83*(3252): 1.

Suganuma, K., Okamoto, T., Koizumi, M. and Shimada, M. (1985). *J. Nucl. Mater., 133–134*: 773.

Sumomogi, T., Kuwahara, K. and Fujiyama, H. (1981). *Thin Solid Films, 79*: 91.

Swain, M. V., Johnson, L. F., Syed, R. and Hasselman, P. P. H. (1986). *J. Mater. Sci. Lett., 5*: 799.

Taira, S. and Inoue, T. (1971). *Proc. International Conference on Thermal Stress and Thermal Fatigue*, Gloucestershire, 1969, Butterworth, Sevenoaks, Kent, England, p. 109.

Tobler, R. L. and Reed, R. P. (1977). *J. Eng. Mater. Technol., 99*(4): 306.

Tochowicz, S., Kaminski, Z. and Szydlo, A. (1976). *Hutnik, 43*(3): 90.

Trietiyachenko, G. N., Kravchuk, L. V. and Semenov, G. R. (1985). *Probl. Prochn.*, (7): 114.

Valli, J., Makela, U. and Mattchews, A. (1986). *Surf. Eng., 2*(1): 49.

Waschkies, E., Hepp, K. and Höller, P. (1986). *Nondestr. Test. Int., 19*(3): 197.

Weisberg, H. and Soldan, H. M. (1954). *Trans. ASME*: 1085.

Weronski, A., and Hejwowski, T. (1987a). Patent pending.

Weronski, A. and Hejwowski, T. (1987b). Patent pending.

Weronski, A. and Hejwowski, T. (1991). To be published in *Przegl. Mech.*

Weronski, A. (1978). Report, Politechnika Lubelska, Lublin, Poland.

Winkler, W. (1983). *Acta Astronau., 10*(4): 189.

Yavari, P. and Langdon, T. G. (1983). *Rev. Sci. Instrum., 54*(3): 353.

Yokobori, T. and Sakata, H. (1979). *Eng. Fract. Mech., 13*: 509.

Zuchowski, R. (1978). Report 194, Instytut Materialoznawstwa i Mechaniki Technicznej Politechniki Wroclawskiej.

Zuchowski, R. (1986). *Analysis of Failure Process in Thermal Fatigue of Metals*, Wydawnictwo Politechniki Wroclawskiej, p. 178.

6

Lifetime Predictions

Depending on the function performed by the element considered, the following failure criteria can be applied:

1. Energy-based criteria, where a material dysfunction is identified with a deterioration of material properties below some acceptable level. The basis of this family of criteria is the assumption of constant energetic capacity of material, irrespective of the way in which the energy is supplied. In the most extreme approach, Ivanova (1963) related energetic capacity to the amount of energy necessary to heat a material to its melting point and to melt it. In a linear damage summation approach, the energy to fracture is taken as mathematically equal to the product of the number of cycles to failure by the energy dissipated in one cycle. Furthermore, this unit portion of energy is identified with the surface confined by the stable hysteresis loop. It was, however, soon recognized that damage does not sum up linearly and that the lifetime relation is a complicated function. The disadvantage of energetic criteria is the necessity to use a number of parameters to describe the hysteresis loop and that in case of thermal fatigue, the material properties change together with the temperature.
2. Strain-based criteria, where the symptom of failure adopted is excessive component deformation.
3. Criteria based on macroscopic coherence.

Consider the latter more explicitly, failure is usually identified with the attainment of a specified crack size or by the attainment of a specified ratio of the cracked surface to the entire component surface.

The majority of lifetime relations applied in thermal fatigue situations have been developed within a low-cycle-fatigue (LCF) data base. These relations are obviously directly applicable to thermal fatigue, provided that:

1. The strain and temperature cycles to which the component is subjected are determined,

161

2. The effect of stress raisers is evaluated,
3. Both of the above enable the maximum effective stress-strain range to be determined, and then the life can be calculated from the life relation chosen.

The second group of lifetime relations comprise those specific for creep, which have already been reviewed in Chapter 2. They are also adopted and used in thermal fatigue situations. The applicability of the Monkman–Grant (1956) criterion was proven by Buba (1983) in thermal fatigue tests on three grades of heat-resistant steel. The coefficient in the Monkman–Grant relation (1956) appeared independent of the test condition and was related to the grade of material. This criterion is considered applicable for asymmetric cycles, that is, where one type of stress—tensile stress—prevails. The third group comprises corrosion-based criteria, also already reviewed.

The most acknowledged strain range–life relation is that originally proposed by Manson (1953) and Coffin (1954),

$$N_f^{1/m} \Delta \varepsilon_p = C$$

where

    m, C = material constants

    N_f = number of cycles to failure

    $\Delta \varepsilon_p$ = plastic strain range in the cycle, which must be precisely known

Applicability of this relation was proved by numerous researchers. The more practical approach where the life relation is written in terms of total strange range, $\Delta \varepsilon_t$, was developed by Manson and Hirschberg (1965). This method is called the *method of universal slopes* since the intercepts of the elastic and plastic lines are related to the tensile properties: σ_u, the ultimate tensile strength, and D, the ductility in a tensile test (i.e., $D = \ln[1/(1 - \psi)]$, where ψ is a reduction in area). This method, presented in Fig. 6.1, is widely used in a preliminary design. It was, however, soon recognized that the method overpredicts lives in tests carried out at elevated temperatures. The short-time tensile properties used in this relation are not affected by oxidation and creep as at elevated temperature exposures. Their effects seem greater in the stage of crack initiation, which in the LCF regime can occupy as little as 10% of element life. For this reason, the life in low-cycle fatigue, assuming cracks to initiate immediately on beginning the test, should be taken equal to 10% of the life calculated by the method of universal slopes. The method, illustrated in Fig. 6.2, is called the *10 percent rule* and has been applied with a fair degree of accuracy.

Yamaguchi and Nishijima (1986) found for creep–fatigue interaction that to reduce the scatter of experimental data, inelastic strain in the Manson–Coffin relation should be normalized by either tensile ductility or creep ductility, depending on the failure mode. Tensile ductility was used in tests with transgranular cracking, the latter with intergranular cracking. The problem encountered here is, however,

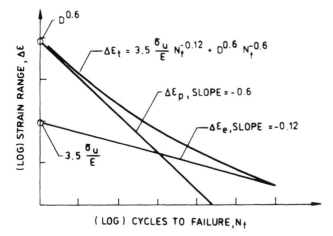

Figure 6.1 Universal slope method for estimating fatigue life.

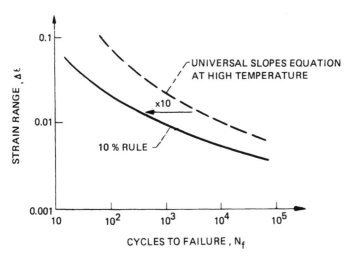

Figure 6.2 The 10 percent rule of Manson and Halford (1967).

that D_c changes remarkably with exposure so that its time relation must be thoroughly evaluated before any life prediction.

Another modification of the Manson–Coffin relation was proposed by Udouchi and Wada (1971):

$$N_f^k \Delta \varepsilon_p F(t) = C_1$$

Here

$$F(t) = \frac{-Q_1}{T_{av}} + \left[1 + C_2 \Delta T \exp\left(\frac{-Q_1}{\Delta T} \right) \right]$$

where

$$\Delta T = \text{amplitude of thermal cycle}$$
$$T_{av} = \text{average temperature in the cycle}$$
$$k, C_1, C_2, Q_1 = \text{material constants}$$

This criterion is recommended for symmetric cycles.

Coffin (1971) proposed a frequency-modified strain-range-life relationship to account for hold-time effects:

$$\Delta \varepsilon_t = \frac{A C_2^n}{E} N_f^{-\beta n} v^{k_1 + (1-k)\beta n} + C_2 N_f^{-\beta} v^{(1-k)\beta}$$

where

the first and second terms represent the elastic and plastic strain range, respectively

C_2 = the fatigue ductility coefficient

β = fatigue ductility exponent

n = cyclic strain hardening exponent

E = Young's modulus

v = frequency

A = cyclic stress–strain coefficient

k, k_1 = coefficients that express the effects of frequency on life

The cycle frequency is calculated according to

$$v = \frac{1}{\tau + \tau_H}$$

where τ is the time for strain reversal and τ_H is the hold time in the cycle.

Kussmaul and Bhongbhibhat (1980) conducted creep–fatigue tests on the cast steel GS-17Cr-Mo-V 5 11 grade at 803 K. Results are plotted and compared with theoretical predictions in Figs. 6.3 and 6.4. The predictive methods used included

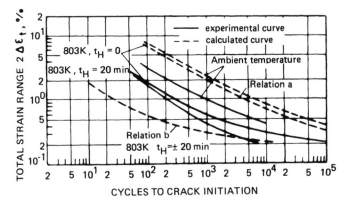

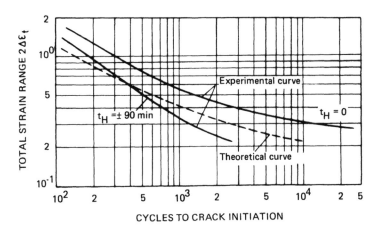

Figure 6.3 Comparison of experimentally determined lives to crack initiation with theoretical predictions .

Figure 6.4 Comparison of experimentally determined lives to crack initiation with theoretical predictions .

(1) the method of universal slopes, (2) the 10 percent rule, and (3) the frequency-modified strain-range-life relationship. It was observed (Fig. 6.3) that theoretical curves obtained by the method of universal slopes and by the 10 percent rule appreciably differ from the experimental data. The 10 percent rule was, however, conservative at lives of about 10^4 cycles. The effect of holding applied in both compressive and tensile half-cycles was also illustrated. Figure 6.4 shows a comparison of the experimental data with predictions made by the third method. The experimental data were generated in two tests with a total cycle time of 180 min. The difference between experiment and theory was more than 300% at $2\Delta\varepsilon_t = 0.2\%$.

The effect of holding times undoubtedly depends on material tested and test conditions; for example, in austenitic steels tensile holds are more damaging than compressive ones, whereas in Cr-Mo steels, case René 80 and IN-738 compressive holds were found most damaging. A satisfactory explanation of these effects was given by Ostergren (1976), who divided the damage introduced by a LCF test into:

1. Strain-induced damage, described by the hysteresis loop and mean stress. The introduction of hold times changes mean stress, as during holds, when the constant strain is maintained, stress partly relaxes. This accounts for the negative effect of compressive holds in some materials.
2. Time-dependent damage, where hold periods can manifest themselves in a changing mode of cracking as it does in austenitic steels.

Furthermore, Ostergren emphasized that most low-cycle fatigue life is spent in crack propagation. For that, the energy supplied to the specimen in the tensile half-cycle is a measure of damage introduced by one cycle and the value of total energy-to-fracture is assumed constant. The cycle frequency is calculated as follows:

1. For austenitic steels, for which compressive holds heal microstructural defects,

$$v = \frac{1}{\tau_0} \qquad \text{for } \tau_t \le \tau_c$$

and

$$v = \frac{1}{\tau_0 + \tau_t - \tau_c} \qquad \text{for } \tau_t > \tau_c$$

2. For Cr-Mo and nickel-based alloys, the frequency is calculated in the usual way:

$$v = \frac{1}{\tau_0 + \tau_t + \tau_c}$$

where

τ_0 = time per cycle spent in continuous strain cycling

τ_t = tension hold time

τ_c = compression hold time

after which, neglecting crack closure effect, the life relationship can be written as

$$\sigma_T \Delta\varepsilon_p N_f^\beta \nu^{\beta(k-1)} = C$$

where σ_T is the maximum tensile stress in the cycle and β, k, and C are constants. A good correlation was found between predictions based on the equation above and the results of experiments on both classes of materials.

If the thermal cycle consists of rapid straining followed by hold time, fatigue and creep contributions can be explicitly distinguished. To compute the N_f number of cycles to failure, the following damage parameter is introduced:

$$D = d_f + d_c = \sum_i \frac{N_i}{N_f(\Delta\varepsilon_{pi})} + \sum_i \frac{t_i}{t_r(\sigma_i)}$$

where

the first term introduces fatigue damage, and the second term is a creep contribution,

N_i = number of cycles applied with a particular plastic strain amplitude $\Delta\varepsilon_i$

$N_f(\Delta\varepsilon_{pi})$ = number of cycles to failure for a particular strain amplitude $\Delta\varepsilon_{pi}$

t_i = total time spent at a particular stress level σ_i

$t_r(\sigma_i)$ = time to rupture corresponding to this σ_i stress level

The D value at failure reveals considerable scatter; for this, the equation above has little predictive capability. For example, in a thermal fatigue test, Kazancev and co-workers (1983) found $d_c + d_f$ within 0.5 and 1.5, depending on test conditions.

In 1971, Manson et al. proposed a new life-prediction method, called the strain-range partitioning method. The method recognized the desirability of dividing the inelastic strain into two components: (1) creep strain, which is time dependent and temperature activated; and (2) plastic strain, which results from crystallographic slip. These two components of strain can be combined in two directions of uniaxial loading, tension and compression, thus producing four different combinations: PP (plastic strain in tension reversed by compressive plastic strain), CP (creep in tension reversed by compressive plastic strain), PC (tensile plastic strain, creep in compression), and CC (tensile creep reversed by compressive plastic strain). The strain-life relationship is determined in a series of laboratory tests that feature these four types of cycles. If all four are drawn in the coordinates inelastic strain range versus number of cycles to failure, the PP curve is usually the highest and the CP curve is usually the lowest (see Fig. 6.5). In the case of a complex load, the resultant hysteresis loop is divided into four basic loops and the predicted life postulated to follow the summation rule:

$$\frac{F_{pp}}{N_{pp}} + \frac{F_{cc}}{N_{cc}} + \frac{F_{cp}}{N_{cp}} + \frac{F_{pc}}{N_{pc}} = \frac{1}{N_{pr}}$$

where N_{pp}, N_{cc}, N_{cp}, and N_{pc} are the cyclic lives determined from the life relationships for each of the strain-range components, and N_{pr} is the predicted life. The strain-range fractions are defined as follows:

$$F_{pp} = \frac{\Delta\varepsilon_{pp}}{\Delta\varepsilon_{in}}$$

$$F_{cc} = \frac{\Delta\varepsilon_{cc}}{\Delta\varepsilon_{in}}$$

$$F_{cp} = \frac{\Delta\varepsilon_{cp}}{\Delta\varepsilon_{in}}$$

$$F_{pc} = \frac{\Delta\varepsilon_{pc}}{\Delta\varepsilon_{in}}$$

where $\Delta\varepsilon_{in}$ is the total inelastic strain range and N_{pp}, N_{cc}, N_{cp}, and N_{pc} are evaluated for the strain range $\Delta\varepsilon_{in}$.

Figure 6.6 shows an arbitrary hysteresis loop. The tensile inelastic strain AD consists of plastic AC and creep CD components. Similarly, the compressive inelastic strain DA can be separated into plastic DB and creep BA components. In

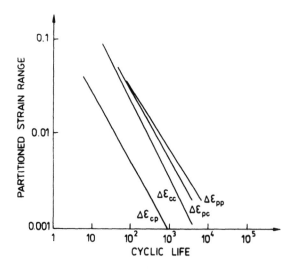

Figure 6.5 Summary of partitioned strain-life relations, type 316 stainless steel, 980 K. (From Manson et al., 1971.)

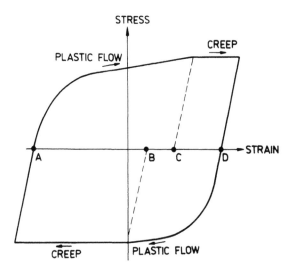

Figure 6.6   Hysteresis loop. (From Manson et al., 1971.)

general, neither plastic components nor creep components equal each other. Obviously, the tensile inelastic strain AD is equal to the compressive inelastic strain DA, as we have a closed hysteresis loop.

Now considering the types of tests generating the strain-life relationships, the following can be taken into account (Manson et al., 1971):

1. N_{pp} can be evaluated in tests run at sufficiently high frequencies to exclude creep components. Frequencies on the order of 0.1 Hz were found adequate for the 2¼Cr–1Mo steel. The corresponding hysteresis loop is depicted in Fig. 6.7a.

2. N_{pc} can be determined in three ways. In the first, tensile straining is applied at a high rate to minimize creep contribution. The compressive portion of the cycle involves a hold period (see Fig. 6.7b), the tensile plastic strain is denoted EB; the compressive plastic strain is BF. In this example, EF can be considered equal to $\Delta\varepsilon_{pc}$, whereas FB equals $\Delta\varepsilon_{pp}$. The second type of test is its modification: Temperature is reduced during tensile straining to make creep contribution negligible. The third type of test involves rapid straining imposed in both tensile and compressive half-cycles and a hold period of constant peak compressive strain. During this hold period stress relaxes and the elastic strain converts into compressive creep strain. The hysteresis loop for this type of test is shown in Fig. 6.7c; $\Delta\varepsilon_{pc}$ equals EF and $\Delta\varepsilon_{pp}$ equals FB.

3. N_{cp}, for which the types of tests available are identical with the above, but the conditions under which the tensile and compressive strains are applied must

be interchanged. As above, $\Delta\varepsilon_{pp}$ is present in these tests. Assuming the linear life fraction rule, we get

$$\frac{N}{N_{pp}} + \frac{N}{N_{cp}} = 1$$

where N is the number of cycles to failure and N_{pp} and N_{cp} are lives corresponding to $\Delta\varepsilon_{pp}$ and $\Delta\varepsilon_{cp}$, respectively. Solving this equation for N_{cp} yields the desired relationship (see Figs. 6.7d and e).

4. N_{cc}, for which the tests considered should involve low stress levels—which, however, makes them impracticable. By raising stress levels we admit plastic components to occur in both half-cycles. These components are denoted $\Delta\varepsilon_{cp}$ and $\Delta\varepsilon_{pc}$. Figure 6.7f shows examples of hysteresis loops obtain in proposed tests. the $\Delta\varepsilon_{cc}$ life relation must be deduced from the equations

$$\frac{N}{N_{pp}} + \frac{N}{N_{pc}} + \frac{N}{N_{cc}} = 1$$

or

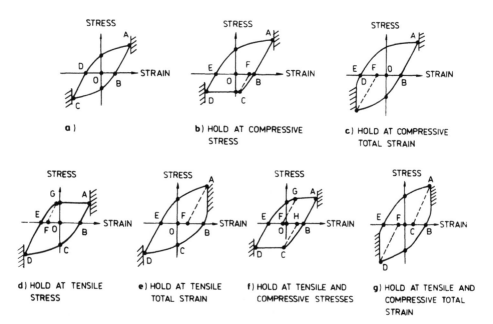

a)

b) HOLD AT COMPRESSIVE STRESS

c) HOLD AT COMPRESSIVE TOTAL STRAIN

d) HOLD AT TENSILE STRESS

e) HOLD AT TENSILE TOTAL STRAIN

f) HOLD AT TENSILE AND COMPRESSIVE STRESSES

g) HOLD AT TENSILE AND COMPRESSIVE TOTAL STRAIN

Figure 6.7 Types of tests applied to reveal relations between cyclic life and each of the basic strain components. (From Manson et al., 1971.)

$$\frac{N}{N_{pp}} + \frac{N}{N_{cp}} + \frac{N}{N_{cc}} = 1$$

depending on test method. In Fig. 6.7f, $\Delta\varepsilon_{cc}$ is given by FB or by the smaller of the two creep strains; $\Delta\varepsilon_{pp}$ is the smaller of the two plastic strains or HB $\Delta\varepsilon_{pc} = EB - HF - GB$. For the loop depicted in Fig.6.7g, $\Delta\varepsilon_{cc} = CB$ and $\Delta\varepsilon_{pp} = EC$; $\Delta\varepsilon_{pc}$ and $\Delta\varepsilon_{cp}$ are neglected.

The SRP method has a number of interesting features. Its major advantage is that it places upper and lower bounds on the cyclic lifetime. It means that the number of cycles to failure should be within a certain limit irrespective of the form of the cycle. The disadvantage is that an accurate partitioning of components at small strain ranges such as those encountered in aircraft gas turbine disks (10^{-4}) was found impractical. Numerous tests showed that the SRP method is not the fully unifying approach for describing the experimental data that was originally hoped. For this reason the validity of the SRP method must be thoroughly checked for the material and temperature range under consideration before any practical use is made of computed lives. To universalize this relationship, Manson (1973) normalized the strain range by dividing it by plastic or creep ductility, depending on which of the basic cycles is considered.

In the case of IN-100, Cailletaud et al. (1983) found poor correlation between the results of thermal fatigue tests and predictions made on the basis of strain-range partitioning. The maximum temperatures of thermal cycles were above 970 K, with

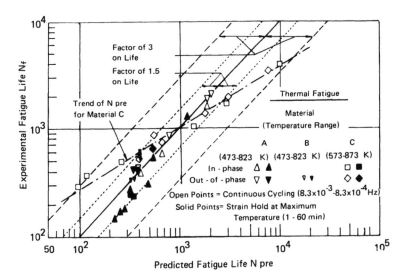

Figure 6.8 Comparison of predicted life with experimental life in thermal-mechanical low-cycle fatigue.

successive stages of γ' precipitation dissolution at high temperatures, and during cooling, new, very fine precipitates were generated. Kuwabara and Nitta (1979) tested three heats of type 304 stainless steel in thermal–mechanical fatigue where both temperature and stress varied independently. Details of test conditions are given in Fig. 6.8. It is seen that thermal mechanical fatigue lives were predicted to a factor of about 1.5 at the lower temperature range 473 to 823 K; at the higher temperature range 573 to 873 K the lives were predicted to within a factor of 3. Two possible sources of inaccuracy were considered: dynamic strain aging and errors associated with the partitioning procedure.

REFERENCES

Buba, N. (1983). Experimental verification of failure criteria applied to thermal fatigue of metals, Ph.D. thesis, Technical University of Wroclaw.

Cailletaud, G., Culié, J. P. and Kaczmarek, H. (1984). Thermal fatigue of a thermally unstable alloy, in *Mechanical Behaviour of Materials*, Vol. IV, *Proc. 4th International Conference*, Stockholm, 1983 (Carlsson, J. and Ohlson, N. G., eds.), Pergamon Press, Elmsford, N.Y.

Coffin, L. F., Jr. (1954). *Trans. ASME, 76*: 931.

Coffin, L. F., Jr. (1971). *Metall. Trans.*, 2: 3105.

Ivanova, W. S. (1963). *Ustalostjoje razrušenie metallov*, Metallurgizdat, Moscow, p. 135.

Kazancev, A. G., Tschemych, A. N., Gusenkov, A. P. (1980). *Zav. Lab. 46:*74.

Kussmaul, K. and Bhongbhibhat, S. (1980). *Stahl Eisen, 100*: 1341.

Kuwabara, K. and Nitta, A. (1979). In *Mechanical Behaviour of Materials*, Vol. 2 (Miller, K. J. and Smith, R. F., eds.) p. 76.

Manson, S. S. (1953). *Behaviour of Materials Under Conditions of Thermal Stress*, NACA TN-2933.

Manson, S. S. (1973). *The Challenge to Unify Treatment of High Temperature Fatigue: A Partisan Proposal Based on Strain-Range Partitioning*, STP 520, ASTM, Philadelphia, p. 744.

Manson, S. S. and Halford, G. R. (1967). A method of estimating high-temperature low-cycle fatigue behaviour of materials, in *Proc. International Conference on Thermal and High-Strain Fatigue*, Metals and Metallurgy Trust, London, pp. 254–270.

Manson, S. S. and Hirschberg, M. H. (1965). *Exp. Mech.*, 5(7): 193.

Manson, S. S., Halford, G. R. and Hirschberg, M. H. (1971). Creep-fatigue analysis by strain-range partitioning, in *Symposium on Design for Elevated Temperature Environment*, ASME, New York, pp. 12–28.

Monkman, F. C. and Grant, N. J. (1956). *Proc. ASTM, 56*: 593.

Ostergren, W. J. (1976). *J. Test. Eval.*, 4: 327.

Udouchi, T. and Wada, T. (1971). Thermal effect on low-cycle fatigue strength of steels, in *Proc. International Conference on thermal Stress and Thermal Fatigue*, Gloucestershire, 1964, Butterworth, Sevenoaks, Kent, England, p. 109.

Yamaguchi, K. and Nishijima, S. (1986). *Fatigue & Fracture of Eng. Mater. & Struc.*, 9(2).

Investigations of the Structure and Properties of Metals in the Course of Thermal Fatigue

7.1 CHARACTERISTIC EXPERIMENTAL CYCLES

The real in-service thermal load conditions of structures and machine components cannot always be simulated precisely in the laboratory, although, of course, simulation of the thermal cycle must be as precise as possible and the model used in testing should resemble the critical features of the actual part. Under industrial conditions, the majority of elements are loaded according to one of the following patterns:

Shock heating, short-duration annealing, and slow cooling
Heating and cooling at high rates
Slow heating and rapid cooling
Heating and cooling at low rates

These patterns are illustrated in Fig. 7.1.

In laboratory testing similarity to conditions of use is required to provide confidence in the answers to questions concerning the service life of the element tested and to identify the factors that influence its durability.

Characteristic testing cycles are shown in Figure 7.2; some of them involve two independent programs, the first for temperature changes and the second for changes in the stress applied. The more complex cycles provide good simulation of industrial conditions and facilitate analysis of the phenomena observed. Coffin's method of investigation is widely used; however, it is modified by some investigators to simplify the testing procedure. During testing the specimen is heated in the specified manner and cooled in various media. Heating is performed by high-frequency current, by electrical resistance heating, in furnaces, in a stream of hot air, in the flame of a gas burner, or by dipping in molten metal. Cooling can be performed in an airstream, water, oil, an oil-and-water emulsion, or in a solution of salt (NaCl) in water.

Specimens used in laboratory testing are of various shapes: cylindrical, in the

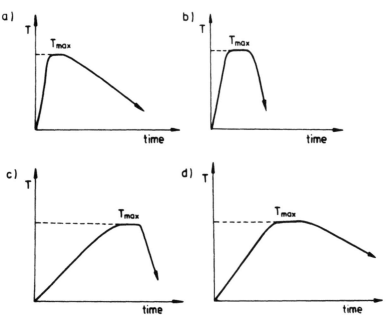

Figure 7.1 Thermal cycle spectra frequently encountered in machine parts working under industrial conditions: (a) rapid heating, annealing, slow cooling; (b) rapid heating, annealing, rapid cooling; (c) slow heating, annealing, rapid cooling; (d) slow heating annealing, slow cooling. (After Weronski, 1983.)

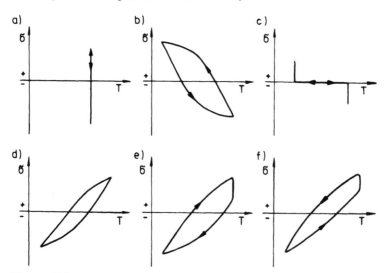

Figure 7.2 Patterns of cyclic stress–temperature variations in characteristic test cycles: (a) isothermal cycle; (b) Coffin's cycle; (c) Torrest–Armstrong cycle; (d) accelerated cycle; (e) relaxation cycle; (f) cycle including creep stage.

form of tubes or rings, cubic, prismatic, or flat. The shape of the test piece should be geometrically similar to the working element.

7.2 MICROSTRESSES IN THE PROCESS OF THERMAL FATIGUE

In conditions of cyclic heating and cooling the process of crack nucleation is determined to a very great extent by the presence of multiphase structures, by the differences in the thermal properties of the various phases, and by their distribution within the matrix. Micronotches can be formed by nonmetallic inclusions of the following types: sulfides, silicates, oxides, or by large coagulations of carbides deposited along grain boundaries. They can also be formed by oxides of the Fe_3O_4 type resulting from surface corrosion, particularly when they are formed in surface recesses. Inclusions can be one of the dominant causes of crack initiation, especially when they are deposited in bands. This is illustrated in simplified form in Fig. 7.3.

Figure 7.3a shows an inclusion notch, of length $2C$, having a coefficient of thermal expansion different from that of the matrix. At a given temperature, zones of stress concentration exist at the notch tips and for the case illustrated are sufficient to produce regions of plastic deformation, indicated by circles of radius r.

Figure 7.3b shows the situation resulting from a rise in temperature. The length of the notch has increased by Δc, producing larger regions of plastic deformation, of diameter $2r + \Delta r$, which overlap the regions of plastic deformation already formed. In the overlapping zones the microstresses may be sufficient to cause decohesion in microvolumes.

Figure 7.3c shows the situation when a number of inclusions constituting micronotches have been deposited in a band. With increasing temperature, the regions of plastic deformation originating at notch tips become larger and progress as described above, the direction of their shift being indicated by arrows. Where two such regions produced by adjacent inclusions overlap, there is a zone of microcrack nucleation, shown as an ellipse.

The foregoing relates to the effects of differences between the coefficients of thermal expansion of adjacent volumes which at any instant are all at the same temperature. This condition is approximated to the extent that the specimen is thin and the rate of change of temperature is low enough.

In other, more common circumstances the temperature will not be uniform throughout the material, and this will lead to additional stresses, even in the absence of variations in the coefficient of thermal expansion within the material. In a multiphase material this effect is enhanced by differences in the thermal conductivity of the various phases. Microstresses will also arise in a multiphase material if, during the thermal cycle, any phase transformations occur and are accompanied by a difference in specific volume.

Depending on the type of material, its history, and the magnitude of the thermal cycle, any or all of these phenomena may occur and the thermal fatigue process may be very complex. Quantitative analysis, which may be confirmed by experi-

a)

b)

c)

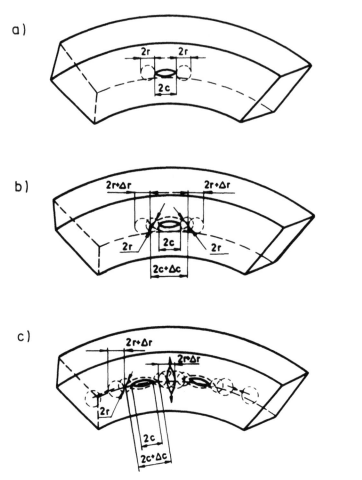

Figure 7.3 Mechanisms of microcrack initiation by inclusions (schematic). (After Weronski, 1983.)

ment, is possible when one mechanism is dominant. This is illustrated in a paper by Kucharski and Wendorff (1975), which presents the results of experiments with carbon steel C = 1.2% and chromium steel 6H15. The authors set out to study the influence of structural microstresses in the vicinity of carbide particles on the progress of thermal fatigue. Suitable structures were obtained by heat treatment, that of the carbon steel being ferrite with spherical inclusions of Fe_3C and that of chromium steel being chromium ferrite with spherical inclusions of $(CrFe)_{23}C_6$.

Metallography was used to find the distribution of the sizes of the carbide particles, which was approximately Gaussian, and for each of the two types of carbide the most frequently occurring diameter was determined for use in the analysis. The spatial distribution of the carbides was assumed, for the purpose of analysis, to be analogous to that of the face-centered cubic lattice of crystallography, but much enlarged. The parameters of these lattices were calculated from the chemical composition of the materials, rather than measured, but the idealized assumptions regarding spatial distribution were supported by metallography.

From the foregoing, the data required for the analytical calculations [Eqs. (7.1) to (7.6) below] were obtained: Most probable diameter of carbide inclusions ($2R_1$): carbon steel, 0.8 μm; chromium steel, 0.45 μm. Minimum separation between adjacent carbide particles: (R_2): carbon steel, 0.77 μm; chromium steel, 0.65 μm.

For the experimental investigation all the specimens were prepared in the form of thin strips, 0.1 mm × 6 mm × 100 to 150 mm, to minimize thermal gradients within the volume of the material and to provide a relatively high electrical resistance, facilitating heating by passing an electric current.

Testing was conducted in an evacuated chamber to minimize the effects due to chemical deterioration of the surface. All the specimens were subjected to constant tensile stress throughout the experiment, 34.3 MPa in the case of carbon steel and 48 MPa for chromium steel. The total permanent elongation of each specimen was taken as the measure of the effect of thermal fatigue. The thermal cycles all consisted of heating at a rate of about 30 K/s followed by free cooling. For each particular specimen or batch of specimens the magnitude and duration of the thermal cycle were unchanged throughout the test, which extended to a substantial number of cycles, but these parameters, and the number of cycles, were deliberately varied between specimens or batches of specimens (particulars of some of the results are given in the notes to Fig. 7.4). The variation in elongation produced by the different thermal regimes is of interest in itself; in addition, the differences between these

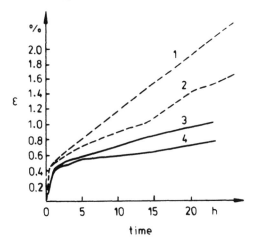

Figure 7.4 Time dependence of elongation of 1.2% carbon steel test pieces exposed to thermal fatigue and of control samples. 1, Test pieces, cycle temperatures 848 to 610 K, cycle period 1 min; 2, control samples for 1 above; 3, test pieces, cycle temperatures 833 to 413 K, cycle period 3 min; 4, control samples for 3 above.

regimes indicate that the general agreement between the predicted and observed phenomena is not the fortuitous result of employing a particular test cycle.

Control specimens identical with those undergoing the thermal fatigue test were subjected to the same mechanical stress but exposed to only one thermal cycle. This cycle was, however, prolonged by a programmed supply of electrical energy during the cooling phase, so that it occupied the same amount of time as the full number of cycles experienced by the corresponding specimens, which were subjected to repeated thermal cycling. The performance of these controls is also shown in Fig. 7.4. It can be seen that their elongation was less than that of the specimens that were cycled repeatedly, although in the case of chromium steel the difference was very small.

At the end of the test, measurements at room temperature showed that in the case of 1.2% carbon steel the permanent elongation of the specimens subjected to repeated thermal cycles was substantially greater than that of the controls. However, in the case of chromium steel no significant differences were observed between the test pieces and controls.

Metallographic examination of the thermally cycled specimens showed that polygonization was more advanced in the carbon steel, but these differences were small and not sufficient to account for the difference in performance of the two classes of steel. The difference is explained by calculating the microstresses in the materials.

The value of the microstress P at the surface of the spherical carbide particles can be calculated from the following expression (Kucharski and Wendorff, 1975; Brooksbank and Andrews, 1969, 1972):

$$P = \frac{(\alpha_2 - \alpha_1)T}{\dfrac{0.5(1 + v_2) + (1 - 2v_2)d^3}{E_2(1 - d^3)} + \dfrac{1 - 2v_1}{E_1}} \tag{7.1}$$

where

R_1 = radius of a carbide particle

R_2 = distance between surfaces of carbide particles

$d = R_1/R_2$

α_1, α_2 = coefficients of thermal expansion of the carbide and matrix material, respectively

v_1, v_2 = Poisson's ratio of carbide and matrix material, respectively

E_1, E_2 = Young's modulus of carbide and matrix material, respectively

T = temperature, Kelvin

In this example the authors choose the value of the microstresses corresponding to the highest temperature of the cycle as the reference level, to simplify calculations. For the same reason, v_1 and v_2 were assumed constant within the range of temperature considered. Variation of yield point with temperature was taken into account, as will be seen later. Following Brooksbank and Andrews (1969, 1972), the radial and tangential components of microstress σ_a and σ_b, respectively, are next found.

Using R to denote distance measured from the center of one carbide particle toward the center of another, we have: for the carbide,

$$\sigma_a = P \quad \text{constant for } 0 \le R \le R_1 \tag{7.2}$$

$$\sigma_b = P \quad 0 \le R \le R_1 \tag{7.3}$$

and for the matrix,

$$\sigma_a = \frac{P}{1 - d^3}\left(\frac{1}{2}\frac{R_1^3}{R^3} - d^3\right) \quad R_1 < R < R_1 + R_2 \tag{7.4}$$

$$\sigma_b = \frac{-P}{1 - d^3}\left(\frac{1}{2}\frac{R_1^3}{R^3} + d^3\right) \quad R_1 < R < R_1 + R_2 \tag{7.5}$$

The stress at any point in the matrix material is the sum of the stresses due to the presence of the carbide particle ($R_1 < R < R_1 + R_2$). The magnitude of the reduced stress is

$$\sigma_{\text{red}} = \frac{3}{2}\frac{|P|}{(1 - d^3)}\frac{R_1^3}{R^3} \tag{7.6}$$

The value of this stress in the space between carbide particles is plotted, as a function of distance from the center of one particle, for carbon and chromium steels in Figs. 7.5 and 7.6. The carbide particles are indicated by arcs. It can be seen that the stress is highest in the immediate vicinity of the carbide particles. The plots of stress distribution are for the lowest temperature in the test cycle, the authors (Kucharski and Wendorff, 1975) having taken the stress at the highest temperature in the cycle as their zero reference on the assumption that stresses were relaxed at this temperature and thereafter built up as the temperature fell. Since the stress at any point in the material is directly proportional to the temperature [Eqs. (7.1) and (7.6)], a plot of stress distribution at any intermediate temperature of interest will have the same form as that shown, but the numerical values of the stresses represented by the ordinates will be reduced. The reduction will be in the ratio that the difference between the temperature of interest and the maximum temperature bears to the difference between the maximum and minimum temperatures.

The regions of the host material that have undergone plastic deformation at the lowest temperature of the cycle are identified after projecting the yield point stress (σ_y) of the host material at that temperature onto the stress distribution shown. This process is repeated with any other temperatures for which the stress distributions have not been plotted by the artifice of projecting the yield point stress at the temperature concerned after it has been increased, by graphical construction, in the

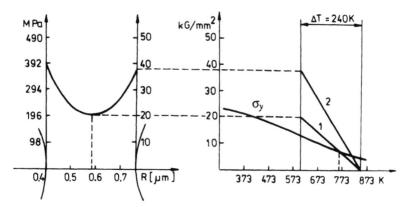

Figure 7.5 Stress distribution in the zone between spherical carbide precipitates for 1.2% carbon steel, cycles 848 to 610 K. σ_{red}, Reduced stress; σ_y, yield point; R, distance from carbide center (μm).

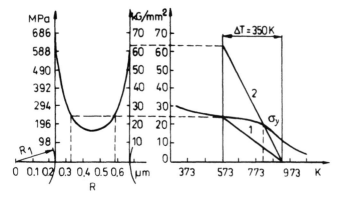

Figure 7.6 Reduced stress distribution in the zone between spherical carbide precipitates for chromium steel, cycles 923 to 573 K. R_1, carbide radius (μm); R, distance from carbide center (μm).

inverse of the ratio by which the stresses in the material have, at that temperature, decreased with respect to the maximum values that have been plotted. Although the absolute values of stress are false, the method is simple and valid for the present purpose.

In Fig. 7.5 it can be seen that the entire volume of the material, carbon steel, undergoes plastic deformation at and below a temperature of about 760 K. This would, of course, be repeated in every cycle (see segments 1 and 2 in this figure).

In contrast, the chromium steel (Fig. 7.6) only starts toward plastic deformation at temperatures below 820 K and throughout the cycle plastic deformation is confined to the vicinity of the carbide particle (see segments 1 and 2 in Fig. 9.6).

The substantial elongation of the carbon steel specimens was no doubt caused by plastic deformation occurring throughout the material, but this by itself cannot account for their elongation being much greater than that of the controls in which plastic deformation would also have occurred in all parts of the material in the course of the long-duration cycle to which they were exposed. This cycle, it will be remembered, exposed the controls to the same range of temperatures as the specimens and occupied the same amount of time as the specimens' many shorter cycles. The explanation lies in the fact that the controls' cycle corresponded to a modified creep test, and they demonstrated normal creep, whereas the specimens demonstrated the effects of thermal fatigue. In the case of chromium steel the elongation of controls and specimens was small and the differences almost negligible; however, insofar as the elongation of the specimen exceeded that of the controls, the same explanation applies.

Regarding the work of Kucharski and Wendorff (1975), we see that their analysis was well supported by experimental results and there is no reason to doubt that the magnitude and distribution of the thermal microstresses in the specimens were as great as their analysis predicted. Nevertheless, their assumption that the stresses were fully relaxed at the highest temperature in each cycle, although probably justified with regard to the controls, seems rather artificial with regard to the specimens.

The controls were subjected to one very slow thermal cycle in which there was time for stresses to be relieved, but the specimens were subjected to cycles lasting for not more than 3 min and the time spent in the vicinity of maximum temperature in each cycle was much less. It therefore seems possible that annealing was not completed during the peak of each cycle. If this was the case, zones of plastic deformation produced in one cycle would have remained in existence, more or less unchanged, at the beginning of the subsequent cycle and the process would have proceeded cumulatively in the manner described earlier and illustrated in Fig. 7.3. The distortion of the carbon steel specimens would have been produced by the overlapping of zones of plastic deformation.

There would have been virtually no overlapping of zones of plastic deformation in chromium steel, as the volume of material in which the plastic limit is exceeded is small and the carbide particles are separated by a much larger volume of material, in which the stress is well below the plastic limit. We consider this a more realistic way of associating the theoretical and experimental results and hence regard the work of Kucharski and Wendorff (1975) as supporting the model shown in Fig. 7.3.

7.3 CRITERIA EMPLOYED IN ESTABLISHING THE RESISTANCE OF COMPONENTS TO THERMAL FATIGUE

Changes in the microstructure of components and equipment exposed to thermal fatigue occur in each thermal cycle and affect the mechanical and physicochemical properties of the material. Although the change occurring in an individual cycle is generally too small to be detected, the effects are cumulative and after some time cause the nucleation of one or more microcracks, which, after reaching critical dimensions, develop rapidly, leading to visible macrocracks and thereafter to fracture.

In the majority of cases some 90% of the life of a component exposed to thermal fatigue is the period of cumulative change in the microstructure, and it is in the last 10% of life that cracks develop. Hence, for reasons of safety and economy, it is important for designers and users to know in advance when the first macrocrack is likely to occur.

Accordingly, laboratory tests are made to estimate, as accurately as possible, the useful service life of a component made of a given material when it is used in the actual industrial conditions for which it is intended. It is therefore necessary to have some practical criterion to identify the end of useful life. The resistance to thermal fatigue can then be expressed as the number of thermal cycles, with the specified parameters, which the component can endure before the end of life, as defined by the criterion, is reached. The choice of the criterion depends on the material, the operating conditions, and the function the component must perform. Ideally, the criterion should be simple to apply in the laboratory and in the field and allow unambiguous interpretation of the results. These factors all point to employing a criterion based on visible changes in the appearance of the outer surface of the test specimen or component (Zmihorski and Zólciak, 1973). It is seldom that much is lost by adopting a simple criterion of this nature since changes occurring below the surface are difficult to detect and also difficult to interpret, whereas the final stage is almost always accompanied by a distinct change at the surface. Accordingly, the number of thermal cycles to the occurrence of the first visible crack is frequently taken as the measure of thermal fatigue resistance, that is, of useful life.

In some materials, especially if the thermal cycles cause substantial deformation, the moment at which a single first crack occurs is difficult to identify, as there is craze cracking in which several cracks may appear almost simultaneously. In such cases the criterion often adopted is the occurrence of a few surface cracks. It is sometimes useful to use an index of surface destroyed (i.e., the ratio of cracked surface area to the entire surface area of the specimen) as the criterion. For components that can be reclaimed, the criterion should be expressed in terms of the maximum length and, especially, the depth of the crack, the number of cracks, consistent with reconditioning. Of course, for vital components, whatever the criterion, examination for cracks must be exceptionally thorough and all of the test specifications rigorously observed.

In general, the resistance of a material to thermal fatigue, expressed by the number of thermal cycles, N, to the appearance of the fatigue sign chosen as a criterion, can be written as (Weronski, 1985)

$$N = f(A, B, C)$$

where

A = set of values describing the properties of the material and test piece

B = set of values characterizing the thermal environment

C = correction factor relating laboratory tests to design and operation

Set A describes the chemical composition and microstructure of the specimen, its mechanical and physicochemical properties, its geometric shape, the influence of stress raisers, and the state of its surface. Set B describes the temperature range of the cycle, the rate of change of temperature during heating and cooling, the duration of the cycles, and the nature of the surrounding media.

In principle the A values are known or determinable in the laboratory; the B values, by the application; and those of C, by the safety factor desired and which takes account of the scale of the laboratory tests, the extent to which they have reproduced the industrial environment, the size and shape of the component, and experience.

These factors influence the result to different degrees. The most important single factor is the chemical composition of the material. If this is unsuitable, treatment aimed at improving the microstructure and protecting the surface will be of little help. The majority of published works relating chemical composition to thermal fatigue performance concern two groups of steels: high-alloy austenitic steels and low-alloy steels. Even in these two cases the number of publications is too small and the differences between the experimental conditions adopted by the various authors are too great to provide unequivocal descriptions of the effect of individual alloying additions in all circumstances.

Largely because of the complicated character of the changes that can occur in the microstructure there is no universal hypothesis or even group of partial hypotheses enabling us to predict the course of thermal fatigue in one alloy used in one set of conditions from the results of tests carried out under different conditions on other alloys of the same elements in different proportions. The explanatory mechanisms currently used are sufficient for reasonably confident prediction of the performance of a material provided that it is substantially similar in chemical composition and microstructure to those for which performance data are available and that the operating conditions envisaged are not widely different from those in which the data were obtained. Although these restrictions are severe, the limited degree of prediction they do permit can often save time and money in planning investigations, as in practice these frequently concern only small departures from known territory. Beyond this, the work reported provides the following general guidelines.

Increasing carbon content decreases resistance to thermal fatigue, the effect becoming pronounced for carbon contents exceeding 0.5% (Gierzynska and Smarzynski, 1979).

Increasing tungsten content also decreases resistance to thermal fatigue, and tungsten content should not exceed 3% (Geller, 1964; Hecht and Hiller, 1959). It has been shown that small amounts of tungsten in conjunction with certain other elements may be beneficial.

Elements such as chromium, molybdenum, vanadium, and niobium increase resistance to thermal fatigue. The effect of the latter three elements is particularly pronounced (Weronski, 1980); additions of only a few tenths of 1% produce a distinct improvement.

The alloying additives have a decisive influence on the microstructure of steel. Molybdenum, especially if accompanied by niobium and tungsten, facilitates the relaxation of stresses and retards further development of cracks.

The thermomechanical properties greatly affect the process of cracking. A number of the processes by which this may occur in a multiphase material were discussed in Section 7.2, particularly those arising from the presence of inclusions having a coefficient of thermal expansion different from that of the surrounding matrix.

The size and shape of a component may affect the initiation and development of cracks in several ways. In thick cross sections the thermal gradients, produced at any but the lowest rates of cycling, increase stress of the first kind because the thermal expansion is not uniformly distributed. Thus we find differences between the results of tests on specimens that are of the same material and shape but of different sizes, and this must be borne in mind when applying the results of tests on small specimens to the design of a large component. Abrupt changes in the profile of a component increase the local concentration of stress (the notch effect) and also affect the distribution of temperature in the material.

The state of the subsurface layer is extremely important, as in the majority of cases thermal fatigue cracks start there. Its structure and properties are affected by rough machining to final dimensions. Surface roughness also increases the effect of a corrosive environment, which will create crack nuclei in the pits of the notches left by machining. Experiments directed to improving the state of stress in the subsurface layer by mechanical treatments such as burnishing, which have been advantageous in the case of mechanical fatigue, show that they are of little advantage in respect to thermal fatigue, as at the temperatures normally encountered in practice, the good effects initially produced are greatly reduced by recovery and recrystallization. It seems more reasonable to carry out suitable thermomechanical treatment, which can provide a considerable increase in thermal strength. Increased resistance to thermal fatigue can also be achieved by the introduction of chromium to the subsurface layer (Weronski, 1983), by chromizing (Pesat, 1972; Restall, 1979), or by cyaniding. Work on the introduction of various elements (e.g., boron) to the subsurface layer is proceeding.

The subsurface layer is also exposed to the processes of adsorption and corrosion. The adverse effects of these processes are intensified at high temperatures and by cooling with media such as wet pressurized air or water. The superposition of these processes on the basic phenomena of thermal fatigue greatly accelerates the process of decohesion. The application of galvanic coatings, especially those of Cr, Ni, and Ni-W, improves resistant to thermal fatigue. The optimum thickness is a compromise between the requirement for protection and the avoidance of excessive stress due to differences in the thermal expansion of coating and substrate (Weronski, 1978).

The range and frequency of the thermal cycle have major effects on the process of thermal fatigue. Obviously, the magnitude of the thermal stress, which is the sum of the stresses due to the temperature gradient in the cross-section of the material and those due to heterogeneity of the material, increases with the amplitude of the cycle, and the thermal gradient itself increases with rate of temperature change and hence is a function of the amplitude, frequency, and waveform of the thermal cycle. The yield point decreases with temperature and this accelerates creep and progression of thermal fatigue. Annealing generally reduces resistance to thermal fatigue to an extent depending on the holding time in each cycle and hence on the amplitude and duration of the cycle.

The factors mentioned in this section are discussed in more detail in Chapter 8. The effect of these factors in a specific practical case can be seen in the study of a steel mold employed in centrifugal casting discussed in Section 7.4.

7.4 STRUCTURE, PROPERTIES, AND BEHAVIOR OF INDUSTRIAL COMPONENTS WORKING UNDER CONDITIONS OF THERMAL FATIGUE

The example considered is a steel mold used in the production of cast iron pipes by the process of centrifugal casting. In this section we are as interested in what actually happens to the mold material as in the theoretical reasons that account for it. Hence, following a brief general description of the centrifugal process, to indicate the nature of the resultant thermal load, we compare the results of approximate solution of the general equation governing the temperature distribution within the mold with some measured values. Similarly, the calculations of the stresses produced by the temperature gradients in the material are compared with those that have been measured.

7.4.1 Characteristics of the Process

Centrifugal casting, introduced over a century ago, is now very widely used to produce castings of many different sizes and shapes in a variety of metals and alloys. In essence, molten metal is poured into a mold which is rotating so that centrifugal force presses the metal against the surface of the mold, from which it takes a good impression. The speed of rotation is usually such that the accelerating field is some tens of times that of gravity. The intensity of the field increases with distance from

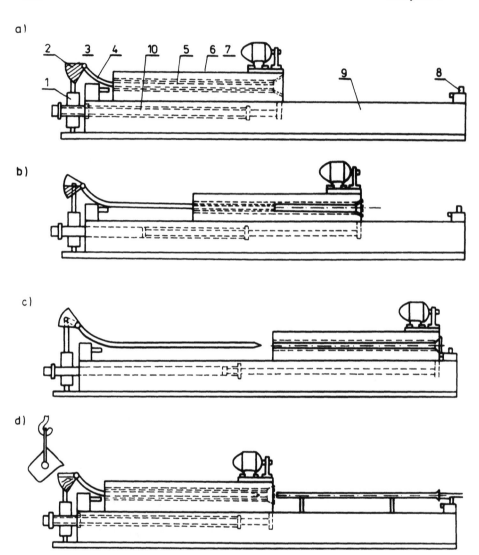

Figure 7.7   Pattern of successive stages in centrifugal casting of pipes by the de Lavaud system. 1, Cylinder operating ladle; 2, cast iron; 3, ladle; 4, chute; 5, trough; 6, water box; 7, motor; 8, puller block; 9, fixed bed; 10, travel cylinder.

the axis of rotation, and the substantial gradient encourages the movement of gases and dross toward the surface nearest the axis of rotation. The casting is designed so that this is a noncritical region. It leads to the production of sound castings. In addition, as the down runners used in gravity casting are not needed, there is a reduction in the amount of liquid metal required to make a given casting.

The process may be performed in many ways: The axis of rotation may be vertical, horizontal, or inclined; molds may be made of various materials but are generally of metal. We are concerned here with metal molds used in the manufacture of cast iron pipes by de Lavaud's method, which is employed extensively for this purpose, being characterized by high productivity of castings free of blowholes and of good structure as a result of the factors mentioned above and the advantageous radial abstraction of heat during solidification (Kolorz, 1980; McIntyre, 1961; Sakwa, 1974).

The equipment used is shown schematically in Fig. 7.7. The mold, in the form of a thick-walled hollow cylinder, is represented by the dashed lines. Its exterior is cooled by water in the surrounding box (6). The mold rotates about a horizontal axis and together with the water box is supported by a fixed bed (9), along which they can be moved by a hydraulic ram operating in the cylinder (10). The operating cycle begins with the mold assembly stationary in the position shown in Fig. 7.7a. The mold starts to rotate, driven by the motor (7), and when the required speed is reached, pouring is begun by tilting the ladle (3). The molten cast iron enters the mold via the chute (4) and the ceramic-covered trough (5). Simultaneous movement of the mold along the fixed bed, as indicated in Fig. 7.7b, ensures even distribution of the metal. Friction between the inner surface of the mold and the molten metal and the viscosity of the latter result in the poured metal acquiring the same angular velocity as the mold and hence being pressed against it by centrifugal action.

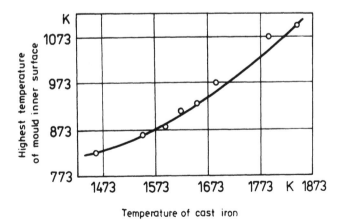

Figure 7.8 Dependence of maximum temperature of mold inner surface on temperature of cast iron in ladle.

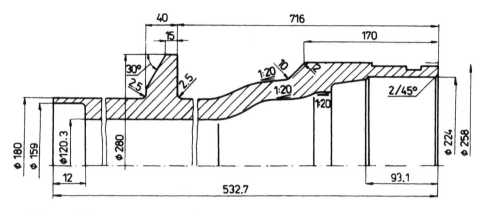

Figure 7.9   View of typical mold for casting pipes centrifugally.

Rotation of the mold is slowed down and stopped after sufficient time for solidification of the casting, which is then secured by internal tongs (8) so that it is removed from the mold when the latter is moved to the ladle end of the track in readiness for the next cycle (Fig. 7.7c and d).

The following data relate to the case to be considered, which is typical of general practice. The articles produced are cast iron water-main pipes 5 to 6 m long and 10 to 40 cm in diameter. The cycle time is approximately 120 s. The temperature of liquid cast iron entering the mold is 1623 to 1673 K. In industrial practice the pouring temperature rises at the inner surface of the mold. Figure 7.8 shows the relation between the maximum temperature of this surface and the initial temperature of the molten cast iron for the case of a mold such as that shown in Fig. 7.9.

The optimum speed of rotation is determined by a large number of factors,

Table 7.1   Adjustments for Producing Centrifugally Cast Pipes

Pipe diameter (mm)	Pipe length (mm)	Weight of cast iron (kg)	Speed of mold movement (m/min)	Mold rotation speed (rpm)
80	4000	70–76	12–13	850–950
100	5000	110–118	20–23	750–850
125	5000	142–150	20–23	650–750
150	6000	212–230	18–21	600–700
200	6000	302–330	14–16	500–600
250	6000	420–450	10–12	400–450
300	6000	525–570	8–10	350–420
400	6000	815–860	6–8	300–370

including the chemical composition, density and viscosity of the casting metal, smoothness of the mold, and gating system. It is usually calculated from the empirical relationship given by Konstantinov:

$$n = \frac{5520}{\sqrt{\gamma r}}$$

where

n = speed, rpm

γ = density of the molten metal, g cm^{-3}

r = inner radius of the mold, cm

Our own recommendations for the production of tube in the sizes commonly used in industry are presented in Table 7.1 and include adjustments based on experimental castings and checked in practice.

7.4.2 Heat Load on the Mold

It follows from the general description presented above that in the course of producing centrifugally cast pipes, the inner surface of the mold is subjected to rapid changes of temperature, whereas the temperature of the water-cooled outer surface is much less variable; in practice it does not exceed 420 K. This can be seen in Fig. 7.10, which shows the results obtained by Königer and Liebmann (1959), who measured the temperature variations at various positions in the cross section of a mold similar to that in Fig. 7.9 throughout successive casting cycles. Later, their results will be used in comparison with calculated temperatures and as a basis in calculating the thermal stresses. At present it is sufficient to note the general form and magnitude of the temperature changes at the inner and outer surfaces of the mold and to observe that they show some differences between the causes of the first casting cycle made in a cold mold and the second cycle, which immediately follows it. The results of subsequent casting cycles are not shown but are in fact very similar to those for the second cycle.

The derivation of an expression for the temperature distribution in the cross section of the mold in the course of a casting cycle may be based on some well-justified assumptions but should also take account of some complicating factors. The first assumption is that the mold and the casting within it may be treated as tubes of infinite length. the second assumption is that the thermal load to which the mold is subjected is axially symmetrical. This relies on the high rotational speed of the mold being sufficient to produce a uniform thickness of the casting material around the inner surface of the mold. The third assumption is that crystallization in the casting begins at the surface in contact with the mold (i.e., that the rate at which heat is abstracted is highest in this region), that the zone of crystallization proceeds radially inward, and that no crystallization is initiated independently at the inner

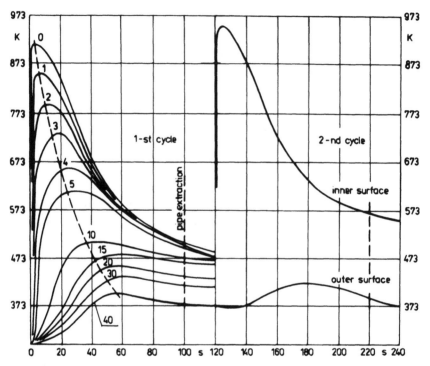

Figure 7.10 Mold temperature variations during centrifugal casting: in particular zones for first cycle, at inner and outer surfaces for second cycle. Numbers indicate distance of zone from inner surface (mm).

surface of the casting which is in contact with air (Königer and Liebmann, 1959; Büchler, 1959; Weronski, 1978).

The complicating factors are that a small gap appears between the inner surface of the mold and the cast material due to contraction of the latter and that oxide films are formed on the surfaces of the casting and mold.

The temperature distribution in the cross section of the mold containing the initially molten metal can be expressed in accordance with Fourier's relation (Skalski, 1973) as

$$q + \frac{1}{r} \frac{\partial}{\partial r}\left[r\lambda(r,T) \frac{\partial T}{\partial r} \right] = C_p(r,t)d(r,t) \frac{\partial T}{\partial t}$$

where

  q = rate of heat emission during solidification of the cast metal and during formation of the thin oxide films on the surfaces of the casting and mold

r = radius of the zone of the mold material that is being considered

$\lambda(r,T)$ = coefficient of thermal conductivity, which is a function of temperature and, because of the axial symmetry of the temperature distribution, is written as a function of both the radius r and the temperature T prevailing in the zone being considered at that radius within the system; the system itself consists of the following isotropic parts: solidifying metal, the oxide layer on it, the air gap, the oxide layer on the mold (the thickness of these three layers is increasing with time), and the mold material itself; the coefficients of thermal conductivity of each of these items are of course substantially different

$C_p(r,t)$ = specific heat and $d(r,t)$ = density, respectively, of each component at radius r from the axis, t being the time measured from the instant at which molten metal first came in contact with the mold

It is virtually impossible to solve this equation without further simplification. Skalski (1973) obtained a solution after making the following further assumptions: Pouring is an instantaneous process; heat is transferred by conduction alone (i.e., heat transfer by radiation across the thin gaseous gap between casting and mold is neglected); heat is not abstracted by the air within the hollow casting; and the two oxide layers and the gaseous gap are treated together as a single thermal insulator whose thickness is time dependent. The values of the thermal and material parameters λ, C_p, and d were treated as temperature dependent, but the dependencies assumed for the mold were those for a steel similar to, but not identical with, that of the mold used by Königer and Liebmann. Initial and boundary conditions and parameters of the materials other than the mold were selected to correspond to the circumstances in which Königer and Liebmann made their experiments.

The initial temperature of the inner surface of the mold was obtained by solving Cauchy's problem for the same system and physical mechanism. The calculated and measured temperature distributions were not in full agreement. The temperature of the inner surface of the mold immediately after the molten cast iron had been poured in was calculated as 948 K and measured as 908 K. It seems probable that the difference was due partly to a slight delay in the response of the thermocouples used to measure the mold's inner surface temperature, since inevitably they could not be in perfect contact with the actual surface itself. In the computation the principal source of error may have arisen from the assumed form of the time dependence of the thickness of the thermal barrier simulating the oxide and air layers between the casting metal and the metallic inner face of the mold. The dependence used would have underestimated the initial thickness.

The calculated temperature distribution in the cross section of the mold, as a function of time, was similar to that obtained experimentally, but the calculated temperatures were lower than those measured. This is accounted for partially by the fact that the temperature of the water surrounding the mold rose appreciably during the experiment, whereas in the calculations it was assumed that the contents of the water box were replaced sufficiently rapidly for its temperature to remain at its low initial value.

Very little work on heat transfer during centrifugal casting has been published and even less on mold oxidation and its effect on heat transfer. Consequently, the two studies discussed above are noteworthy. The measurements of temperatures at the positions selected for the thermocouples must be considered to be more accurate than the computation of the temperatures at these positions, performed with the aid of many simplifying assumptions. However, the computation sought to provide an overall picture of conditions throughout the mold and, being based on the dimensions of the mold and the properties of its material, may indicate their influence on the temperature distribution. The agreement between the two sets of results, although far from perfect, is reassuring, as the time-varying temperature distribution is an essential basis for considering the resultant stresses and thermal fatigue.

7.4.3 Thermal Stresses in the Mold Cross Section

It is apparent from the temperature distributions shown in Fig. 7.10 that the mold material will be subject to thermal stresses and that as the distributions are time dependent, these stresses are likely to vary with time. Thus, considering a thin annulus of material surrounding and containing the inner surface of the mold, it is obvious that at the moment when casting begins and its temperature rises very rapidly, its expansion will be resisted by the mold material surrounding it, in which the temperature rises less rapidly. Hence at this moment the innermost annulus will be subject to compression and that surrounding it to tension. Clearly, these stresses will diminish as the temperature of the surrounding material rises, and inspection of the temperature distributions at successive stages in the casting cycle (Fig. 7.10) strongly suggests that they will change sign. However, a quantitative description of the thermal stresses in the mold is necessary to account for the thermal fatigue that occurs, and this is provided by the work of Königer and Liebmann (1959), who used strain gauges to measure the tangential stress at selected points in the mold cross section. Their findings may be said to be supported by the work of Skalski (1973), who calculated the theoretical stresses on the basis of Königer and Liebmann's (1959) measurements of the temperature distributions. Both approaches showed that stresses occur in excess of the elastic limit. Precise description and calculation of the distribution of the thermal stresses is difficult because of the form and time dependence of the thermal distribution and the temperature dependence of the mechanical and physical properties of the material. The mold material is also subjected to mechanical stresses due to its mass and charge and the high speed of rotation. In principle, the superposition of mechanical stresses on the thermally produced stress in the nonlinear system may alter the position of the boundary between zones of elastic and plastic deformation. However, the form of the mold and the way in which it is supported make this factor sufficiently small to be neglected in what follows.

With the same assumptions as in Section 7.4.2 (including the fact that the material is isotropic) and the further assumptions that Young's modulus and Poisson's ratio are independent of temperature, the thermal stress within a region of elastic deformation can be expressed in a system of cylindrical coordinates (r, ψ, l), in which the axis l is parallel to the longitudinal axis of the mold, as

$$\sigma_r = -\frac{E}{1-v}\frac{1}{r^2}\int_{r_1}^{r}(\alpha_0 + \alpha_1 T)Tr\,dr + \frac{E}{1+v}\left(\frac{X_1}{1-2v} - \frac{X_2}{r^2}\right)$$

$$\sigma_\psi = \frac{E}{1+v}\frac{1}{r^2}\int_{r_1}^{r}(\alpha_0 + \alpha_1 T)Tr\,dr - \frac{E}{1-v}(\alpha_0 + \alpha_1 T)T + \frac{E}{1+v}\left(\frac{X_1}{1-2v} + \frac{X_2}{r^2}\right)$$

$$\sigma_l = -\frac{E(\alpha_0 + \alpha_1 T)T}{1-v} + \frac{2vEX_1}{(1+v)(1-2v)} + \sigma_c$$

where

$$E = \text{Young's modulus}$$

$$v = \text{Poisson's ratio}$$

$$r_1 = \text{radius of the inner boundary of the region of elastic deformation}$$

$$r = \text{radius of the zone being considered}$$

$$T = \text{temperature of the material at radius } r$$

$$\alpha = (\alpha_0 + \alpha_1 T) = \text{coefficient of linear thermal expansion}$$

$$X_1, X_2 = \text{constants that can be calculated from the boundary conditions}$$

$$\sigma_c = \text{correction factor that may be applied to take into account the free ends of the mold in the interest of greater accuracy}$$

It can be proved that the tangential stress is the greatest of those in the region of elastic deformation; however, it is not the cause of mold failure. The service life of the mold is limited by the thermal stresses in the region of plastic deformation.

Increased temperature in the zone considered reduces the yield point of the material, and this together with the stresses corresponding to the new temperature distribution results in the elastic limit being exceeded. In this case the components of stress can be expressed as

$$\sigma_r = \pm\int_{r_1}^{r}\frac{\sigma_y(T)}{r}\,dr + X$$

$$\sigma_\psi = \pm\sigma_y(T) + \sigma_r$$

$$\sigma_l = \frac{\sigma_r f(r) - \sigma_\psi}{f(r) - 1} + \sigma_c$$

where

σ_y = yield point

$f(r) = d\varepsilon_\psi / d\varepsilon_r$

ε_ψ = tangential deformation

ε_r = radial deformation

σ_c = correction factor taking into account the free ends of the mold

X = constant of integration

The progress of plastic deformation is obtained by evaluating these expressions for successive values of temperature corresponding to small increments of time. When the temperature begins to fall, the reduction of stress and increase in yield point in the zones that have been subjected to elastic deformation eventually results in reversion to elastic deformation.

The calculation of strain is thereafter based on Hooke's law, taking the dimensions resulting from the previous plastic deformation as the entry point. More generally, once the maximum values of stresses, deformations, and temperature have been found, they become the reference levels for subsequent calculation.

Skalski (1973) was able to show that the greatest stresses arise in the immediate vicinity of the inner surface of the mold and that the compressive stress in this subsurface zone, reaching its maximum 5 s after pouring begins, exceeds the elastic limit of the material and causes substantial plastic flow. He also showed that these stresses would be reduced if the wall of the mold were less thick. His consequent recommendation is generally applicable, particularly if the rotating mold can be given some form of additional support (Weronski, 1978). His second recommendation, concerning the application of thermal insulation to the inner surface of the mold, is not universally applicable. Such insulation is sometimes provided, for example, by introducing a small quantity of ferrosilicon before the molten metal is poured in, the rotation of the mold spreading this in an even layer around the inner surface. However, the metallurgical properties of the casting required for many applications arise from the rapid extraction of heat by contact between the molten metal and the cooler metal of the mold, and this precludes the interposition of a thermal barrier. In general, the use of such insulation, although beneficial to the mold, by increasing the solidification time slows down the casting cycle and may also cause undesirable chemical segregation in the casting.

The measurements made by Königer and Liebmann (1959) also showed that the thermal stress is greatest in the subsurface zone of the inner surface of the mold, that is, reaches its maximum, compressive, value 5 s after pouring begins, and that it exceeds the elastic limit of the mold material. There is, however, a difference between the measured and calculated values of the maximum compressive stress, which were 580 and 539 MPa, respectively. The measured and calculated stress distributions within the entire mold throughout the casting cycle were similar in form, but the absolute values of the stresses obtained by the two methods diverged

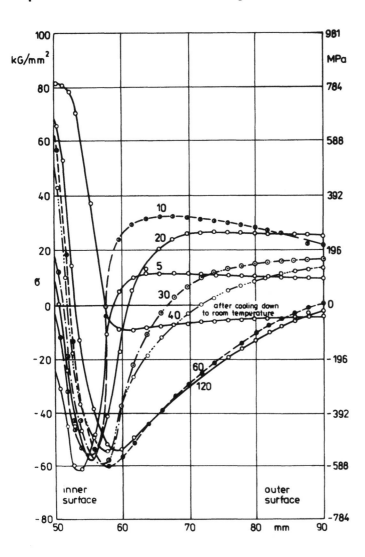

Figure 7.11 Distribution of tangential stress in mold cross section. Numbers indicate time in seconds from start of operation.

with distance from the inner surface and with time from the beginning of casting. Despite the considerable difficulties in the experiment, the measured values should be considered the more accurate, and further discussion is based on these results, as presented in Fig. 7.11.

Königer and Liebmann (1959) presented their measurements of tangential stress distribution in the mold cross section by means of this diagram, in which the abscissa is the distance from the mold axis in millimeters, 50 corresponding to the inner surface zone and 90 to the outer surface. Each curve shows the instantaneous magnitude of the stresses at a time in seconds measured from the beginning of the casting cycle indicated by the adjacent number. It can be seen that the stress in the subsurface zone of the inner surface exceeds 580 MPa compressive within 5 s from the beginning of pouring but become tensile after 30 s. At the end of the cycle, after 120 s, the tensile stress in the innermost zone is 680 MPa.

Further very interesting information is contained in the same authors' diagram of measured stress and strain, a portion of which is shown in Fig. 7.12. The stresses

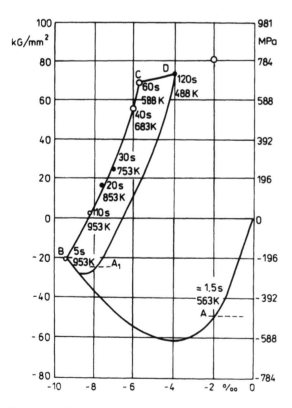

Figure 7.12 Stress and strain at mold inner surface at various stages of the casting process. Numbers indicate time in seconds from start of operation and corresponding temperature.

and corresponding strains at the inner surface of the mold during the first casting cycle are represented by the curve *OABCD*. The straightness of the portion *OA* shows that in the initial stage of the casting cycle stress and strain obey Hooke's law; after 1.5 s the limit of proportionality (343 MPa) is exceeded. Point *A* on the curve represents the yield point of the mold material.

A maximum compressive stress of about 600 MPa is reached later and produces a strain of –4‰. Thereafter, as a result of conduction of heat to adjacent zones and their consequent expansion, the compressive stress falls.

The maximum strain of approximately 10‰ occurs after 5 s, the temperature of the inner surface then being 953 K (point *B*). The *BC* portion of the curve shows intensive reduction of the deformation of the inner surface. At point *C* the yield point is again exceeded and the reduction of the strain originally produced by compression proceeds rapidly for the remaining 60 s of the cycle (i.e., to *D*). The strains and stresses in the next cycle are shown by the curve DA_1BCD. The region within the curves A_1BCD can be associated with the energy stored in the innermost subsurface of the mold, and as it accumulates in successive cycles it can produce cracks.

7.4.4 Chemical Composition, Structure, and Mechanical Properties of Mold Materials

It is obvious that because of their size and the critical technical requirements they must satisfy, the molds for centrifugal cast iron water-main pipes are expensive. Despite successive and efficient reconditioning, the total mold service life is only some 2000 cycles. To the cost of providing a new mold must be added the costs incurred through interruption of production each time a mold has to be reconditioned or replaced. Accordingly, the durability of the mold is of great importance. In this section the materials and processes commonly used in making molds in various countries and the properties obtained are reviewed briefly prior to describing our own experiments with mold steels and consequent recommendations.

The structure and behavior of heat-resistant steels is influenced significantly by the following elements if they are present to a sufficient extent: Cr, Mo, Ni, and Si. The specifications below omit elements which, if present are in amounts so small that their effect is negligible.

In the Soviet Union molds are usually made of 12H, 15H, or 35H steels containing, respectively, 0.12%, 0.15%, and 0.35% C together with 0.8 to 1.1% of Cr and small but significant amounts of molybdenum. In Great Britain the steel used contains 1.6% Cr together with 0.6% Ni. In the United States molds are manufactured from steels containing 0.25% Mo together with chromium and nickel. In Germany two groups of steels are used: the first contain 0.8 to 1.2% Cr with more than 0.3% C, together with molybdenum in the range 0.4 to 0.6% Mo. Steels in the second group contain = 2% Cr and 0.2 to 0.23% C, with molybdenum in the range 0.4 to 0.6% Mo.

Investigations carried out by Büchler (1959) showed that molds had good durability when made from steels containing 1.8 to 2% Cr, C ≤ 0.28%, together with 0.3% Mo and that the carbon content of steels containing more than 2% Cr should be less than 0.2%. All the German molds, after forging, drilling, and hardening, are tempered at a high temperature. Their tensile strength is in the range 588 to 833 MPa and their impact strength is in excess of 137×10^4 J/m^2. In Poland molds are made of 20H2M steel, which has a chemical content similar to that of the German mold steel. It contains 0.15 to 0.25% C, 0.17 to 0.37% Si, 0.5 to 0.8% Mn, 1.7 to 2% Cr, and 0.45 to 0.65% Mo. After quenching and tempering the mechanical properties are: hardness 207 to 241 HB, tensile strength σ_u 686 to 814 MPa, and impact strength $\geq 98 \times 10^4$ J/m^2.

The performance of the steels used for molds in various countries is nearly the same and the differences in the service life of the molds made from them arise from differences in the industrial practice of various factories in regard to their use rather than from their chemical composition.

In considering the most important properties of the mold material and how they may be optimized, we start with the facts established in Sections 7.4.2 and 7.4.3: that the zone adjacent to the inner surface of the molds is subjected to very rapid changes in temperature and stress and that this is where cracking starts. This leads to obvious conclusions: first, that the condition and mechanical properties of the material in this region have a fundamental influence on the service life of the mold, and second, that because of the rapid changes in temperature and stress, the impact strength must be a very important mechanical property of the mold material. To this may be added the fact that as differences between the mechanical properties of adjacent regions of material will produce local concentrations of stress, uniformity of hardness should be regarded as one indication that the material is in a suitable condition.

It is conventional industrial practice for the measurements of the mechanical properties of the mold material to be made at room temperature, but in seeking to identify optimum mechanical properties and the mechanism of mold failure, the present author has preferred to measure the mechanical properties at temperatures in the region of those at which a mold works in practice. These experiments and the study of the microstructure produced by various treatments were carried out of 20H2M steel, but following Büchler (1959), the conclusions may be expected to apply to most of the steels currently used in the production of molds. The 20H2M steel belongs to the group of bainitic steels and the considerable differences can be produced in its microstructure, depending on the treatment.

Figure 7.13 shows the TTT (time–temperature–transition) diagram for continuous cooling of this steel at various rates. Its characteristic feature is a wide region of bainitic transformation, B, displaced with respect to that of ferrite–pearlite transformation, F-P. In the figure the temperature variation with time for each specimen is shown by a continuous line. The percentage of the material that underwent a given transformation is shown by the number below the region of transformation and to the left of the cooling curve concerned. The circled numbers are the hardness of the material at room temperature; the three on the left of the figure (310, 302, 265) are Vickers hardness numbers, the remainder are Rockwell.

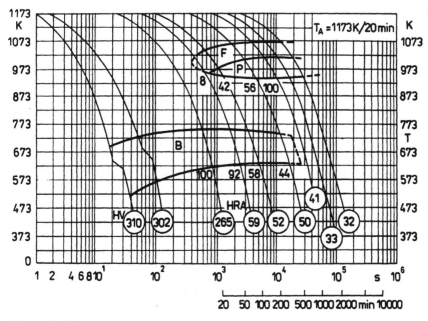

Figure 7.13 TTT diagram for 20H2M steel.

Thus in the case of the fourth cooling curve from the left, 8% of the material underwent ferritic–pearlitic transformation and 92% bainitic transformation, the hardness at room temperature being 59 on the Rockwell scale.

Cooling at rates within the range 0.2 to 12 K/s results in a uniform homogeneous bainitic structure. It is notable that a wide range of cooling rates has little effect on the room-temperature hardness, which is substantially constant and close to 300 HV. The ferrite–pearlite transformation occurs to varying extents with cooling rates of less than 0.1 K/s, and at cooling rates in the range 0.1 to 0.02 K/s it is followed by bainitic transformation. The variation in the percentage content of pearlite–ferrite and bainite structures with cooling rate can be seen in the figure.

As the structural properties of 20H2M steel are very similar to those of the steel used for the mold studied by Königer and Liebmann (1959), it is justifiable to consider the cooling rates they measured (Fig. 7.10) in conjunction with the TTT diagram shown in Fig. 7.13. From this it can be seen that only the material adjacent to the inner surface of the mold undergoes a transformation and that this is between ferrite–pearlite and bainite.

The difference between the mechanical properties of these phases intensifies stress at the boundary between transformed and untransformed material, a boundary which of course moves as the temperature distribution changes during the casting cycle. The effect is cumulative over a number of casting cycles and is one of the causes of mold failure.

Similar investigations were made on other steels that are promising materials for molds: 26H2MF, 12H1MF, and 15H1MF. Metallographic examination of 26H2MF steel after austenitizing at 1290 K and oil hardening showed a martensitic structure. Cooling from 1290 K at the much lower rate of 120 K/min produced a structure consisting of upper and lower bainite accompanied by retained austenite. The structure corresponding to cooling rates of 3 to 50 K/min consisted primarily of upper bainite and retained austenite. The 12H1MF and 15H1MF steels behaved similarly, but the cooling rates to produce the corresponding structure were somewhat smaller. It should be noted that lower cooling rates, in the range corresponding to bainite transformation, enhance the diffusion of carbon from ferrite to austenite, enriching it with carbon and thus accelerating its transformation into ferrite and carbides.

Investigation of the creep resistance of all four of these steels showed that the bainitic structure provided the greatest durability, the martensitic structure medium durability, and the pearlite–ferrite structure the lowest durability. All these structures can, of course, be obtained by austenitizing at temperatures above A_{c_3} and appropriate choice of cooling rate. The creep resistance depends to a great extent on distribution of the alloying additives between the ferrite and the carbides and on the spatial distribution and particle size of the latter. It also depends on the match between the crystalline lattice of the carbides and that of the matrix material. In these steels the hardness after tempering is increased by the molybdenum additive due to precipitation of the Mo_2C carbide.

The influence of austenitizing temperature, quenching medium, and tempering on the microstructure, hardness, and impact strength was studied in further detail using 20H2M steel. The microstructures produced by austenitizing at temperatures of 1123, 1173, and 1223 K followed by quenching in water or in oil are shown in Figs. 7.14 to 7.19. Optical microscopy did not reveal distinct differences between the microstructures due to austenitizing at each of these temperatures and quenching in either medium. The structures were in each case composed of bainite similar in configuration and size of spines. This indicates the facility with which this steel can be hardened under industrial conditions in which the hardening temperature cannot always be controlled with great precision.

Examination under an electron microscope at a magnification of 17,000× did, however, reveal minor differences between the structures produced by hardening in water and in oil. The microstructure of specimens hardened in water was composed of α matrix having a distinct needle-shaped form, a relatively high density of dislocations, and carbide precipitates of varying size and shape, mostly deposited at the boundaries of grains and grain blocks (see Fig. 7.16). The microstructure of specimens hardened in oil was composed of α matrix having a needle-shaped form only in some areas and fewer carbide precipitates; careful examination revealed the presence of tempered martensite in some regions. This martensite probably came from retained austenite.

In mechanical tests the hardness of the specimens was approximately 300 HV

Figure 7.14 Microstructure of 20H2M steel austenitized at 1173 K for 90 min and hardened in water (500×).

Figure 7.15 Microstructure of 20H2M steel austenitized at 1173 K for 90 min and hardened in oil (500×).

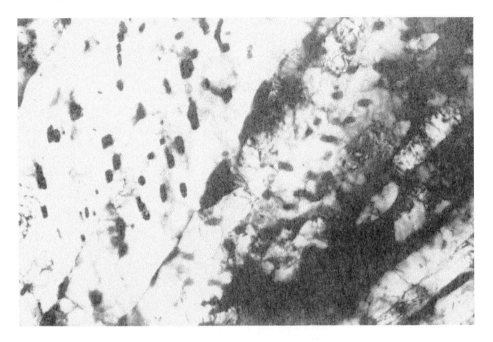

Figure 7.16 Microstructure of 20H2M steel austenitized at 1173 K for 90 min and hardened in water, electron microscope, thin film (17,000×).

and impact strengths in the transverse direction were in the approximate range 60 to 100×10^4 J/m². The impact strength of this still depends, of course, on the austenitizing temperature from which it was quenched. Increasing this temperature from 1123 K to 1173 K raised the impact strength from about 80×10^4 J/m², which resulted from enlargement of the austenite grain size and higher quenching stresses. The impact test fractures of water-hardened specimens are shown in Figs. 7.17 to 7.19. Examination of the fracture of the specimens austenized at 1123 K revealed mainly ductile fracture together with some regions of cleavage fracture with characteristic oval zones. Plastic deformation of the specimens was clearly evident. In regions where parting fractures occurred, jogs on the cleavage planes could be observed (Fig. 7.17).

Raising the austenitizing temperature to 1173 K increased the ductility of the fracture, but a further increase to 1223 K reduced the plastic deformation, increased the size and number of oval zones of fracture, and made the zones of parting fractures more apparent. Formation of the oval zones observed under the electron microscope was probably due to the local concentration of microstresses, which either created cracks or facilitated the progress of cracking during impact testing.

All hardened articles such as molds are tempered to stabilize their mechanical properties and reduce quenching stresses. The mechanical properties of 20H2M

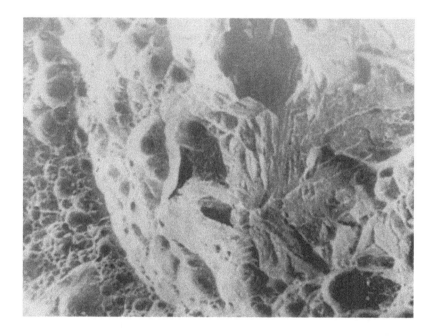

Figure 7.17 Fracture appearance of 20H2M steel austenitized at 1123 K for 90 min and hardened in water, scanning electron microscope (1000×).

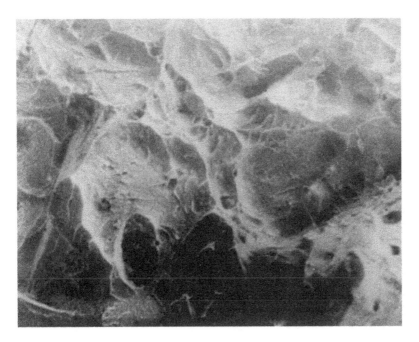

Figure 7.18 Fracture appearance of 20H2M steel austenitized at 1223 K for 90 min and hardened in water, scanning electron microscope (3000×).

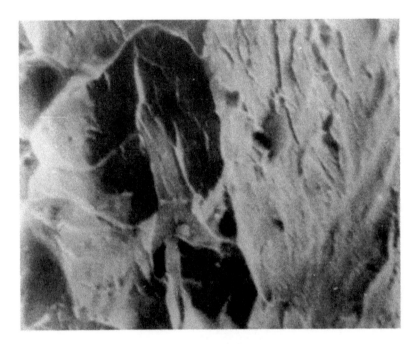

Figure 7.19 Fracture appearance of 20H2M steel austenitized at 1173 K for 90 min and hardened in water, scanning electron microscope (3000×).

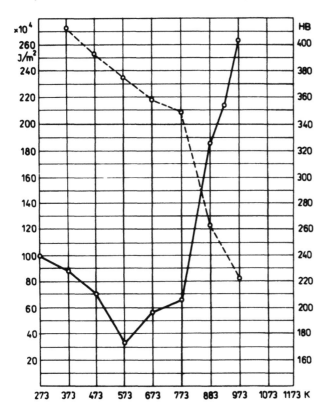

Figure 7.20 Influence of tempering temperature on impact strength and hardness of 20H2M steel austenitized at 1173 K for 90 min and hardened in water. Solid line, impact strength; dashed line, hardness.

steel austenized at 1173 K, hardened in water, and tempered at various temperatures are shown in Fig. 7.20.

The hardness of the structure obtained is indicated by the dashed line and the impact strength by the full line. It can be seen that as the temperature of tempering increases, the hardness decreases gradually from an initial, untempered, value in the region of 420 HB to 350 HB for material tempered at 773 K. Thereafter the hardness falls more rapidly with rise in the tempering temperature.

The impact strength has a local minimum of 35×10^4 J/m^2 at 573 K and rises rapidly with temperatures above 773 K. It should be mentioned that the rate at which the tempered samples were cooled did not affect their impact strength; this was probably due to the molybdenum additive and the small cross section of the specimen.

Examination of samples tempered in the range below 573 K did not reveal significant differences in their microstructure. Although the transformation of retained austenite to tempered mortensite was to be expected, it was not observed. However, this transformation is very difficult to detect and may in fact have occurred, in which case the increase in specific volume would have increased the state of stresses. Further increase in the temperature of tempering to 873 K distinctly reduces the needle-shaped structure of tempered bainite, and this reduction becomes more pronounced for higher temperatures, from 873 to 973 K. The microstructure produced by tempering at the highest temperatures of the test series were composed of reconstructed α phase, a large number of evenly distributed carbide coagulations, and some fine submicroscopic carbide particles.

Further examination of the impact test fractures of hardened and tempered specimens were made under the electron microscope and examples are shown in Fig. 7.21 to 7.23. The cleavage fracture of the specimen tempered at 373 K for 5 h (Fig. 7.21) was typically transcrystalline with a characteristic uniform distribution of deformed oval zones. Tempering at a higher temperature of 473 K increased the size of the plastically deformed zones. Examination of the specimens tempered at 573 K revealed cleavage fracture and large zones in which parting fracture had taken place, which could be caused by transformation of the retained austenite increasing the state of stresses. Clusters of carbides are visible adjacent to these zones (Fig.

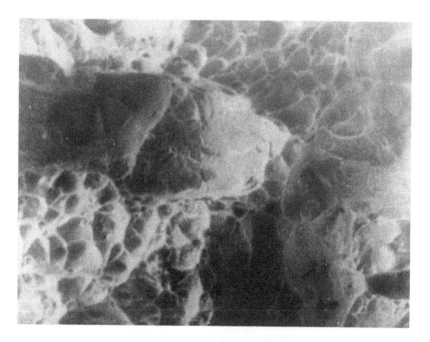

Figure 7.21 Fracture appearance of 20H2M steel austenitized at 1173 K for 90 min, hardened in water and tempered at 373 K for 5 h, scanning microscope (1000×).

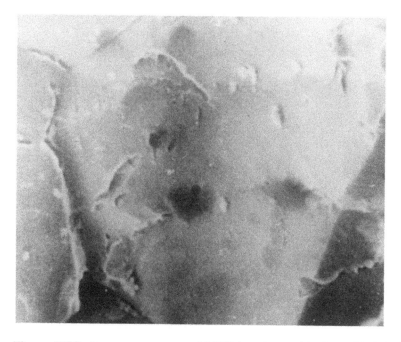

Figure 7.22   Fracture appearance of 20H2M steel austenitized at 1173 K for 90 min, hardened in water and tempered at 573 K for 5 h, scanning microscope (1000×).

7.22. Increasing the temperature of tempering to 673 K produced indications of a somewhat increased degree of plasticity, manifested by the occurrence of small zones of plastic deformation. Still further increase in the tempering temperature resulted in a mixed ductile and parting mode of fracture. Figure 7.23 shows the fracture of a specimen tempered at 873 K. It can be seen that transcrystalline–ductile fracture had taken place. Fine carbide precipitations are visible in the oval zones of the fracture. The appearance of the fracture after tempering at higher temperatures is similar.

These results of the examination of microstructure are in accord with the plots of hardness and impact strength shown in Fig. 7.20; that is, they provide a coherent description of the results within the framework of conventional heat treatment of the steel concerned. It will, of course, be realized that in the case of a mold, the microstructure and mechanical properties will depend on the heat treatment it has received during manufacture and also, especially in the region of the inner surface, on the further heat treatment it receives in the course of use.

Accordingly, further investigations were carried out on specimens austenitized at 1173 K for 90 min, hardened in water, tempered at 923 K or at 973 K, and additionally annealed for 10 min at various temperatures. Three micrographs from these specimens are shown in Figs. 7.24 to 7.26. The austenitizing temperature of

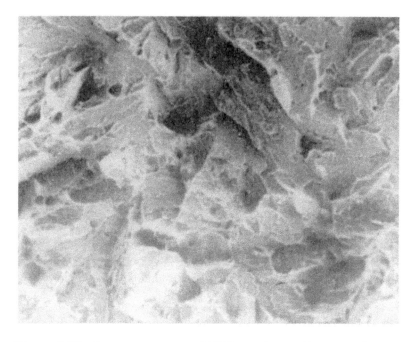

Figure 7.23   Fracture appearance of 20H2M steel austenitized at 1173 K for 90 min, hardened in water and tempered at 873 K for 5 h, scanning microscope (1000×).

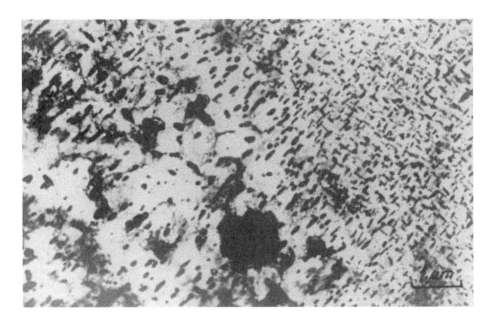

Figure 7.24 Microstructure of 20H2M steel austenitized at 1173 K for 90 min, hardened in water, tempered at 923 K for 5 h, additionally annealed at 673 K for 10 min, electron transmission microscope, thin film (10,000×).

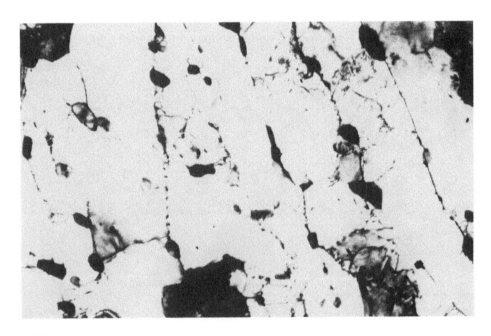

Figure 7.25 Microstructure of 20H2M steel austenitized at 1173 K for 90 min, hardened in water, tempered at 973 K for 5 h, additionally annealed at 573 K for 10 min, transmission microscope, thin film (33,000×).

Figure 7.26 Microstructure of 20H2M steel austenitized at 1173 K for 90 min, tempered at 923 K for 5 h, additionally annealed at 673 K for 10 min, transmission microscope, thin film (27,000×).

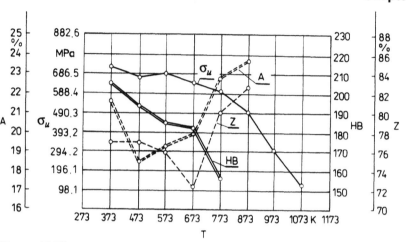

Figure 7.27   Temperature dependence of mechanical properties of 20H2M steel.

1173 K is that from which mold materials are usually quenched. The temperatures used for tempering (923 or 973 K) are the lower and upper limits of the maximum temperature attained by the mold inner surface when used in accordance with normal casting practice. The temperatures at which the additional annealing was carried out cover the range encountered in the casting cycle and at the same time the range in which a reduction in mechanical properties is observed (see Fig. 7.10 and 7.27). The time for which the material was held at the annealing temperature approximates the cumulative total of the periods in which the inner surface zones are at their maximum temperature in the course of production runs extending from a new or reconditioned mold to the first appearance of cracks.

Investigations of the effect of the holding temperature of this additional 10-min anneal were made with the transmission microscope. These showed that holding temperatures of less than 473 K did not have any substantial effect on the micro-structure, although there was a slight increase in the number of carbide particles; these were deposited chiefly at grain boundaries and grain block boundaries. Holding temperatures in the range 573 to 673 K lead to the appearance of fine precipitations of M_3C-type carbides and the larger $M_{23}C_6$-type carbides.

In some parts of the specimen, cross-section mesh-like structures of carbide precipitates could be seen. The holding temperature of 773 K produced specimens in which Mo_2C carbides were observed. In some places these formed pine-tree-like structures in which the branches were generally at right angles to each other. Such carbide precipitations are known to be one of the major causes of the reduction of the impact strength of steel in the temperature range 773 to 873 K.

A higher holding temperature (approximately 973 K) caused dissolution of Mo_2C carbides and intensive coagulation of M_7C_3 carbides and their distribution at subgrain boundaries, sometimes immediately adjacent to each other. A difference in the density of dislocation was observed in subgrains, especially in regions adjacent to carbides, which may point to blocking of dislocation motion.

Investigations of the mechanical properties of the steel (i.e., tensile test, impact test, and Brinell hardness measurement) were carried out at elevated temperatures. The test pieces were austenitized at 1173 for 90 min, quenched in water, and tempered at 923 K for 5 h. Thus their heat treatment was the same as that used in the manufacture of a mold. The results obtained are shown in Fig. 7.27, in which Z is the percentage reduction of area of specimen, A the strain, σ_u the ultimate tensile strength, and HB the Brinell hardness. It can be seen that as the temperature increases, the plasticity of the test piece diminished to a minimum at 673 K. It should be stressed that both the ultimate tensile strength and the hardness vary continuously and almost monotonically with temperature, in contrast to Z and A, which have local minima.

Micrographs of two fractures produced in the impact test are shown in Figs. 7.28 and 7.29. Fractures produced in tests at the relatively low temperature of 473 K were of a ductile nature; those produced at 873 K were also ductile and had a well-developed plastically deformed fracture surface, as shown in Fig. 7.28. At the test temperature of 973 K fracture was ductile and the fracture surface consisted of

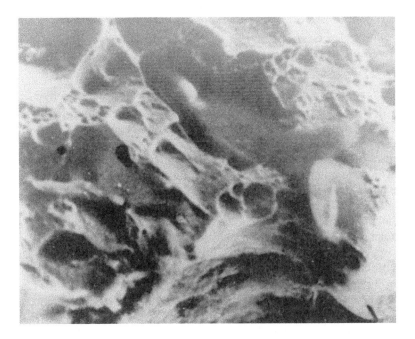

Figure 7.28 Fracture appearance of 20H2M steel austenitized at 1173 K for 90 min, hardened in water, tempered at 923 K for 5 h. Produced in impact test at 873 K, scanning microscope (3000×).

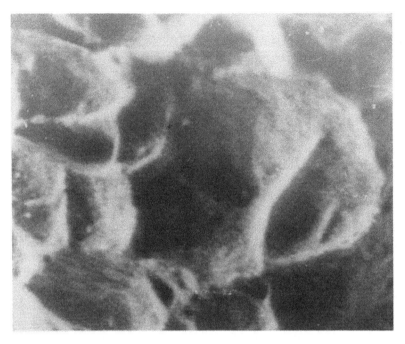

Figure 7.29 Fracture appearance of 20H2M steel austenitized at 1173 K for 90 min, hardened in water, tempered at 923 K for 5 h. Produced in impact test at 973 K, scanning microscope (3000×).

uniformly distributed regions of plastic deformations, as shown in Fig. 7.29. The irregularities on the deformed surfaces could have been created by carbides or nonmetallic inclusions.

The results of the impact tests are shown in Fig. 7.30. It can be seen that the impact strength has minima at 473 K and 873 K and reaches a maximum at 973 K. At temperatures above 973 K the impact strength decreases as the A_{c_1} temperature is exceeded.

The laboratory tests were based on two approaches to simulating the thermal regime of the casting cycle in order to study its effects on the mold material. The first approach comprised the work described above, in which the effects of particular segments of the cycle, such as the annealing period, were studied individually in isolation from the remainder of the cycle. This had among its advantages the fact that the essentially static temperature of an isolated segment of the cycle could be reproduced very accurately.

The second approach was more direct. The specimen was a ring cut from a mold and hence accurately simulating an actual mold in all respects save axial length. The test equipment and procedure employed were described in Chapter 5. Recapitulating briefly, the outer cylindrical surface was cooled by a water jacket and the

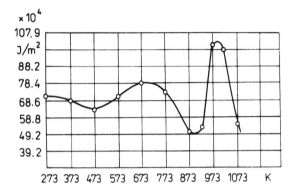

Figure 7.30 Temperature dependence of impact strength in transverse direction of 20H2M steel austenitized at 1173 K for 90 min, hardened in water, tempered at 923 K for 5 h.

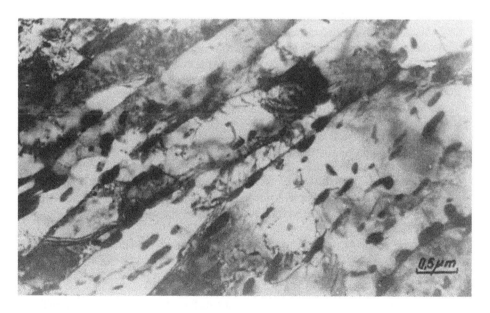

Figure 7.31 Microstructure of 20H2M steel austenitized at 1173 K for 90 min, hardened in water, tempered at 923 K for 5 h after 400 cycles of 293 to 973 K, zone extending 1 mm beneath mold inner surface, transmission microscope, thin film (20,000×).

inner cylindrical surface was subjected to programmed heating by an induction coil mounted within it. The form of the cycle was similar to that shown in Fig. 7.1a. For a variety of reasons, including enhancement of the end effects due to the relatively short axial length of the ring, it was not possible to simulate very precisely the conditions within full-scale molds, and the results should be considered as complementary to, but not overriding, those of the first approach. The cyclic variation in surface temperature was between 293 and 973 K, and the heating rate was sufficient to raise the temperature of the surface zone from minimum to maximum in 5 s (compare Fig. 7.10).

After 400 cycles the microstructure of the zone extending to 1 mm beneath the inner surface was examined, this being the zone in which cracks usually start. A micrograph of this microstructure is shown in Fig. 7.31. It can be seen that carbides are distributed within the grains and along the boundaries, impairing their cohesion. The density of dislocations is high, as is the size of the carbides, especially those deposited along grain boundaries. The carbides M_3C and $M_{23}C_6$, together with a small amount of M_7C_3, can be seen in the microstructure. These carbides have a tendency, enhanced by thermal cycling, to dissolve in the matrix material, supersaturating it locally, which produces microstresses through the change in specific volume. This restricts the movement of dislocations, and the carbides also act as barriers to dislocations. Both the microstresses and blocking of dislocation motion facilitate crack initiation.

The information presented in this chapter indicated the nature of the processes that occur in a centrifugal casting mold and lead to failure of its inner surface. We described some classical experiments and computations which reveal the magnitude of the thermal stresses involved and demonstrate underlying mechanism of failure. We also described laboratory work that illustrates the microstructures and properties produced in the manufacture of a mold and their subsequent modification by the simulated conditions of use. Inevitably, only the salient aspects were described and a large number of data from experiments and field experience could not be set down, although their nature was implied. All this, together with material contained in earlier chapters, leads to the following recommendations, which should be of service to designers and users of molds for centrifugal casting.

1. Design should tend toward a thin-walled mold. Alternatively, to minimize the overall costs of providing molds and maintaining production with them, a two-part construction should be considered. In this the relatively thin replaceable liner is supported by the water-cooled outer member, the combination providing adequate rigidity.

2. The wall thickness of conventional molds should be checked regularly, especially after reconditioning. Nonuniform thickness causes differences between the stress distributions in adjacent regions, which accelerates failure. The additional stresses caused by dynamic unbalance also reduce mold life.

3. To ensure uniform hardness of the inner surface, the mold material should be homogeneous and easily hardened. he difference in hardness measured at any points on this surface should not exceed 12HB.

4. The molds should be made from ingots cast in vacuum, or the steel used for

their production should be blown with argon or another inert gas to minimize the amount of inclusions and metallurgical impurities. The process of manufacture and reclamation should include multipoint burnishing or burnishing accompanied by surface hardening of the inner surface (Weronski, 1980). Multipoint burnishing provides more consistent thermal contact between the molten cast iron and the mold surface, which helps to control the deformation process. The temperature at which the mold is tempered during its manufactures should not be such that the plasticity is reduced. For the widely used steels similar to 20H2M, this means that tempering in the region of 673 K should be avoided. In production use the temperature of the mold's inner surface should not exceed 973 K.

5. Rebuilding and refacing the inner surface with an automatic welder is a technique to be recommended, as it extends the life of the mold many times.

Finally, we should not be afraid to consider triple-layer molds composed of an inner layer of tungsten carbide embedded in cobalt matrix attached to an intermediate layer made of a ferrous alloy with good plastic properties and supported by the outer layer, which provides the system with good rigidity. This seems promising, as tungsten carbides are very resistant to rapid temperature variations, as shown by Merten and Steinhurst (1984).

REFERENCES

Brooksbank, D. and Andrews, K. W. (1969). *J. Iron Steel Inst.*, 207: 474.

Brooksbank, D., and Andrews, K. W. (1972). *J. Iron Steel Inst.*, 210:246.

Büchler, F. (1959). *Stahl Eisen*, 79: 1722.

Geller, J. (1964). *Izw. Czorn. Miet.*, 7.

Gierzynska, M., and Smarzynski, A. (1979). *Mechanik*, 52(3): 149.

Hecht, H. and Hiller, H. M. (1959). *Werkstattstechnik*, 10.

Königer, A. and Liebmann, W. (1959). *Stahl Eisen*, 79: 1730.

Kolorz, A. (1980). *Giesserei*, 20: 652.

Kucharski, K. and Wendorff, Z. (1975). *Arch. Hutn. PAN*, 2(1): 50.

McIntyre, J. B. (1961). *Foundry Trade J.*, 459.

Merten, C. W. and Steinhurst, W. R. (1984). *J. Eng. Ind.*, 106: 325.

Pesat, V. (1972). *Hutn. Listy*, 11.

Restall, J. E. (1979). *Metallurgia*, 46: 676.

Sakwa, W. (1974). *Cast Iron*, Slask, Katowice, Poland.

Skalski, K. (1973). Thesis, Technical University, Warsaw.

Weronski, A. (1978). *Improvement of Resistance of Steel Moulds to Thermal Fatigue*, report.

Weronski, A. (1980). *Improvement of Resistance of Steel Moulds to Thermal Fatigue*, report.

Weronski, A. (1983). *Thermal Fatigue of Metals*, WNT, Warsaw, pp. 178–186.

Weronski, A. (1985). *Materials Science: Experimental Methods*, Technical University, Lublin, pp. 254–268.

Żmihorski, F. and Żólciak, T. (1973). *Metalozn. Obrobka Cieplna*, 3: 2 .

8

Practical Remarks and Advice Concerning Operational Use of Some Industrial Components Working Under Conditions of Thermal Fatigue

8.1 MOLDS FOR CENTRIFUGAL CASTING

In Chapter 7 the preceding chapters were summarized in a way that enabled the reader to find the proper method of solving a thermal fatigue problem in a specific industrial component relatively easily. In addition, working conditions of steel molds for the centrifugal casting of cast iron pipes were comprehensively reviewed together with results of experimental work carried out by the authors in seeking causes for their premature retirement. Similar or even more extensive experimental procedures are standard for solving problems associated with thermal fatigue.

In this chapter we describe briefly problems associated with the dysfunction of industrial components, their causes, and methods of prevention. We are limited to such a treatment because of space limitations and by the fact that detailed analyses of the performance of industrial components are almost unavailable in the literature. Moreover, the published data deal with specific components working under specific conditions, and therefore only very general conclusions can be drawn. Of course, the choice of proposed preventive methods and technologies depends on the working conditions of the component considered, and thus readers themselves should choose among the methods recommended.

Working conditions of molds used for the production of centrifugally cast iron pipes are summarized so as to be consistent with Chapter 4. The influence on service life of the following factors has been established:

1. Cyclic temperature changes in the approximate range 500 to 950 K at the mold's inner surface, with a thermal cycle length of approximately 120 s
2. High centrifugal forces caused by the fact that the mold's rotational speed approaches 950 rpm
3. High stresses in the zone adjacent to the mold's inner surface (i.e., about 200 MPa in compression and 780 MPa in tension (maximum)
4. Corrosion caused by air and gases released from solidifying metal

In addition, scratch parks parallel to the mold axis are formed during the extraction of the cast pipe. These micronotches enhance the progress of thermal fatigue. Naturally, the degree to which these factors participate in the process of thermal fatigue and the ranges of temperatures, stresses, and centrifugal forces depend on the casting conditions, the material being case, and the mold material. This is illustrated by the following:

1. Slusarjev (1975) quoted cycle times for centrifugal casting of cast iron pipes as 100 to 120 s for pipes of 100 mm diameter and 120 to 144 s for 150-mm-diameter pipes.
2. According to information supplied by Wisconsin Centrifugal Inc., bronze alloys, stainless steel alloys, and chromium–nickel alloys are produced centrifugally. The shapes and sizes of castings depart appreciably from those that could be expected on the grounds of Chapter 7. For example, a 9-ton hub for a controllable-pitch marine propeller, a 330-cm-diameter ball mill bearing, and a 10.4-ton spherical bearing were produced centrifugally.

The usual geometry of the horizontal casting permanent mold (i.e., taper and axially symmetric shape) enables the production of simple-shaped products only. Therefore, the producers develop their own innovative casting technologies and mold designs. Among these, the process developed and patented by Charles Nobel, which involves replacement of the permanent mold by a thin-walled tube lined with zirconia seems very profitable (Burden, 1982). As the zirconia being in direct contact with the molten iron constitutes a thermal barrier, this eliminates thermal fatigue as a factor affecting the life of the mold shell. Zirconia lining is collected and recycled like core materials in a foundry. This method enables complex shapes to be cast and requirements of dimensional accuracy to be met. Moreover, the outlook for zirconia availability is good (Vingas, 1982).

Data presented later in this section were gathered by the senior author during work in industry several years ago and relate to the centrifugal casting of cast iron pipes. Working conditions of the mold were presented in Chapter 7. Molds were made of 20H2M steel by conventional casting and forging, with subsequent thermal treatment carried out in accordance with the conclusions of Chapter 7.

8.1.1 Performance Tests

The first cracks were observed by the naked eye on a mold's internal surface after about 200 casting cycles. Cracks were aligned in directions almost parallel to the mold axis. Most were located 1 m from the smooth end of the mold (for a mold 6 m long) or, less often, at the mold faucet. The typical pattern of cracks is depicted in Fig. 8.1. Observation of cracks in a plane perpendicular to the mold axis revealed that the cracks penetrated to a depth of a few tenths of a millimeter and were branched (see Fig. 8.2). Branches usually ran along prior austenite grains. Cavities were found along the path of the crack or in the immediate vicinity. These were probably formed at a border of a few grains or resulted from coagulation of

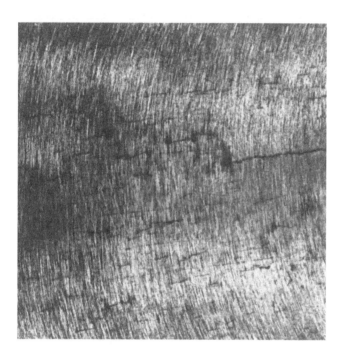

Figure 8.1   Typical pattern of cracks (2×).

vacancies, although this seems less likely. During further casting cycles, cracks grew into the mold material and joined, forming the frequently observed pattern of fire cracks. A gradually progressing corrosion was also observed.

Taking into account the mutual orientation of cracks to the mold axis, it can be noted that the first stage of cracking is influenced predominantly by circumferential stresses and scratch marks. The first factor was discussed in detail in Chapter 7, but it should be mentioned that during successive casting cycles, mechanical properties at the mold inner surface gradually decrease. This is distinctly suggested by experimental results (see Figs. 7.24 and 7.25). Almost half of the cracks were formed at inclusions or linked them following the pattern of crack initiation given in Chapter 7.

In the second phase of cracking (i.e., joining cracks and forming fire cracks), the following factors are important: (1) weakening of the mold structure by cracks already formed, (2) the continuing corrosion process, and (3) the formation of plastic zones at crack tips. A decision on mold reclaiming, involving the removal of a thin layer by machining and grinding, is made when the operator realizes that there are problems with pipe extraction caused by pouring liquid metal into the crack. This moment correlates roughly to a crack depth of 1 to 2 mm and the length of a biggest crack of 100 mm. A mold is scrapped when the internal diameter is enlarged to such a degree that pipes produced can no longer meet tolerance limits. The total number of cast pipes is 1500 to 2000 per mold.

The data on mechanical properties and microstructure of mold material that have been presented were obtained from tests carried out at ambient or elevated temperatures or after the specimen was exposed to a series of thermal cycles. The real working conditions of the mold are more complex, as the inner material of the mold surface is in contact with molten metal, mold powders, and coatings used during casting. In the case mentioned, graphite was applied to the chute.

Measurements of chemical composition and hardness were done on scrapped molds, and these results are shown in Fig. 8.3 and 8.4. Nonhomogeneous distribution of elements in the cross section of molds is caused by diffusion and is additionally enhanced by stresses varying cyclically. However, the distribution of carbon is affected by graphite applied to the chute so that the expected influence of diffusion is masked. The change in the chemical composition of the material of molds and microstructural changes cause the distribution of hardness to be nonhomogeneous (see Fig. 8.4). The most rapid hardness change is seen to occur in the immediate vicinity of the mold's internal surface. At depths exceeding 15 mm, changes in hardness are connected with random effects rather than with the influence of thermal cycling. As these results were obtained from a scrapped mold exposed to 1500 to 2000 thermal cycles, these changes are expected to be less for molds in service.

The second cause of mold retirement is twisting of the mold. A performance test has demonstrated that this phenomenon is always caused by the eccentricity of the internal bore. However, these instances, which are related to the fault of the operator, were omitted in the discussion of mold durability.

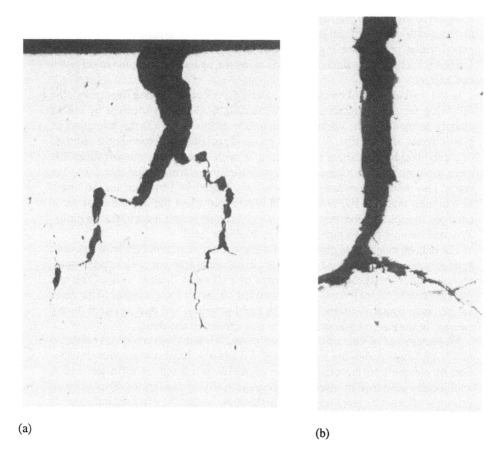

(a)   (b)

Figure 8.2   Crack observed in transverse cross section of mold: (a) (20×); (b) (100×); (c) etched (500×).

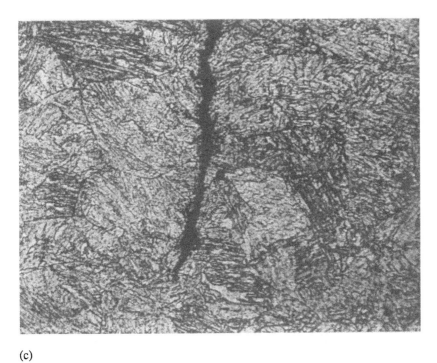

(c)

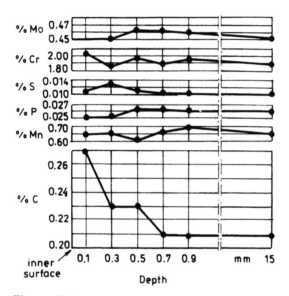

Figure 8.3   Distribution of chemical composition in the scrapped mold. (From Maciejny and Weronski, 1979.)

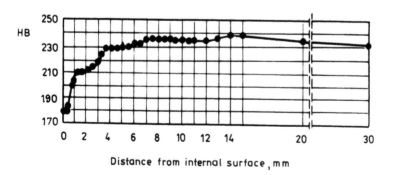

Figure 8.4   Distribution of hardness in the interior of scrapped mold. (From Maciejny and Weronski, 1979.)

8.1.2 Technological Recommendations

Performance tests have proved that there is no connection between the mechanical properties of mold material (i.e., the results of tensile tests and hardness measurements) and the service life of the mold. This is supported by two facts: first, that the mechanical properties undergo gradual degradation under the influence of temperature cycling; and second, that mold working conditions make contradictory demands on mechanical properties (i.e., high hardness), so that premature crack initiation should be avoided and good plasticity maintained to delay crack development. However, impact strength, characterized as the stability to withstand rapid loads, must be higher than 110×10^4 to 120×10^4 J/m^2 (this is shown clearly in Fig. 8.5). The data given were obtained from a test carried out at room temperature on specimens cut transverse to the mold axis.

It has also been demonstrated (Weronski, 1983) that the smoothness of the internal mold surface is an important factor influencing mold durability. The decrease in surface finish from $R_a = 2.5$ µm to $R_a = 10$ µm (where R_a is the average value of surface irregularity) limited the life of the mold by 25%. The influence of smoothness should be considered with regard to the following three aspects: (1) notch effect, (2) presence of an air film between the notch bottom and a metal being cast, and (3) an effect of surface irregularities on scratches formed during pipe extraction. The superposition of these effects produces a decrease in the life of the mold. Of course, this necessitates very thorough mold reclamation. Mold material should not be overheated during grinding because it could cause phase transformations and a buildup of excessive internal stresses.

Heats of which ingots for molds are cast should either be vacuum degassed or

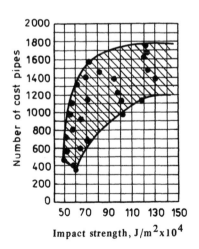

Figure 8.5 Mold life as a function of impact strength measured on transversely cut test pieces. (From Weronski, 1979.)

blown in ladle with inert gas to displace inclusions to the ingot head. The senior author recommends reducing the sulfur and phosphorus contents to a maximum of 0.025% each by dephosphorization and desulfurization in the first and second phases of refining, respectively. An effective way to achieve this is by electroslag remelting, and this is also recommended as an alternative method.

Slusarjev (1975) has proved that centrifugal casting of ingot for molds followed by thermal treatment enables about 40% of the metal to be saved and enables the production of molds with properties comparable to those produced by common technologies involving casting and forging. This method seems to be very profitable.

Protective coatings and mold powders are very promising means of increasing mold service life. This problem is discussed in Chapter 9 and in Section 8.2. The Noble process is an advantageous alternative.

Olejarski (1986) demonstrated that the resistance of 20H2M steel to thermal fatigue can be increased by 80 to 150% through the introduction of a small amount of niobium and vanadium and by increasing the molybdenum content to 2%. The elements mentioned maintain the distribution of hardness in the mold cross section at an almost constant level. The high thermal fatigue resistance of 20H2M steel of altered chemical composition is connected with the presence of simple MC carbides instead of the M_3C, M_6C, and $M_{23}C_6$ found for nominal chemical composition.

The recommendations given in this section, together with those mentioned in Chapter 7, offer feasible amendments that may be introduced in the future. The second cause of mold retirement, axis twisting, can easily be eliminated through controlled measurements of wall thickness in the mutually perpendicular planes. The admissible deviation is 0.2 mm in the case of molds of size similar to that depicted in Fig. 7.9. The eventual eccentricity of the mold bore should be corrected by machining and grinding to meet this requirement.

8.2 INGOT MOLDS

8.2.1 Working Conditions

Working conditions for ingot molds (i.e., temperatures to which they are exposed and stresses occurring in them) cover a wide range. This is caused by a diversity of production technologies and by the fact that pouring and cooling conditions for molds are controlled with limited accuracy. For example:

1. Die-casting dies employed in casting aluminum alloys are exposed to temperatures of approximately 850 to 900 K and the time spent in the die by an ingot is short, typically a few tens of seconds.
2. The temperatures of the internal surfaces of molds used for continuous casting are lowered by intensive water cooling. Additional factors are mechanical fatigue due to mold oscillatory movement and more serious erosion than for die-casting dies. The time of the working cycle is determined in this case by the heat size and "capacity" of the caster.

3. Working conditions for molds used to cast steel ingots differ widely in temperature (typically, 970 to 1170 K) and in the length of the casting cycle period (from a few to more than 10 h).

Broadly speaking, mold working conditions depend on the following technological factors of the casting process:

The temperature of liquid metal while being poured into the mold and the mold temperature before casting begins
The grade of material being cast
The way in which the mold is filled up with molten metal (i.e., from the top or by uphill pouring) and the rate of pouring
Cooling conditions (i.e., the rate of temperature drop and the manner of cooling)
The quality of the ceramics of which the bottom plate of the mold is made and the type of mold powders
The manner in which ingots are stripped from the mold

The large number of technological factors influencing the durability of molds makes an analysis of their failures difficult, and any extension of a mold's life can only be realized through the change in their working conditions performed in successive steps. Troubleshooting and limiting economical results of failures often resemble a successive approximations procedure, as a detailed analysis of working conditions has to take into consideration all of the operational aspects and the quality of the final product. Of course, such an analysis cannot ignore mold design elements: (1) mold material, weight, overall dimensions, and capacity; (2) cross-sectional shape and thickness of sidewalls; and (3) taper value and location and shape of trunnion.

8.2.2 Mold Failures and Their Causes

Practical experience from the die-casting industry shows that thermal fatigue is a severe problem in brass die-casting dies, is less important in aluminum die-casting dies, and is of minor importance in zinc die-casting dies. The likelihood of thermal fatigue failure, neglecting cycle shape and duration, is dependent on the magnitude of the thermal cycle involved and on its maximum temperature. This rather rough estimation supports field experience. A more detailed attempt to find rules governing mold life was undertaken by Danzer and Sturm (1982). The case in question was pressure-die casting of the aluminum alloy GD-Al-Si, which has a pouring temperature of 950 to 980 K. The corresponding temperatures of the die-casting die surface are 850 to 900 K and with the die preheat temperature of 530 K constitutes the thermal cycle to which the die-casting die material is exposed. The durability was evaluated utilizing Spera's (1969) method. Factors influencing die life were identified with mechanical fatigue and creep. An evaluation of die life necessitated an estimation of stresses and deformations in the subsurface layer of the material by the method of finite elements, on the basis of temperature measurements, and on knowledge of the mechanical properties of the die material (i.e., the hot-work tool steel X-40Cr-Mo-V51).

Buchmayr et al. (1982) examined the creep performance of this steel and came to the following conclusions:

1. Creep rate at temperatures below 720 K is negligibly low, so Danzer and Sturm (1982) omitted this range of temperatures in their considerations.
2. In the range of temperatures between approximately 720 K and the α–γ transition temperature, the creep rate obeys the following empirical formula:

$$t_{f,c} = \left[\exp\left(-28.3 + \frac{62,000}{T + 273} \right) \right] \sigma^{-6.06} \times 10^{8 \times 10^{-3}\sigma - 9.5 \times 10^{-6}\sigma^2}$$

where

$t_{f,c}$ = time until fracture

σ = is the stress to which the specimen is exposed

T = temperature

The Manson-Coffin law for this steel can be formulated as follows (Williams et al., 1960):

$$N_{f,f} = 0.1(\Delta\varepsilon_{In})^{-2}$$

where $N_{f,f}$ is the number of cycles until failure and $\Delta\varepsilon_{In}$ is the range of plastic deformation.

The dependence of Young's modulus on temperature can be described by the equation obtained by numerically fitting the polynomial to the experimental points on the assumption that the Poisson's ratio $\nu = 0.31$:

$$E = 2.04 \times 10^5 + 30.5T - 0.148T^2$$

and the coefficient of thermal expansion is given by the formula

$$\alpha = 1.15 \times 10^{-5} - 9.8 \times 10^{-10}T + 2.9 \times 10^{-11}T^2 - 3.5 \times 10^{-14}T^3$$

Apart from that, the tensile tests were performed at temperatures of 770, 870, and 970 K. Ludwig's law was applied to interpolate the results:

$$\sigma = K(\varepsilon_0 + \varepsilon_{In})^m$$

where

σ = stress

k,m = coefficients

ε_0 = elastic strain

ε_{In} = plastic strain

Through the numerical analysis of data, the following dependences of K, ε_0, and m on temperature were found:

$$K = 343 + 9.5T - 0.01T^2$$

$$\varepsilon_0 = 4.1 \times 10^{-3} - 1.2 \times 10^{-5}T + 9.4 \times 10^{-9}T^2$$

$$m = 0.65 - 2.2 \times 10^{-2}T + 2.1 \times 10^{-6}T^2$$

As mentioned in Section 8.2.1, the maximum temperature of the die's inner surface is in the range 850 to 900 K, which is determined mainly by the temperature of molten metal. The temperature distribution in the inner die surface is depicted in Fig. 8.6. The maximum temperature is reached a few seconds after pouring molten metal. The subsequent drop in die temperature results from opening the die and spraying parting compound into the die. Before beginning the next casting cycle, the die is preheated to 530 K. The temperature distribution in the zone adjacent to the die inner surface is shown in Fig. 8.7 for four different stages of the casting cycle.

Stresses and deformations were calculated by means of the finite element method assuming a two-dimensional state of deformation. The estimation of die life was performed according to Miner's rule:

$$\frac{1}{N_f} = \frac{1}{N_{f,c}} + \frac{1}{N_{f,f}}$$

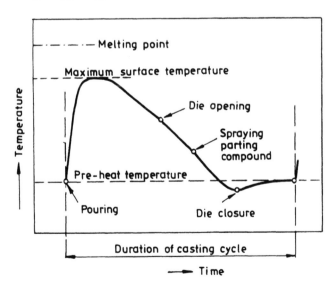

Figure 8.6 Die-casting die surface thermal cycle. (From Danzer and Sturm, 1982.)

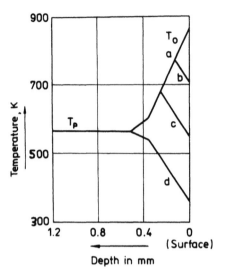

Figure 8.7   Schematic representation of temperature distribution in the subsurface layer of a die-casting die. a, During solidification of the casting; b, during cooling of the casting; c, on die opening; d, during spraying with parting compound. Tp, preheating temperature; T_0, surface temperature. (From Danzer and Sturm, 1982.)

where N_f is the number of casting cycles until die failure. The first summand on the right side of the equation describes the contribution of creep in accord with the formula

$$N_{f,c} = \int_0^{\Delta t} \frac{dt}{t_{f,c}(\sigma,T)}$$

where Δt is the duration of casting cycle and $t_{f,c}(\sigma,T)$ is the time to rupture under the stress and at temperature T. The second summand, describing the contribution of mechanical fatigue, can be connected with deformations calculated by means of the finite element method through the Manson-Coffin law.

It follows from the calculations that mechanical fatigue contributes only 10% of the damage; the remaining part of the damage is caused by creep. The lines of constant life calculated are shown in Fig. 8.8. It is seen that the most favorable working conditions of the die are when the mold is heated to a relatively high temperature before the next casting cycle is begun.

The data quoted were obtained using model calculations and thus should be considered only approximate. In practice, the influence of corrosion, which was taken into consideration only indirectly through the results of creep tests in the aforementioned work, is evident. Filling of cracks by oxides causes their wedging, and thus producing tensile stress at the tips of cracks promotes crack propagation.

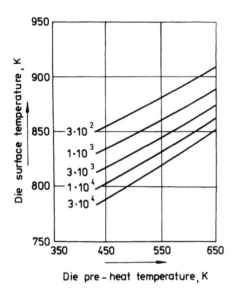

Figure 8.8 Calculated lines of constant life (in working cycles). (From Danzer and Sturm, 1982.)

Mirtich and co-workers (1981) attempted to protect aluminum die-casting dies made of H-13 steel from thermal cycles produced by heating in an aluminum-alloy bath and cooling with an air–water–lubricant spray. Laboratory tests conditions resembled industrial working conditions. The criterion used to evaluate the performance of candidate coatings employed in this study was the value of the cracked surface, which rendered demands of dimensional accuracy and surface quality of the product. The thickness of the ion-beam-sputtered coatings was about 1 μm and had practically no effect on the magnitude of thermal stresses. The experimental tests proved that the superior W and Pt coatings increased durability severalfold. A detailed analysis of coating performances showed that the tungsten coating remained almost intact throughout the tests; there were fewer and larger cracks than in uncoated specimens. The thermal fatigue resistance afforded was attributed to the corrosion protection. In Pt-coated specimens many shallow cracks resulted, alleviating thermal stresses.

In continuous casting the problem is quality of billets produced rather than mold durability. A mold is made of copper covered with protective coating (usually chromium) and is cooled intensively with water. During casting, a mold undergoes distortion caused by thermal stresses induced by temperature differences in a particular part of the mold (Brimacombe et al., 1984). This affects the quality of the billet, of course, causing rhomboidity and corner cracking.

Ingot molds are still used to an appreciable degree in the iron and steel industry. The major cause of premature retirement of molds are cracks located either at the bottom of sidewalls or in mold corners. Both are due to the accumulation of damage contributed by successive casting cycles. Analyzing the working conditions of an ingot mold, it can be said that: (1) the temperature of the internal mold surface is 970 to 1170 K, depending mostly on mold design and the grade of steel being case, and (2) heat flux from steel to mold is affected by mold distortion and drops from the initial value of approximately 4.2 GJ/m^2 to 0.2 GJ/m^2 when a gap is formed between the ingot and the mold wall. Apart from that, it should be noted that initiation of cracks and heat flux are influenced by corrosion.

Figure 8.9 shows the temperature of mold internal surface and its deformation as functions of time. Measurements were made on a typical industrial bottom-poured 12.8-ton ingot mold of rectangular cross-section with a side-to-side ratio of 1.5. Measurements related to the half-height occupied by the ingot.

A more detailed experiment was carried out by Tlenkler and Czech (1962) on an 18-ton mold with a side-to-side ratio of 1.62. Measurements were conducted throughout the entire casting cycle and the results obtained are depicted in Fig. 8.10 together with locations at which they were taken. The results are shown in Fig. 8.11, where the deformations have been exaggerated. It is seen that:

In the first stage of the casting cycle the broad sidewall becomes concave and the narrow one, convex; maximum concavity occurs in the seventh minute (see Fig. 8.10).

After 40 min the concavity diminishes and maximum deformation appears after 120 min.

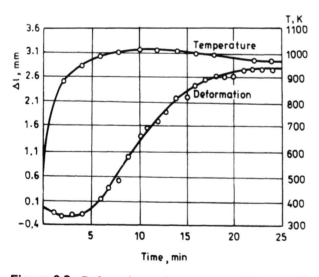

Figure 8.9 Deformation and temperature of ingot mold internal surface. (From Jefimov, 1960.)

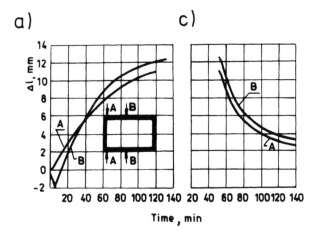

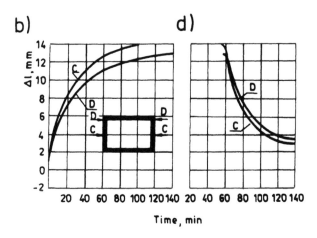

Figure 8.10 Ingot mold deformation throughout casting cycle measured at half height: (a) and (b) on pouring; (c) and (d) after stripping.

a)

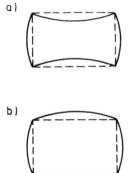

b)

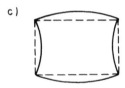

c)

Figure 8.11 Schematic representation of ingot mold deformation: (a) on pouring, (b) just before stripping; (c) after stripping.

Certainly, these changes in the mold's shape cause the corners to work in repeated bending. This can be referred to as mechanical fatigue at elevated temperatures. It was found that apart from fire cracks, corner cracking is the most serious problem. In seeking preventive measures the thermally activated processes should be discussed (i.e., diffusion, creep, phase transitions, corrosion). Undoubtedly these factors can be considered for a given mold and the specific technological process. Later in this section an attempt to cure this problem is described; however, limited space dictates only a brief discussion.

Stachurski (1972) measured the permanent deformations of 18-ton ingot molds made of pig iron having internal dimensions of 1640×695 mm at the bottom and 1575×625 mm at the top and a wall thickness of 200 mm. Ingot molds were used in 10- to 16-h cycles in open-hearth and converted plants to produce, by top casting, ingots of killed and semikilled steels. Pouring molten metal into mold lasted 3 to 5 min. Ingots were removed from molds after 3 to 4 h and molds were allowed to cool freely for some time, practically to the end of the casting cycle, when water spray was applied. Measurements of permanent deformations of ingot molds were taken after 20 to 60 cycles, with an average value of 30.

An ingot mold is shown in Fig. 8.12 together with locations of measuring points. The changes found in dimensions were:

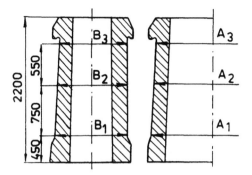

Figure 8.12 Ingot mold and locations of measuring points.

A_1 = +2 to +11 mm, A_2 = +2 to +8 mm, A_3 = −1 to +6 mm

B_1 = −16 to −2 mm, B_2 = −1 to −15 mm, B_3 = −8 to +2 mm

A plus sign indicates an increase in a dimension, a minus sign, a decrease. The increase in wall thickness measured along their centerlines was 1 to 2 mm for a narrow wall and 3 to 6 mm for a broad wall. The change in mold dimensions is caused by permanent deformation, whereas that in the thickness of the side walls is due to swelling of the mold material resulting from cyclic variations in temperature. Temperature cycling causes decomposition of cementite. The second competitive cause are cracks formed in mold walls, impairing their rigidity. A change in mold dimensions reduces the taper of the sidewalls which often causes a seizure of the ingot and mold failure while stripping it. It also follows from the measurements that these changes in dimensions grow larger with an increasing number of casting cycles, but it is associated with large variations in their magnitudes, even between molds that were in service for the same length of time. Most resistant to deformations are ingot molds that are circular in cross section, then square ingots, then rectangular ones. In the latter two cases it is advantageous to round off passages between neighboring sidewalls.

Apart from corner cracking and permanent deformation there is also a problem with fire cracks. Their origin is caused by thermal stresses resulting from differences in temperature. Even in sophisticated designs employing materials of good thermal conductance, the differences in temperature between the center of the sidewall and the corner are about 100 K. The second source of thermal stresses is the restriction of deformation of the inner mold surface by cooler, outer layers of material. The measure that makes it possible to reduce the magnitude of thermal stress is the choice of proper material. Cast iron is comparatively often employed for high thermal conductance of graphite. Maszkov and Zawgorodniev (1973) investigated the influence of thermal cycles resembling those occurring in ingots on the stresses induced in cast iron. The material investigated was cast iron (C = 2.9%, Si = 2.1%, Mn = 1.26%, S = 0.01%, P = 0.098%, Mg = 0.02%) having, after graphitizing, the following mechanical properties: ultimate tensile strength 500 to 550 MN/m^2 and elongation 1.5 to 2%. It was found that after a few thermal cycles a nearly constant state of stresses developed that favored cracking.

Deilmann and Schreiber (1974) attempted to extend the service life of 5-ton molds. The factors examined as affecting their performance were the shape of the ingot mold (see Chapter 10) and the chemical composition of the mold material. Generally speaking, a mold's material has to meet two contradictory demands: It should be soft to avoid fire cracks and hard to eliminate corner cracking. The material commonly employed for production of ingot molds is hematite pig iron having a silicon content of up to approximately 2% and carbon mostly in a form of graphite, which makes it possible to avoid the risk of fire cracking through reduction of the thermal gradient. Deilmann and Schreiber (1974) optimized the chemical composition of steelmaking pig iron by providing a pearlitic matrix in the mold and a higher amount of combined iron. This grade of material can withstand higher loads than hematite pig iron and proves useful in casting ingots of killed steel. The results of the field trials on 6.3-ton molds can be summarized as follows:

For Mn/Si ratios approaching 2, ingot molds should be soft annealed at 1270 K before being placed into service.
For Mn/Si < 1, ingot molds withstand more than 100 pourings for Si contents below 1.5% and the carbon content has practically no effect on the mold's durability.
For Mn/Si > 1, the Si content should be below 0.8%.
For high Mn/Si ratios exceeding 1, the C/Mn ratio should be above 4.
For C/Mn < 3.2, the durability of molds is low and independent of the Mn/Si ratio.

The results obtained made it possible to find a better combination of carbon, manganese, and silicon contents, thus prolonging the service life of molds. More explicit chemical contents of these elements are given by Schmidt (1979). Values recommended for molds used in converter plants are 3.5 to 3.6% C, 1.4 to 1.5% Si, and 0.7 to 0.9% Mn.

It can be expected that nodular irons could offer far higher durability than other currently used ingot mold materials for more advantageous forms of graphite and better stability of the microstructure in service at high temperatures. However, it is also essential in this case to optimize the chemical composition to avoid permanent deformation of the ingot mold after a long run.

Cracked ingot molds are repaired by welding preceded by grinding to the sound material. This method is capable of at least doubling the service life. Moreover, trying to find a more general preventive measure, the application of protective coatings and alteration of pouring practice should be considered. An example of the former method was given by Chladek (1980). The Cu-Ni protective coatings applied to the surfaces of the cast iron ingot molds used to produce grate bars increased the durability by as much as 80%. The latter method was evaluated experimentally and practically by Droniuk et al. (1982). It was found that the application of exothermic mold powder including 10 to 12% aluminum, 9 to 12% soda niter, 19 to 21% soluble glass, and other ingredients retards corrosion progress in an ingot mold made of cast iron. The corrosion protection was attributed to the diffusion into the working layer of the ingot mold of elements constituting mold powder and formation of composite oxides on their basis. The coagulation of these oxides and filling gaps left by oxidized graphite restored the thermal fatigue resistance of the subsurface layer.

An extension of the service life of ingot mold necessitates an estimate of the usefulness of the potential remedial measures reviewed in this section: that is, optimizing the microstructure and the related chemical composition of ingot mold material, the application of protective coatings and mold powders, maintenance by welding, and redesign of the ingot mold to achieve better fatigue resistance and application of water spray cooling systems as described in Chapter 10.

8.3 TURBINES

8.3.1 Working Conditions

It is generally accepted by responsible experts that until the end of the century most of our electrical energy will be produced by turbogenerators. In addition, the high power output-to-weight ratio of turbines deserves attention. Turbines are conventionally divided into two groups: steam and gas. Each includes a variety of turbines of different designs working under different conditions. In addition, turbines have to meet specific requirements. For instances, economical aspects are most important in the case of steam turbines utilized to produce electrical energy, including their availability and efficiency, whereas high power output is demanded of aircraft turbine engines, which gives rise to a pronounced reduction in durability. The life expectancy of large turbosets is 100,000 h whereas that of aircraft turbine engines is between 5000 and 10,000 h. Diesel engines and steam turbines have also reached a well-established state of development, and any improvement in their performance is difficult to achieve. In contrast, the inlet gas temperature in open-cycle combustion turbines can be raised relatively easily. A 100-K increase in this temperature results in an increase in net power output of about 20%. This causes higher working temperatures to be involved and new material to be applied.

Steam turbines currently utilized in Europe to produce electric energy are usually between 600 and 800 MW, and in Japan and the United States they are even above 1000 MW. The trend to increase the rating of turbosets is because the installation of one highly rated turboset is 10 to 20% cheaper than the installation of two turbosets which together produce the same net power output. In steam turbines at present, inlet temperatures reach 838 K and upper values of steam pressures are between 20 and 35 MPa. An analysis of their working conditions shows that they have to meet the requirement of high operational flexibility. This is caused by changes in the consumption of electric energy, the rapid drop during weekends being particularly obvious (Fig. 8.13). Consequently, steam turbosets operate under changing-in-time conditions, frequently referred to as *cold starts*, *warm starts*, *hot starts*, and *overspeedings*. The first term refers to the situation when the turboset is

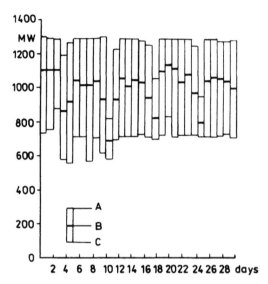

Figure 8.13 Load range of the 1300-MW turboset Gavin 2 (U.S.) for September 1977. A, Maximum load; B, average load; C, minimum load. (From Hohn, 1980.)

switched on after a period of rest. The following two terms describe a load change such that no condensation of steam can take place; the difference between them lies in the extent of load change. The final term relates to the case when the rotational speed of the rotor exceeds a nominal value. According to Kumeno et al. (1977), the approximate numbers of operations that have to be taken into account by the designer are 30 overspeedings, 100 cold starts, and 10,000 warm or hot starts. Of course, all changes in working conditions must be carried out gradually according to routine procedures, but it follows from field experience (Somm and Reinhardt, 1979) that breakdowns generally occur during load changes. These failures are caused by thermal stresses induced in turbine components. In this section failures and their causes will be considered for rotors, disks, and blading, as almost any damage to them decreases the performance of the turbine and often causes the failure of other components. The forces acting on these components are as follows:

1. Forces caused by steam pressure differences exerted on the surfaces of a component: steady and varying due to rotating blades, fluctuations, or separation of flow
2. Forces caused by rotary movement and vibrations due to unbalance of the rotating mass
3. Forces caused by restrained thermal expansion
4. The weight of the component.

The resultant stresses acting within components considered can be divided into the following:

1. Stresses due to steady external or inertial forces causing creep.
2. Stresses resulting from thermal gradients existing during steady operation. At working temperatures these stresses can relax and partly vanish; the effect produced by them can, however, be referred to as thermal fatigue due to cyclic mode of turbine operation.
3. Stresses resulting from temperature gradients occurring during cold starts, warm and hot starts, and shutdowns causing thermal fatigue.

From the above it can be concluded that working conditions include creep, mechanical fatigue, and thermal fatigue. To complete the picture it should be mentioned that steam and water droplets cause erosion of guide vanes and blades. Working temperatures of industrial gas turbines are in the range 1120 to 1470 K. In aircraft turbine engines the inlet gas temperature is near this upper limit. The temperature in the automotive gas turbine engine must be even higher (1570 to 1670 K) to achieve competitive fuel consumption. In some instances, hot parts can be cooled by air applied to the combustor to cool combustion gases from the temperature of the flame, and stationary blades can be cooled by air flowing through the slot designed in the trailing edge. Parts exposed to the highest temperatures are often made of ceramic (e.g., the rotor of automotive turbine engines) (Lines, 1974).

Working conditions of gas turbine components can be listed as for steam turbines, including, however, the much higher temperatures involved and specific features of fossil combustion, such as (1) highly aggressive products of combustion and the presence of trace elements (V, K, S, Na, Cl) which greatly contribute to the process of corrosion; and (2) severe thermal gradients in blades, causing damage to the already formed protective scale and frequently causing blade cracking.

8.3.2 Damage to Turbine Components and Methods of Mitigation

This section deals with damage to rotors, disks, and blades. Damage to these components in steam turbines is very important, as the very high rotational speed of these components may lead to their projection as "missiles," subsequently penetrating the turbine casing or even causing damage to a steam supply system. Economically, any repair or replacement of these components is time consuming and expensive, as the cost of power to replace the lost capacity is added to the cost of repair. These problems have been a focal point since the occurrence of a series of catastrophic failures of steam turbine rotors.

At least 16 of the failures or near-failures of steam turbine rotors occurring in the 1950s can be attributed to the brittleness of rotor material (Viswanathan and Jaffee, 1983). In the 1950s, ingots for rotor forgings with a chemical composition of 1% Cr, 1% Mo, 0.25% V were melted in open-hearth furnaces. Segregation bands of manganese sulfides, tramp elements, and alloying elements were reported in the region of the rotor bore. Ingots were contaminated with high concentrations of H, P, Sb, Sn, As, and S, and these elements are known to have a detrimental influence on mechanical properties. Typical FATT (ductile-to-brittle fracture appearance transition temperature) values were about 420 K. One of the high-pressure/intermediate-pressure rotor of this vintage failed in 1974 at the TVA Gallatin station in the United States. The fracture was found to have nucleated in the bore region. It was concluded from an analysis that the rotor failed after about 4 h from the beginning of the warm-up period after synchronous speed was reached. At the moment of failure there was a dangerous combination of temperature, tangential stress, and fracture toughness (K_{Ic}) of rotor material in the bore region. Under these circumstances, the already formed crack developed in an avalanche fashion as it reached critical size. According to fracture mechanics, the critical crack size (a_c) that leads to unstable growth is as follows (Greenberg et al., 1969):

$$a_c = \begin{cases} \dfrac{K_{Ic}^2 \cdot Q}{\pi \sigma^2} & \text{for internal crack} \\[4mm] \dfrac{K_{Ic}^2 \cdot Q}{1.21 \pi \sigma^2} & \text{for surface crack} \end{cases}$$

where

σ = stress at the point of interest,

Q = flaw shape parameter, which is a function of crack geometry and the ratio of applied stress to yield stress of the material,

K_{Ic} = fracture toughness of material at the point of interest under the temperature and stress conditions considered

The a_c value for this particular rotor was estimated as 7.3 mm (Cook et al., 1978). It is obvious that such a large crack was formed by flaw growth. When causes of flaw growth are considered, the following points must be noted:

1. Corrosion cannot play any role, as the steam turbine rotor was operating in dry steam.
2. The creep phenomenon seems unlikely as a potential cause, as the fractured part was in the cooler part of the rotor.
3. The last and most probable alternative is mechanical failure caused by superimposing the transient stresses resulting from the cold start on the stresses related to the rotary motion of the turbine rotor.

In conclusion, the preexisting flaw developed during cold starts, load changes, and shutdowns, and its progress can be described by Paris's equation (see Chapter 2).

At the end of 1950s and the beginning of the 1960s cracks were also reported to occur at the rotor surface in the region adjacent to the high-pressure turbine impulse chamber, where the temperature variations during load changes were the greatest (Berry and Johnson, 1964). The method applied to solve this problem involved close control of temperature rise rates during major cycles of load changes. It was studied by Hoyle and Mahabir (1963) in relation to marine steam turbines using computer simulation in the work concerned with the emergency starting of marine turbines from cold. In this case a heavy condensation of steam occurs and the heat flux to the rotor material exceeds values corresponding to a routine major load change. The most acceptable procedure, occupying 53 min, was recalculated by Chow (1971). The reduced thermal stresses induced in the region of the first stage are shown in Fig. 8.14. Other factors affecting rotor cracking are FATT and fracture toughness. The effect of these factors can be limited by the use of rotors with an axial center bore. This enables the following:

1. Removal of segregates and nonmetallic inclusions.
2. Evaluation of mechanical properties at the bore region.
3. Performance of ultrasonic and magnetic testing of the bore.
4. Promotion of hydrogen diffusion.

The other design approach is the welded rotor, and this is shown in Chapter 10.

The typical FATT values for Cr-Mo-V steel rotors dropped from 420 K in the 1950s to about 360 K in the 1970s. The ASTM A470 class specification limits the FATT value to 393 K. It is noteworthy that the FATT value of the turbine can change during service as the P, S, Sn, As, and Sb elements can diffuse toward grain boundaries, causing the material to become brittle. This is enhanced by the presence of Mn and Si as alloying elements. The amount of hydrogen can be reduced by vacuum ladle treatment and vacuum degassing. The alternative method is anti-snowflake heat treatment (i.e., long-term annealing at 870 to 930 K). The holding time should be 20 to 30 h per 100 mm of component diameter. Good prospects of quality improvement are offered by vacuum carbon deoxidation (VCD), electroslag remelting (ESR), and low-sulfur silicon deoxidized conventional (LSC) technologies. Ladle chemical analyses are listed in Table 8.1 and compared with those of other ingots. It can be seen that all three methods provide lower S contents, LSC,

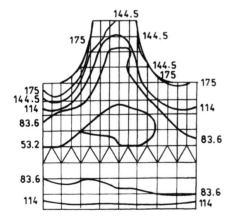

Figure 8.14   Reduced thermal stresses in marine steam turbine rotor. (From Chow, 1971.)

Table 8.1 Chemical Composition of Cr-Mo-V Steels[a]

	C	Mn	P	S	Si	Ni	Cr	Mo	V	Sn	Sb	As	Nb	V	Al	H
ASTM A470 Class 8	0.25–0.35	1.0 Maximum	0.015 Maximum	0.018 Maximum	0.15–0.35 Maximum	0.75 Maximum	0.9–1.5	1.0–1.5	0.2–0.3	—	—	—	—	—	—	—
Conventional[b]	0.32	0.83	0.009	0.009	0.27	0.23	1.07	1.17	0.25	0.008	0.001	0.01	—	—	0.003	NR
Advanced Technology Rotors																
LSC	0.31	0.78	0.007	<0.001	0.28	0.33	1.13	1.15	0.23	0.002	0.0012	0.003	—	—	0.004	NR
VCD	0.28	0.76	0.004	0.001	0.05	0.40	1.18	1.21	0.26	0.010	0.0015	0.006	—	—	0.005	NR
CSR	0.31	0.78	0.009	0.002	0.19	0.27	1.18	1.18	0.26	0.003	0.001	0.011	—	—	0.009	NR
1950 Vintage Rotors																
Buck	0.31	0.72	0.028	0.036	0.29	0.14	1.01	1.25	0.22	0.0903	NR	NR	—	—	NR	NR
Joppa	0.30	0.81	0.013	0.015	0.32	NR	0.97	1.26	0.23	0.014	NR	NR	—	—	NR	NR
Gallatin	0.36	0.84	0.012	0.011	0.23	0.06	1.07	1.34	0.26	0.03	NR	NR	—	—	NR	0.00036

Source: Viswanathan and Jaffee (1983).
[a]NR, not reported.
[b]Mean value of 29 conventional rotors purchased by Westinghouse Corporation during the period 1971–1977.

VCD, and also phosphorus levels compared to those of the conventional forgings. The soundness of forgings in terms of porosity, segregation of sulfur, and piping is excellent. Moreover, VCD and ESR technologies provide lower FATT values and all three methods mentioned above provide higher impact strength and fracture toughness (Viswanathan and Jaffe, 1983).

Kumeno et al. (1977) investigated two large forgings made in 1970. One was a 1% Cr-Mo-V steel forging for an intermediate-pressure turbine rotor for a 350-MW generator with a weight of 31.6 metric tons. The other was a 3.5% Ni-Cr-Mo-V steel forging for a low-pressure rotor of a 600-MW steam turbine. Their chemical compositions and average mechanical properties are listed in Table 8.2 together with thermal treatment conditions. The results of a non-destructive test are shown in Table 8.3. The defect in the Cr-Mo-V forging was a sulfide inclusion, and in the Ni-Cr-Mo-V forging the defect was microporosity. The allowable size of the sulfide inclusion located at the rotor bore surface was estimated to be about 1.6 mm in length. This was based on the assumption that the rotor has to withstand 10,000 cycles of major load changes, which corresponds to the real working conditions. However, as this estimation neglects the influence of adjacent flaws and a change in fracture toughness with temperature, the critical size quoted should be considered only an approximation.

McCann and co-workers (1973) investigated two IP steam turbine rotors composed of 1% Cr-Mo-V steel. They attempted to correlate chemical composition, microstructure, and mechanical properties. The first rotor was withdrawn from production because of porosity found at the bore, and the second was rejected because of hardening cracks. Both ingots were arc melted, silicon deoxidized, and vacuum poured. Forgings were heat treated in accordance with the following procedure: rotor A—oil hardened from 1248 K, tempered at 970 K for 18 h, furnace cooled; and rotor B—oil hardened from 1298 K, rehardened from 1248 K, tempered at 973 K for 19 h, furnace cooled. Chemical analyses revealed a little segregation of elements in rotor A. Results of the chemical analyses for rotor B are shown in Table 8.4. Normal test positions are shown in Fig. 8.15 and positions from which additional test material was selected are indicated in Fig. 8.16. It can be seen that the carbon and molybdenum contents are within the range 0.27 to 0.37% and 0.91 to 1.08%, respectively. Microstructural examinations revealed the structure of upper bainite developed by thermal treatment. This distribution of carbides was a little less homogeneous at the surface of the rim of rotor B, perhaps because of the higher content of mid-bainite. In some regions a lack of fine precipitates of VC was found, probably because of the formation of coarse alloy carbides. In a similar way, the microsegregation of molybdenum causes the precipitation of M_2C carbides instead of the formation of fine V_4C_3 precipitates. The tensile tests on specimens taken from different locations on rotors A and B showed that 0.2% proof stress and ultimate tensile strengths were fairly constant regardless of test location or test direction. Elongations obtained in tensile tests carried out at room temperature and at 783 K were nearly independent of specimen location for rotor A. This is in contrast to the results obtain for rotor B, where elongations varied from 6 to 25% at room temperature and from 15.7 to 28% at 783 K. It can be concluded that the

Table 8.2 Chemical Compositions and Mechanical Properties of the Rotor Forgings Tested

	Chemical composition (wt % ladle)									
	C	Si	Mn	P	S	Ni	Cr	Cu	Mo	V
Cr-Mo-V forging	0.28	0.28	0.71	0.010	0.009	0.35	1.15	—	1.31	0.25
Ni-Cr-Mo-V forging	0.24	0.06	0.30	0.009	0.011	3.68	1.80	0.06	0.45	0.12

	Mechanical property[a]				
	Tensile strength (MPa)	Yield strength (MPa)	Elongation (%)	Reduction in area (%)	FATT (K)
Cr-Mo-V forging	819(R) 816 (C)	636 (R) 615 (C)	22.1 (L) 18.7 (C)	63.0 (L) 52.2 (C)	360 (R) 372 (C)
Ni-Cr-Mo-V forging	857 (R) 872 (C)	730 (R) 724 (C)	22.8 (L) 24.5 (C)	71.5 (L) 68.2 (C)	165 (R) 212 (C)

	Heat treatment	
Cr-Mo-V forging:	Quenched: 1240 K × 21 h	Fan cooled
	Tempered: 940 K × 54 h	Furnace cooled
Ni-Cr-Mo-V forging:	Quenched: 1110 K×36 h	Water spray
	Tempered: 870 K × 74 h	Slow cooled

Source: Kumeno, et al. (1977).
[a]R, radial body; C, longitudinal center core; L, longitudinal.

Table 8.3 Nondestructive Tests Results of the Forgings

	Cr-Mo-V Forging	Ni-Cr-MO-V Forging
Periphery ultrasonic test	$B_1 = 100\%$[a]: no indication High sensitivity[b]: noise echo	$B_1 = 100\%$[a]: no indication High sensitivity: 194 indications, max. 20%
Bore-sonic test[c]	No indications	152 indications, max. 30%
Bore magnetic particle test[d]	53 indications, max. 1.6 mm	162 indications, max. 3.0 mm
Bore-scope test	305 flaws, max. 2.3 mm (Bore diameter 72 mm)	89 flaws, max. 1.5 mm diameter

Source: Kumeno et al. (1977).
[a]Setting the first back reflection to 100% on the oscilloscope.
[b]Sensitivity so as to get 10% reflection from the 1.6-mm-diameter penny-shaped crack at the center bore.
[c]Sensitivity: Set the reflection from a 10-mm-diameter 100-mm-length axially drilled calibration hole centered 61 mm apart from the bore surface equal to 100%.
[d]Magnetizing current is 39 A/cm bore diameter.

Table 8.4　Chemical Analysis Survey Throughout Rotor B

Sample position[a]	Chemical composition (wt %)								
	C	Si	Mn	S	P	Ni	Cr	Mo	V
Cast analysis	0.30	0.20	0.80	0.011	0.009	0.67	1.20	1.08	0.37
Standard test positions									
Bottom shaft	0.31	0.20	0.82	0.019	0.013	0.63	1.13	1.02	0.35
Top shaft	0.33	0.19	0.79	0.019	0.013	0.62	1.11	1.01	0.34
Top—body rim	0.29	0.20	0.78	0.011	0.010	0.63	1.10	1.06	0.32
Bottom—body rim	0.30	0.20	0.79	0.011	0.010	0.64	1.18	1.08	0.36
Center—body rim	0.30	0.20	0.81	0.010	0.010	0.64	1.13	1.00	0.31
Center—body rim	0.30	0.20	0.80	0.011	0.009	0.65	1.12	1.02	0.31
Midcore	0.26	0.19	0.79	0.008	0.007	0.62	1.13	0.97	0.34
Bottom core	0.24	0.19	0.74	0.009	0.007	0.60	1.09	0.91	0.32
Top core	0.37	0.21	0.83	0.016	0.012	0.66	1.15	1.05	0.35
Extra tests									
Top body T2	0.36	0.24	0.84	0.021	0.013	0.69	1.21	1.17	0.30
Top body T2	0.33	0.24	0.83	0.011	0.010	0.67	1.18	1.01	0.34
Top body T1	0.34	0.30	0.84	0.011	0.011	0.67	1.18	1.01	0.33
Top body T1	0.35	0.26	0.84	0.011	0.011	0.69	1.20	1.01	0.34
Top body T1	0.36	0.20	0.84	0.019	0.012	0.68	1.18	1.14	0.34
Mid body M2	0.34	0.20	0.84	0.011	0.012	0.74	1.30	1.08	0.33
Mid body M1	0.31	0.28	0.83	0.012	0.010	0.66	1.17	0.86	0.34
Bottom body B2	0.31	0.27	0.83	0.011	0.011	0.65	1.17	0.85	0.34
Bottom body B1	0.27	0.19	0.79	0.010	0.010	0.62	1.09	0.98	0.31

Source: Journal of the Iron & Steel Institute.
[a]Positions marked "Top," and so on, correspond to the position of the forging with respect to the original ingot. Positions marked "1" correspond to test positions 2 in away from the core. Positions marked "2" correspond to test positions 10 in away from the core.

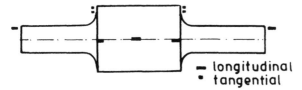

Figure 8.15 Positions of normal production tests. (Reprinted with permission from Journal of the Iron & Steel Institute.)

segregation of elements in rotor B and differences in the rate of cooling for particular layers of material during quenching caused variations of bainitic microstructure which affected the mechanical properties.

The problem of optimizing the chemical composition and heat treatment of rotors was studied by Norton and Strang (1969). It was found that to obtain a predominant structure of upper bainite in the body of a large steam turbine rotor, which provides the most satisfactory dispersion of fine vanadium carbides and thus good creep properties, the chemical composition of rotor material has to be altered so that the proper balance between carbide-forming elements will be maintained. The resultant chemical composition of 1% Cr-Mo-V steel includes:

Manganese and nickel; the upper limit for both should be 0.75%, as at higher percentages vanadium carbides can become coarse.

A molybdenum content of 0.6 to 0.9%; a higher content reduces the amount of V_4C_3 carbide, and a lower content is not recommended because molybdenum stabilizes fine vanadium carbide, maintaining its coherency with the matrix.

Chromium should be kept below 1%, as a higher content reduces creep properties.

The vanadium-to-carbon ratio should be 1.5:1 to 2:1, to balance the creep rupture properties and ductility.

The steady creep rate of optimal-composition rotor forgings was lower by some tens of percentage points than those of conventional composition. However, it should be mentioned that optimization of the microstructure can also be achieved through redesign of a rotor, thus reducing its diameter or gashing disks before final heat treatment or by the application of a welded rotor (see Chapter 10).

The other alternative is to produce rotors with an axial gradient in properties. This

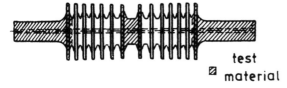

Figure 8.16 Schematic diagram of rotor B and positions of extra tests. (Reprinted with permission from Journal of the Iron & Steel Institute.)

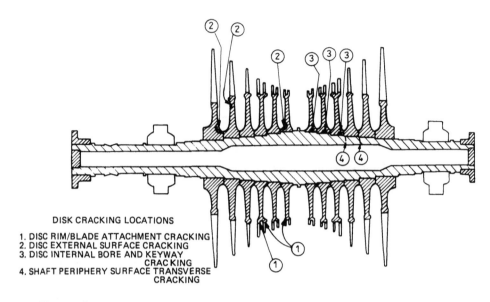

DISK CRACKING LOCATIONS

1. DISC RIM/BLADE ATTACHMENT CRACKING
2. DISC EXTERNAL SURFACE CRACKING
3. DISC INTERNAL BORE AND KEYWAY
   CRACKING
4. SHAFT PERIPHERY SURFACE TRANSVERSE
   CRACKING

Figure 8.17 Low-pressure turbine configuration and disk cracking locations. (From Lamping, 1980.)

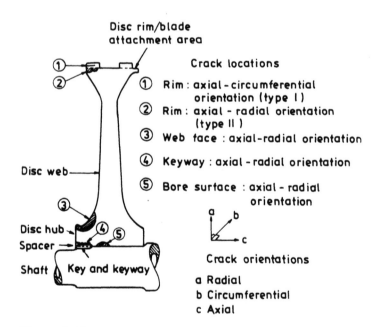

Disc rim/blade
attachment area

Crack locations

① Rim : axial - circumferential
   orientation (type I)
② Rim : axial - radial orientation
   (type II)
③ Web face : axial-radial orientation
④ Keyway : axial - radial orientation
⑤ Bore surface : axial - radial
   orientation

Disc web

Disc hub
Spacer
Shaft   Key and keyway

Crack orientations

a Radial
b Circumferential
c Axial

Figure 8.18 Locations and orientations of cracks in low-pressure turbine disks in U.S. power plants. (From Lyle et al., 1985.)

enables a balance to be kept between fracture toughness and creep properties, as the former should gradually increase from the high-pressure section of turbine to the low-pressure section, and the latter should decrease (Sawada et al., 1981).

From the above it can be concluded that the life of integral design rotors can be prolonged by:

Optimizing the chemical composition

Elimination of nonmetallic inclusions and segregates

Redesign, which makes it possible to avoid excessive stress concentrations and obtain the desired microstructure

Optimizing thermal treatment and forging; field experience indicates that the degree of forging should be two to three

Accurate control of material properties (i.e., creep performance and fracture toughness) and through nondestructive examinations of forgings produced

Routine inspection of working steam turbines

Selection and evaluation of materials of which components affecting turbine availability are made (e.g., steam piping)

Turbine disks transmit torque from the blades to the shaft. In modern turbine rotors there are 10 to 16 disks. Steam turbine rotors are often produced by shrink-fit of disks to the shaft. To prevent a disk from slipping, one or more "keys" are placed into a keyway between the rotor shaft and disk. In the United States, disks are made of quenched and tempered 3.5 Ni-Cr-Mo-V steel. In the United Kingdom disks are frequently composed of 3% Cr-Mo steel. The U.S. 3.5 Ni-Cr-Mo-V steel has to meet the requirements set down in ASTM A471: ultimate tensile strength 965.3 MPa minimum; yield strength, 862 to 1000 (0.02%); elongation, 15% minimum; reduction, 43% minimum.

In 1969 the catastrophic failure of a forged shrunk-on steel disk in a low-pressure turbine rotor at the Hinkley Point nuclear power plant was reported. This stimulated the inspection of disks and initiated investigations of their failures. The British examined 810 disks from 102 turbine rotors, and 124 disks on 50 rotors were found to be cracked. The U.S. experience corresponds well with British findings. The cross section of a low-pressure turbine with shrunk-on disks is depicted in Fig. 8.17, and the crack locations are also marked. In analyzing the 1969 failure, it was found that the crack originated in the keyway. The regions of the disk in which cracking was revealed are shown in Fig. 8.18. In Fig. 8.19 the location of cracks on disk rim/blade attachment devices is shown for some steeple configurations (Lamping, 1980). For the fir tree type of attachment, cracks occur in serrations of steeples and propagate tangentially to the disk rim (type I crack) or propagate from the innermost serration in the radial direction (i.e., parallel to the disk web). For the second and third attachment configurations, cracks propagate axially (parallel to the shaft axis) and tangential to the disk rim. Apart from these occurring in rectangular keyways, grooves growing from the corners of keyways were also found. The largest groove was 9.4 mm. It is believed that they are related to erosion by condensed water, as they have been found when steam entering the turbine was not reheated. The disk cracking described previously was associated with stress-corrosion cracking. Cracks

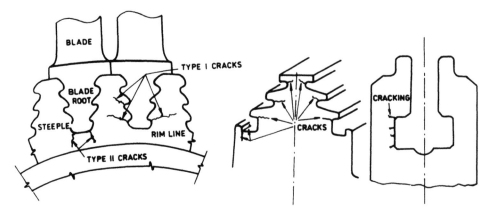

Figure 8.19 Disk rim/blade attachment cracks.

found in disks were predominantly intergranular, branched, and filled with iron oxides (Lyle et al., 1985).

A correlation between cracking incidents and the following factors has been established (Lyle et al., 1985):

1. The magnitude of stresses in the disk rim/blade attachment device. The higher the stress, the more likely it is that a crack will occur.
2. The number of cold starts and condenser leakages. Both of these introduce air into the turbine and thus produce steam contamination. As it is shown by data obtained in industry and from experimental results that most cracks grow under conditions of pure steam, steam contamination is probably important in crack initiation.
3. The presence and width of crevices between keys and keyways. A laboratory test with pure steam passing through a crevice revealed that the critical gap width is of 0.05 mm or less (Wigmore, 1979).
4. Temperature. Cracks developed faster at higher temperatures.

It is recommended that:

1. Reheaters be operated properly to maintain the moisture line in the same location as that where disk cracking has occurred (this pertains only where a liquid phase is present). Any change in the location of the moisture line increases the number of disks liable to stress corrosion.
2. Crevices between keys and keyways be avoided.
3. Steam contamination be prevented.
4. Materials more resistant to stress-corrosion cracking be used (i.e., more pure materials of lower yield strength), although this is difficult to do in a manufactured turbine rotor, as the casing offers limited space.
5. Routine nondestructive examinations be carried out, and failed components, on which turbine availability is dependent, be replaced promptly.

It seems obvious from our earlier review of the working conditions of gas turbines that turbine blades are the parts of the gas turbine most liable to wear. Turbine blading works under creep conditions, gas corrosion, and often subject to severe thermal stresses. The latter was the cause of the leading blade edge cracking observed in a J47 aircraft engine. In laboratory examinations it was found that the maximal temperature difference between the leading edge and the midchord was 467 K. The problem was solved by designing the special fuel distribution valve supplying fuel to each combustor nozzle in sequence, which enabled simultaneous ignition of all combustors to be avoided and thus thermal gradients in the blade to be reduced (Johnston et al., 1959).

The problem of thermal stresses occurs to a large extent in air-cooled airfoils. Cooling is applied to reduce blade temperature and therefore to meet creep resistance requirements. The typical temperature reduction of that is of about 300 K at the expense of a buildup of thermal stresses. The solution to this problem is either the application of ceramic blades, which has been studied extensively, or the unidirectional solidification of blade material. According to Spera and co-workers (1971), the columnar-grain Ni-based MAR-M200 superalloy is 100 times more resistant to thermal fatigue than that conventionally cast. Of course, the axis of preferred growth has to be close to the axis of principal stress. The increase in thermal fatigue resistance can be attributed to two factors: realignment of grain boundaries parallel to the direction of principal stress, and to lower elastic modulus for the preferred growth direction of unidirectionally cast material. A further increase can be accomplished by the application of monocrystalline structure blading.

Vanes in aircraft turbines are repaired by standardized procedures recommended by the producer. As they are made of cobalt alloys, they are easily welded, whereas turbine blades made of high-strength Ni-based superalloys, having a much lower response to welding, have to be repaired either by brazing, solid-state diffusion, or transient liquid-phase (TLP) bonding. Full rejuvenation of noncracked service-exposed turbine blades can be achieved by the hot isostatic pressing (HIP) process. This process involves simultaneous application of high temperature and high pressure, which causes creep cavities to close and diffusion bonding of cavity surfaces (Koul et al., 1984).

Sudden increases in fuel oil prices have led to the use of lower-grade fuels. As such fuels have a much higher content of sodium and vanadium compared to other metallic species, the use of these fuels is suspected to contribute to the corrosion process. For example, 4-GT fuel can contain up to 500 ppm vanadium. Vanadium forms a molten slag layer on the blade surface which fluxes the oxide layer and thus enhances the corrosion process. The deposit on the blade surface can contain a mixture of V_2O_5 and Na_2SO_4. It is known that such a mixture is particularly aggressive. There are two feasible ways to deal with this problem: to reduce the blade temperature to 965 K (i.e., below the melting point of V_2O_5), which is not profitable economically; or to add magnesium to the fuel, which forms compounds (e.g., magnesium vanadates) of relatively high melting point. Stevens and Tidy (1983) recommend an Mg/V weight ratio of at least 1.

In the case of aircraft turbines running on distillated fuel, this phenomenon does not occur. However, corrosion enhancement has been observed, and it was found that salt mist ingested with intake air can be responsible for this.

The methods by which protective coatings are applied to airfoils are described in Chapter 9 together with some results of operational tests. It is worth noting that the coated blade has to possess adequate creep strength, corrosion resistance, resistance to erosion by products of combustion (e.g., hydrocarbons, ashes, and oxides), and must be able to withstand the impact of foreign objects on the turbine, which often happens in aircraft turbines.

The nominal chemical compositions of rotor materials are given in Chapter 3. It should be noted that the spread of mechanical properties within particular ingots of superalloys is larger than that for steel. This causes powder processing to be a competitive production method of gas turbine components. However, it seems obvious that a radical change in the technology of materials needs great experimental exertion.

A temporary method that enables the problem of turbine safety to be solved to some extent involves the inspection and retirement of components based on a routine schedule. This can be illustrated by the case of the 485-CAN 40 turbine engine used in CF-114 trainer aircraft. An analysis performed on the grounds of fracture mechanics enabled safe life limits to be specified and the determination of an inspection procedure. The crack-prone region of this engine was a fifth-stage compressor disk, and based on the analysis mentioned, this component was used well beyond the safe life limits recommended by the producer (Koul et al., 1984).

8.4 EQUIPMENT FOR HEAT TREATMENT

The term *heat treatment* covers a variety of processes carried out at elevated temperatures and frequently also in carefully controlled atmospheres. The processes consist of the following principal stages: heating to a required temperature, annealing for a specified periods, and cooling down. The precise characteristics of the thermal cycle and the atmosphere involved are suited to a particular application, such as the chemical composition of a component being heat treated, its size, and the resultant properties required.

The variety of cycle characteristics and the use of equipment of different shapes and sizes destined either to manipulate heat-treated parts or to create and maintain necessary temperature and environment frequently necessitates a specific, individualized approach. A routine procedure leading to cost reduction in such instances consists of:

An analysis of equipment damage: causes and results,
An estimation of working conditions,
Laboratory tests to simulate the effect of working conditions on the element under consideration,
Field trials

While considering candidate materials for particular applications, it should be remembered that components made of inexpensive materials may require frequent repair, increasing the cost to an unacceptable level and that, for example, an increase in the heat-treating cycle temperature is, in general, profitable, as it enhances efficiency. For this reason, laboratory tests should be aimed at estimating relatively vast group of materials, and test conditions should enable us to evaluate purposefulness of alteration in cycle characteristics. However, laboratory tests provide only approximate information on materials performance, since an estimate of working conditions made on the basis of a user's observation usually differs from real conditions, due mainly to limited controllability of heat-treating processes and the omission of seemingly marginal factors. Moreover, the data regarding performance of materials provided by the producer are obtained in routine tests and therefore are of rather limited usefulness. For example, oxidation tests are usually carried out in isothermic conditions in still air with periodic (e.g., once a day, once a week, or after completion of the test) withdrawal of specimens from the furnace to obtain weight-increase data and to assess depth of internal penetration by metallographic examination. In contrast to this, the duration of the thermal cycle employed in a heat-treating plant is usually of some hours, and the oxide layer is easily impaired, with the consequent formation of new oxides in unprotected zones. Because of this, routine laboratory tests and even tests developed purposefully to simulate real working conditions provide data of limited value, so that what is required is final verification of experimental results in the field. Complexity of oxidation tests can be illustrated by the work of Rundell (1985a,b), who found that oxidation rates in isothermic tests carried out below 1270 K, estimated in terms of internal penetration, are similar to those obtained in conditions of cyclic exposure, whereas at temperatures exceeding 1370 K, the oxidation rates in cyclic tests are a few times greater than for isothermic tests. In addition, the oxidation rate of AISI 309 steel in cyclic tests carried out at 1423 K with a cycle duration of 20 h depends strongly on the chromium content of the specimen tested and can differ by the factor of 4 for compositions consistent with the materials specification. The oxidation rate in cyclic tests showed marked instabilities connected with spalling of formed oxides, which demonstrated itself as repeated sequences of rapid oxidizing followed by a recovery period. This, of course, makes an extrapolation of test results difficult. The other noteworthy finding was that the fraction of adherent oxides depends on material and temperature, as well as on test conditions, and can be as low as a few percent. Examples of effective selections of material for heat-treating equipment supported by field trials are given below.

Fluck et al. (1985) demonstrated a procedure undertaken to select a material for a flame hood used in a heat-treating plant. The chemical compositions of alloys considered are given in Table 8.5. The oxidation resistances of all of them, except for Cabot alloy 214, depend on the formation of chromium-rich scale. At temperatures exceeding 1353 K, however, Cr_2O_3 is converted into volatile CrO_3, which does not provide proper protection from oxidation. Alloy 214 forms aluminum oxide having good stability and adherence at elevated temperatures as well as chromium oxide of limited efficiency. Results of a static oxidation test are shown

Table 8.5 Nominal Compositions of Alloys

Alloy (UNS No.)	Ni	Fe	Co	Cr	Mo	W	Al	Ti	Other
					Composition (%)				
Cabot alloy 214	75	2.5	2.0[a]	16.0	0.5[a]	0.5[a]	4.5	—	Y (present)
Hastelloy alloy X (N06002)	47	18.5	1.5	22.0	9.0	0.6	—	—	
Alloy 600 (N06600)	72	8.0	1.0[a]	15.5	—	—	0.35[a]	0.3[a]	
Inconel alloy 601 (N06601)	Balance	14.0	—	25	—	—	1.4	0.3	
Cabot alloy 625 (N06625)	Balance	5.0[a]	1.0[a]	21.5	9.0	—	0.4[a]	0.4[a]	Cb+Ta = 3.5
Haynes alloy 230	57	3.0[a]	3.0[a]	22.0	2.0	14.0	0.3	—	0.03 La
Hastelloy alloy S	67	3.0[a]	2.0[a]	15.5	14.5	1.0[a]	0.25	—	0.05 La
Waspaloy alloy (N07001)	58	2.0[a]	13.5	19.0	4.3	—	1.5	3.0	0.05 Zr
RA 333 (N06333)	45	18.0	3.0	25.0	3.0	3.0	—	—	1.25 Si
Haynes alloy 188 (R30188)	22	3.0[a]	39.0	22.0	—	14.0	—	—	0.07 La
Alloy 800H (N08810)	32.5[a]	44	2.0[a]	21	—	—	0.4	0.4	
Haynes alloy 556	20.0	31	18.0	22	3.0	2.5	0.2	—	0.8 Ta, 0.02 La, 0.02 Zr
Multimet alloy (R30155)	20.0	30	20.0	21	3.0	2.5	—	—	Cb+Ta = 1.0
RA 330 (N08330)	35.0	43	—	19	—	—	—	—	1.25 Si
Type 304 stainless steel (S30400)	9.0	Balance	—	19	—	—	—	—	2.0 Mn, 1.0 Si
Type 310 stainless steel (S31000)	20	Balance	—	25	—	—	—	—	2.0 Mn, 1.5 Si
Type 316 stainless steel (S31600)	12	Balance	—	17	2.5	—	—	—	2.0 Mn, 1.0 Si
Type 446 stainless steel (S44600)	—	Balance	—	25	—	—	—	—	1.5 Mn, 1.0 Si

Source: Institute of Metals
[a]Maximum.

Table 8.6 Static Oxidation Behavior of Alloys

Alloy	Metal loss + continuous internal penetration (μm per side)[a]			
	1253 K	1368 K	1423 K	1478 K
Alloy 214	5	3	8	18
Alloy 600	23	41	74	213
Alloy X	23	69	147	(672 h)[b]
Alloy 556	28	66	295	(168 h)[b]
Type 310	28	58	112	262
Alloy 800H	46	188	226	345
Type 446	58	368	>584	(1000 h)[b]
RA330	109	170	221	211
Type 304	206	>584	>584	(336 h)[b]
Type 316	363	>1753	>2642	(168 h)[b]

[a]1008-h exposures in flowing air. Cycled to room temperature once a week.
[b]Alloy consumed in test
Source: Institute of Metals.

in Table 8.6. The hood made of Cabot alloy 214 lasted five times longer than that of nickel-based alloy. Selection of material for this application was based on the observation that contaminants are absent in the working environment of the flame hood.

Pachowski (1970) analyzed the durability of baskets and cover plates employed in the heat-treating shop. The components mentioned were used in quenching, carbonitriding, and annealing steel elements. The laboratory tests involved estimation of the deformation of thermally fatigued specimens and were verified by field trials. The candidate materials were cast steels of grade H18N35, H18N38, and H7N60, equivalent, respectively, to the HT, HV, and HW cast steels developed by Alloy Casting Institute in the United States. The specimens employed in this study were of L shape and resembled a critical aspect of the component design (i.e., its corner). The range of cyclic temperature changes (323 ↔ 1223 K) corresponded to the most severe thermal cycle that occurred in the shop considered. The cycle duration, however, was much shorter (580 s) than the quickest one employed in industry. The laboratory tests demonstrated that specimens made of H18N35 were the most resistant to permanent deformation. Field trials verified this conclusion.

In his study evaluating heat-resisting alloys for bar baskets used in heat-treating job shops, Rundell (1982) demonstrated an approach that differed from those mentioned above. Because of that, the laboratory tests are of only limited validity, as they are unable to render such factors as basket loading and mechanical abuse; the four bar baskets were fabricated from 13-mm-diameter rods of candidate materials and evaluated after 9 months of service. The floor of the baskets was made of a series of long bars (referred to as runner bars) aligned in one direction, welded to short bars (called crossbars) oriented crosswise to the former. Runner bars at the

Table 8.7 Chemical Composition and Location of Materials in Bar Baskets

Alloy	Condition[a]	Location in basket	Composition (%)[b]							
			C	Si	Mn	Ni	Cr	Ti	Al	Other
B	HRAP	Crossbar	0.02	0.17	0.2	61	21.5	0.34	1.30	—
RA 330	ACG	Crossbar	0.05	1.3	1.5	36	19.4	—	—	—
D	ACD	Crossbar	0.05	1.4	1.3	35	14.9	—	—	—
330-1	ACG	Crossbar	0.07	0.4	1.4	36	18.5	—	—	—
330-2	ACG	Crossbar	0.09	0.4	1.4	36	18.5	—	0.22	—
330-4	ACG	Crossbar	0.11	0.4	1.4	36	18.5	—	0.74	—
330-5	ACG	Crossbar	0.10	1.2	1.4	36	18.5	—	1.1	—
330-13	ACG	Crossbar	0.08	3.2	1.4	36	18.5	—	—	—
B	HRAP	Runner bar	0.02	0.17	0.2	61	21.5	0.34	1.30	—
A	HRAP	Runner bar	—	—	—	(75)	(15)	—	—	—
C	HRAP	Runner bar	0.04	0.25	0.95	31	21.3	0.42	0.42	—
RA 330	HRAP	Runner bar	0.07	1.1	1.9	35	18.8	—	—	—
RA 333	ACG	Runner bar	0.06	1.4	1.7	45	25.5	—	—	3.0 W, 3.0 Mo, 3.0 Co
RA 330	ACG	Runner bar	0.05	1.3	1.5	36	19.4	—	—	—
D	ACD	Runner bar	0.05	1.4	1.3	35	14.9	—	—	—

[a]HRAP, hot rolled, annealed, and pickled; ACG, annealed and centerless ground; ACD, annealed and cold drawn.
[b]Balance Fe.
Source: Institute of Metals

bottom of the baskets were exposed to sliding wear. The chemical compositions of alloys considered are given in Table 8.7, together with their position in the baskets. The first heat-treating job shop did most of its carbonitriding at 1118 to 1143 K, whereas the temperature range employed by the second shop was 1088 to 1253 K. Carburizing, a minor factor, was carried out at 1198 to 1228 K. In both shops, thermal cycles were terminated by oil quenching from slightly reduced temperature (i.e., 1118 K). Each shop was furnished with a pair of bar baskets fabricated from candidate materials which were subsequently taken out of use to evaluate the extent of damage. Runner bars in baskets removed from service were far more damaged than were crossbars, due to sliding wear with consequent detachment of protective scale. Experimental alloys 330-1 through 330-5 experienced heavy damage, estimated in terms of the extent of carburization and the number of cracks formed. The fifth that contained over 3% silicon suffered less damage. The fact that silicon increases resistance to carburizing is well acknowledged (e.g., Harrison et al., 1979). Nickel, although providing good heat resistance, does not form stable carbides, and therefore a suitable amount of chromium-limiting inward diffusion of carbon has to be included in the chemical composition of the alloy. According to Andrzejewski (1980), the ratio of chromium to nickel content should be 2.0 to 2.3, as an excess of chromium causes grain coarsening and reduces heat resistance. From the alloys considered by Rundell (1982), alloy RA 333 was most resistant to carburizing and alloys A and C were least resistant to it. Judging from material performance during service, alloy RA 333 was rated as the best material, alloy A and RA 330 ACG were rated the next highest, the poorest performance was shown by alloy C.

Another example of material selection was given by Fluck et al. (1985). In the study quoted, material performance was estimated with regard to resistance to molten salts. On the basis of preliminary laboratory tests, alloy 214 was considered the most resistant to $BaCl_2$–KCl–NaCl molten salts. Field trials showed, however, that pots made of this material, operating at temperatures up to 1280 K, lasted only

Table 8.8 Resistance to Molten Chloride Salts at 1118 K

Alloy	Metal loss + continuous internal penetration (mm per side)
Alloy 188	0.69
Multimet alloy	0.76
Alloy X	0.97
Alloy S	1.0
Alloy 556	1.1
Waspaloy alloy	1.7
Alloy 214	1.8
Type 304	1.9
Type 310	2.0
Alloy 600	2.4

Source: Institute of Metals.

a few days. Metallographic examinations revealed severe scaling on the outside surface of the pot, as well as cracking. The damage to the material was due to the vapor present in the heat chamber. Repeated tests carried out in more realistic conditions included the exposure of alloy coupons to mixed chloride salt baths at 1118 K for 30 days. The results obtained are given in Table 8.8; see Table 8.5 for the chemical compositions of the alloys.

Some problems with material performance cannot be solved simply by replacing the material used by another. An example was given by Fluck et al. (1985) in which fans running in a carburizing furnace in an endothermic atmosphere containing 40% H_2, 20% CO, 0.1 to 0.2% CO_2, and the balance N_2 at 1033 to 1173 K failed prematurely, whereas other furnace internals made of the same Multimet alloy remained unaffected. The problem was cured by a minor design change to improve gas flow, as the low clearance between the fan and the heat-resistant lining formed a dead space with stagnant gas that favored an increase in carbon activity with consequent severe carburization and pitting of the fan material.

8.5 MILL ROLLS

8.5.1 Working Conditions

Mill work rolls are in direct contact with processed material, and thus any deterioration of their surfaces affects the quality of the product. Demands for higher efficiency and product quality require that more attention be paid to the problem of the service life of mill rolls. Apart from this, new grades of material, frequently more difficult to roll, are put into production. All these factors cause higher roll loads and higher temperatures at roll surfaces. In general, the working conditions of rolls depend on the applied technology of rolling. However, the working conditions of upper and lower work rolls differ somewhat. There are also differences in working conditions between adjacent stands, related to the loads applied, cooling of the material as it passes across the mill line, and so on. Determining the causes of roll failure and describing methods for their mitigation will be simplified and focused on a particular stand.

Wear resistance of mill work rolls demands high hardness of working surfaces, whereas the transmission of variable-in-time loads contradicts this. Mill work roll materials and technologies employed are balanced so that both requirements are fulfilled. In general, the life of mill work rolls depends on:

1. Working conditions: applied load and variable-in-time loads, and range and gradient of temperatures,
2. Mechanical and structural properties of mill roll material,
3. Configuration of backup and work rolls and assembly accuracy

The magnitude of a roll load depends on the degree of reduction in material thickness, the rolling speed, and the processed material itself. The load of mill rolls changes rapidly when the processed material is entering or leaving a roll gap and is also dependent on the roughness of the rolling material surface and the difference in the rolling speed of adjacent stands.

The range and gradient of temperatures in the mill roll material is a function of applied technology: the mass of the mill roll, the rolling speed, and the efficiency of the cooling system. Rolling technologies are conventionally divided into two principal groups: (1) cold rolling, where the temperature of the rolling material is low, usually ambient; and (2) hot rolling, where the temperature is above the recrystallization temperature of the processed material.

The factors influencing the service life of mill work rolls are quite different in these two technologies. Hot rolling will be discussed first. The surface of a mill work roll is rapidly heated when the rolling material enters the roll gap, and is subsequently cooled by the spray of coolant. After some time, the interior of the

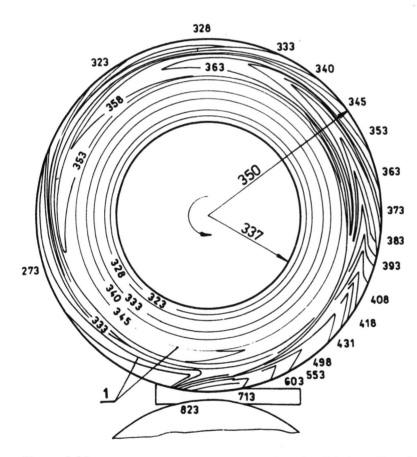

Figure 8.20 Temperature distribution in a hot mill work roll during rolling of a steel strip. (From Jalovoj, 1973.)

Table 8.9 Compositions of Roll Samples

Sample number	Roll material	C	Mn	P	S	Si	Ni	Cr	Mo
						Composition (%)			
1	Steel	1.65	0.53	0.010	0.014	0.62	1.68	0.03	0.57
2	Steel	1.67	0.55	0.012	0.010	0.73	1.66	0.98	0.56
3	Nodular iron	3.42	0.47	0.050	0.013	1.85	3.10	0.73	0.68
4	Nodular iron	3.42	0.47	0.050	0.013	1.85	3.10	0.73	0.68
5	Grain iron	3.38	0.48	0.080	0.061	0.84	4.39	1.67	0.34
6	Chrome-iron	2.39	0.70	0.045	0.056	0.62	0.63	18.1	1.42
7	Chrome-iron	2.35	0.67	0.045	0.060	0.64	0.64	18.2	1.47
8	Chrome-iron	2.49	1.46	0.021	0.030	0.52	1.06	14.5	0.57
9	Chrome-iron	2.68	1.82	0.050	0.041	0.79	0.76	13.9	0.58
10	Chrome-iron	2.70	1.50	0.020	0.030	0.72	0.90	16.0	0.45

Source: AISE Year Book (1981).

work roll has a stabilized temperature of about 350 K. Cyclic variation of subsurface temperature causes nucleation and the development of cracks. The pattern of temperature changes is shown in Fig. 8.20. It is somewhat similar to that discussed in detail in Chapter 7 for the centrifugal cast mold. The mechanism of mill roll cracking has been described by Weronski (1977).

Apart from the above, the temperature of the roll surface causes its corrosion, and this is greatly accelerated in surface irregularities because of the concentration of stresses occurring in these regions. Friction, an inherent condition of rolling, enhances the corrosion process by removing the oxide layer already formed. It is noteworthy that roll loads employed in this technology are not large, hence the hardness of roll surfaces is usually between 280 and 450 HB. To simplify this, we can say that rolls are working under thermal fatigue conditions in a hostile environment.

Work rolls employed in cold reducing mills transmit very high loads, typically 400 to 3000 metric tons. The surface temperature during the rolling of structural or low-alloy steels stabilizes after about 30 min at 350 K. This temperature rise is caused by high loads and friction between the roll and the processed material. Therefore, we can say that the mill roll is working under conditions of mechanical fatigue. The effective state of stresses acting on roll material is the result of internal stresses in the mill roll material and the applied load. The hardness of the surface layer is higher than that in the previous cases and is usually 90 to 130 HS.

8.5.2 Damage to Roll Surfaces and Methods of Mitigation

Visual inspection of hot mill work roll surfaces usually reveals thermal fatigue cracks and oxide buildup some time after the beginning of rolling. These factors cause abnormal wear of work rolls. This problem has been reviewed by de Barbadillo and Trozzi (1981). Examination of scrapped cast rolls (see Table 8.9) showed that only in the case of chrome-iron was the pattern of thermal fatigue cracks regular (nearly square); for other roll materials it formed an almost random structure. The origin of this pattern seems to be service conditions and internal stresses rather than the microstructure. The much finer secondary pattern was observed to bear a close relationship to the cell structure of the roll material. It was found that cracks forming both patterns had initiated and grown in a carbide phase or preferred the directions along carbides. Steel rolls had fewer thermal fatigue cracks than did rolls made of other materials. Examination of the cracks indicated that those parallel to surface cracks were filled with oxide, which was identified as iron oxide. The subsurface cracks had a scale consisting of oxides of alloying elements. No softening of roll material, decarburization, or phase transition was revealed. The model of hot mill work roll deterioration (de Barbadillo and Trozzi, 1981) presented below, based on the assumption of a sound roll material, involves gradual progressive changes in the structure of the roll material as follows:

1. On heating the subsurface layer of the roll material through contact with the hot, processed element, the yield stress is exceeded and the material undergoes local plastic deformation.

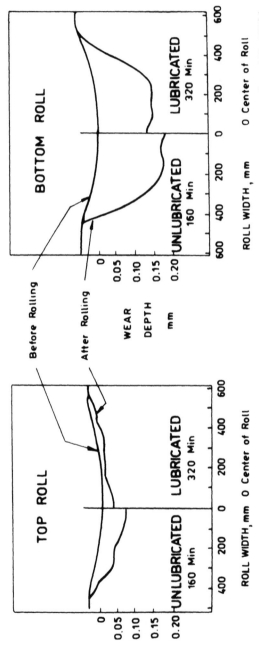

Figure 8.21   Wear of hot strip mill work rolls. Time indicated is strip-in-stand (min). (From AISE Year Book, p. 103, 1980.)

2. On cooling of this layer by contact with a backup roll and by a coolant stream, tensile stress develops, as the yield stress in tension is not exceeded at low temperature.
3. The residual compressive stress introduced during the production process is reduced after one roll revolution and ultimately disappears locally.
4. The change in the state of effective stresses causes a situation where the surface layer of roll material can no longer withstand thermal cycles. However, thermal cracks cannot develop into deep layers of roll material because of the residual compressive stress existing in the interior of the roll.
5. Thermal fatigue cracks begin to join up forming a network that makes the roll pressure a more important factor, and then the difference in coefficients of expansion between carbides and matrix becomes an operative mechanism, resulting in microcracking transverse to the roll surface, referred to as a secondary pattern.
6. Subsurface cracks push up the surface of the material; heat conductivity is locally reduced by underlying cracks.
7. Corrosion products of the strip adhere to the roll surface. Corrosion products are no longer removed from the entire roll surface because of the presence of surface irregularities.
8. Joining of subsurface cracks greatly enhanced by progressive corrosion, and fragments of the roll surface become dislodged.

With regard to the requirements, these have to be met by work roll material. It can be said that apart from resistance to thermal fatigue and corrosion, the hardened surface layer should be of even thickness to provide slow wear, and the roll core must be ductile to withstand dynamic loads.

The roll temperature range should be maintained at a low level by surface cooling to avoid the drop in mechanical properties. Hostetter and Vyas (1973) have shown that an application of lubricant to mitigate friction between the work and the backup roll is also advantageous and can reduce downtime by 10% and increase work roll life by a factor of 2 without affecting strip quality. Lubricant can be applied either in neat form or as a mixture. A comparison of wear profiles between lubricated and nonlubricated work rolls is shown in Fig. 8.21.

The mechanism of damage to cold mill work rolls is of a different nature, caused by the fact that mechanical loads are greater than in the previous case and thermal stresses can appear while the processed material is wrecked between rolls. Industrial observations have shown that damage to cold mill work rolls is located in the center of the working area if wide strips are rolled, and in peripheral regions of the working area if strips of narrow width are rolled. This seems to be caused by the fact that narrow strips are processed by randomly chosen parts of the roll surface and that less worn parts of the roll surface work harder and therefore undergo progressive wear. In the case of wide strips, the central part of the roll surface works as strips enter the stand in almost the same way.

There are two causes of work roll failure. The first is wear caused by friction between the roll and the processed material. The second cause, spalling, is the

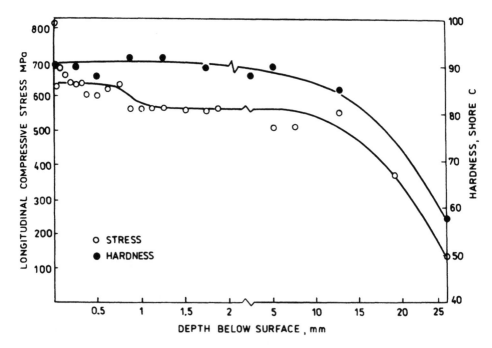

Figure 8.22 Stress and hardness profiles for a new roll. (From Chilton and Roberts, 1981.)

principal cause of premature, sometimes unexpected failure of rolls. According to Chilton and Roberts (1981), it can be caused either by mechanically induced damage or localized overheating. Examination of spalls showed that they were preceded by cracks nucleated at or near the work surface. These cracks propagated into the roll material at an oblique angle and subsequently grew parallel to the surface.

In an investigation to determine the cause of spalling, Chilton and Roberts (1981) conducted microstructural examinations of a number of scrapped rolls made of steel containing about 0.8% C with small amounts of carbide-forming elements. Rolls produced from forgings were surface hardened and low-temperature tempered. The results obtained evidenced local changes in the microstructure of the subsurface layer. Light-etching regions of oval shape were found that extended 0.07 mm below the roll surface in most cases. It was proved by microhardness measurements that these regions were harder than the surrounding matrix. Therefore, these could be potential stress risers, and in fact, some cracks were found associated with them. Similar microstructure was revealed in a specimen of which the surface was struck with a ball peen hammer. Hence it can be concluded that these light-etching regions are produced by localized overloading or by deformation with a high strain rate. Electron micrographs indicated an extremely fine microstructure of light-etching regions and an electron microprobe did not show any significant differences in chemical composition between these regions and the adjacent matrix. Bedford et al. (1974) suggested that they can consist of untempered martensite resulting from reaustenitizing by local rapid heating. The other possibility, partially confirmed by laboratory results, is that they consist of heavily deformed tempered martensite.

In the case where rolls scrapped after mill wreck with subsequent welding of sheet metal to the surface were examined, the roll material adjacent to the weld was found to be reaustenitized and quenched. Therefore, this roll material became harder than the surrounding matrix, of which the temperature was above the tempering temperature but no high enough to cause reaustenitizing. These microstructural changes are able to cause roll failure after some time, although in severe incidents, cracks can be produced instantly by induced thermal stresses.

Figure 8.22 shows the distribution of a longitudinal component of stress and hardness versus depth below the surface for a new 52.5-cm-diameter roll of Shore 90 hardness. Figures 8.23 and 8.24 show stress and hardness profiles for used rolls. It is seen that the magnitude of surface compressive stress can be increased during normal service. However, stress–depth profiles taken on rolls severely damaged by the mill wreck indicated that compressive stress can be reduced locally, and in some cases tensile stress was observed in the damaged zone. The additional conclusion that can be drawn from Fig. 8.23 is that rolls should be reclaimed by removing a thin layer to retain a profitable state of stress at the surface.

It follows from the discussion above that the requirements of cold reducing mill roll material can be listed in almost the same order as those for hot mill work roll material, with the exception of resistance to hot corrosion, which does not play any role in this case.

Backup rolls are not exposed to severe thermal shocks and overloadings. The desired quality of the surface is much lower than that of work rolls and therefore it

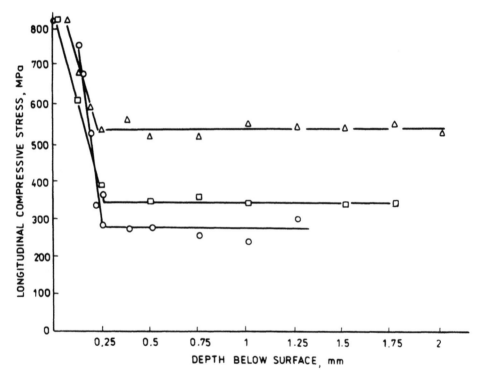

Figure 8.23   Stress–depth profile for used rolls. (From Chilton and Roberts, 1981.)

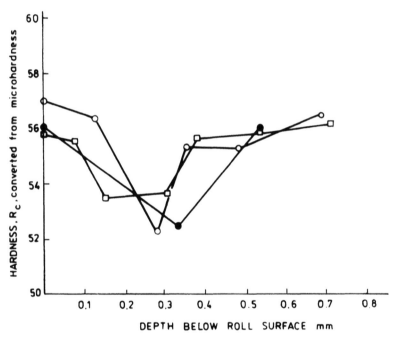

Figure 8.24 Hardness–depth profile for used rolls. (From Chilton and Roberts, 1981.)

Table 8.10 Distribution of Elements in a Volume of Grooved Roll Forging

Sample location	Chemical composition							Degree of forging
	C	Mn	Si	P	S	Cr	V	
A_1	0.50	0.62	0.23	0.021	0.025	0.55	0.19	
A_2	0.48	0.62	0.23	0.022	0.026	0.56	0.19	
A_3	0.47	0.62	0.24	0.021	0.023	0.57	0.19	9
A_4	0.51	0.62	0.23	0.022	0.024	0.56	0.19	
A_5	0.69	0.67	0.24	0.033	0.036	0.62	0.21	
B_1	0.49	0.61	0.26	0.022	0.028	0.57	0.20	
B_2	0.48	0.61	0.24	0.021	0.026	0.57	0.21	
B_3	0.50	0.62	0.26	0.023	0.027	0.57	0.22	5
B_4	0.53	0.63	0.28	0.024	0.026	0.60	0.23	
B_5	0.61	0.66	0.28	0.028	0.027	0.61	0.25	
B_6	0.65	0.67	0.26	0.030	0.030	0.63	0.27	
C_1	0.49	0.62	0.24	0.022	0.025	0.60	0.20	
C_2	0.49	0.61	0.24	0.021	0.022	0.61	0.20	
C_3	0.48	0.62	0.23	0.020	0.022	0.57	0.20	3
C_4	0.46	0.60	0.22	0.019	0.022	0.57	0.21	
C_5	0.46	0.61	0.23	0.020	0.023	0.58	0.20	
C_6	0.44	0.60	0.22	0.021	0.023	0.57	0.20	
D_1	0.48	0.61	0.19	0.021	0.024	0.55	0.19	
D_2	0.49	0.61	0.21	0.022	0.025	0.56	0.20	
D_3	0.48	0.61	0.18	0.021	0.022	0.56	0.19	
D_4	0.47	0.61	0.19	0.024	0.027	0.57	0.20	
D_5	0.49	0.53	0.17	0.021	0.024	0.56	0.20	1.5
D_6	0.52	0.58	0.20	0.022	0.025	0.57	0.20	
D_7	0.55	0.62	0.23	0.024	0.026	0.59	0.21	
D_8	0.56	0.63	0.23	0.025	0.028	0.59	0.21	
D_9	0.56	0.63	0.24	0.026	0.028	0.56	0.22	
D_{10}	0.60	0.64	0.29	0.028	0.033	0.60	0.21	
E_1	0.49	0.62	0.26	0.021	0.025	0.57	0.19	
E_2	0.49	0.62	0.24	0.021	0.028	0.59	0.20	
E_3	0.49	0.62	0.24	0.020	0.027	0.59	0.19	
E_4	0.48	0.60	0.24	0.016	0.027	0.57	0.19	4
E_5	0.51	0.61	0.24	0.016	0.027	0.59	0.22	
E_6	0.59	0.65	0.26	0.024	0.030	0.60	0.22	
E_7	0.64	0.60	0.26	0.028	0.033	0.62	0.25	
F_1	0.48	0.62	0.23	0.019	0.028	0.58	0.18	
F_2	0.48	0.62	0.23	0.016	0.029	0.58	0.19	
F_3	0.46	0.61	0.23	0.018	0.028	0.58	0.19	5
F_4	0.45	0.60	0.22	0.022	0.024	0.57	0.20	
F_5	0.45	0.60	0.23	0.018	0.023	0.54	0.18	
G_1	0.50	0.62	0.25	0.023	0.024	0.57	0.20	
G_2	0.49	0.62	0.23	0.022	0.026	0.59	0.20	11
G_3	0.46	0.61	0.24	0.022	0.024	0.58	0.19	
G_4	0.45	0.60	0.24	0.023	0.024	0.58	0.18	

is unnecessary to give any details, bearing in mind that some of the factors mentioned can affect service life.

Methods to increase the service life of work rolls are discussed next. According to Bradd (1961), oxide inclusions can be a predominant cause of premature failure. This is evident if their properties (i.e., Young's modulus and a coefficient of thermal expansion) are compared with those of steel. However, good casting practice can greatly reduce the amount of inclusions. This is especially important for such large ingots as those of which rolls are made. The tendency of inclusions to form aggregates is enhanced by slow cooling of an ingot and by forging. Inclusions can easily come from refractory linings of the furnace and ladle while molten metal is in contact with them. Oxide inclusions are formed when molten metal is poured from the furnace to a ladle and from the ladle to an ingot mold, or are products of oxidation of Mn, Al, and Si. The solubility of sulfides is better in molten metal, and thus they tend to precipitate while the ingot is solidifying. The number of inclusions can be reduced by removing the central part of an ingot and its skin. Apart from this, vacuum casting and especially electroslag remelting are frequently applied and reduce the number of inclusions very effectively. The additional profit of applying the latter two procedures is a reduction in the content of dissolved gases. The choice of the degree of forging is a compromise between the high value demanded by spoiling the dendritic structure and welding internal flaws, and the low value necessary to avoid a band structure of inclusion aggregates. It also affects the distribution of elements in a volume of forging. This is shown in Table 8.10 for 50 HMV steel. Locations at which compositions were measured are shown in Fig. 8.25.

The second factor influencing the service life of rolls is the microstructure of the roll material. Contradictory demands of roll material (i.e., good wear resistance and resistance to varying-in-time loads) have led to the development of rolls with a tough core and a hard surface. This type of microstructure can be obtained either by differential heat treatment of rolls or by centrifugal casting. The first of these (Linhardt, 1972) involves low-frequency induction heating of the barrel surface, which enables heat treatment of the subsurface zone above the critical temperature to be performed and to retain the fine-grain microstructure of a core developed by prior heat treatment. In the United States, the use of differentially treated bloomers and slabbers has doubled the service life of work rolls made by conventional methods. Sleeving the worn-out rolls enables additional profit to be gained. Differentially heat-treated sleeves with a hardness of 60 to 62 scleroscope C performed twice as well as did solid backup rolls of 46 to 50 scleroscope C hardness (see Fig. 8.26). However, sleeved rolls can fail as a result of splitting caused by cracks propagating from the roll journal or through slippage when the shrink joint is too loose. These problems can be solved by the application of epoxy cement to the joint, which equalizes temperatures of the sleeve and core, thus eliminating splitting of the sleeve, and provides sufficient mechanical bonding of the components. Sleeved rolls are more liable to deform during rolling than are solid rolls (Kobayashi, 1972). However, this can be solved by the concept of a pair cross mill (Nakajama et al., 1985), in which the backup and corresponding work roll are kept parallel to each

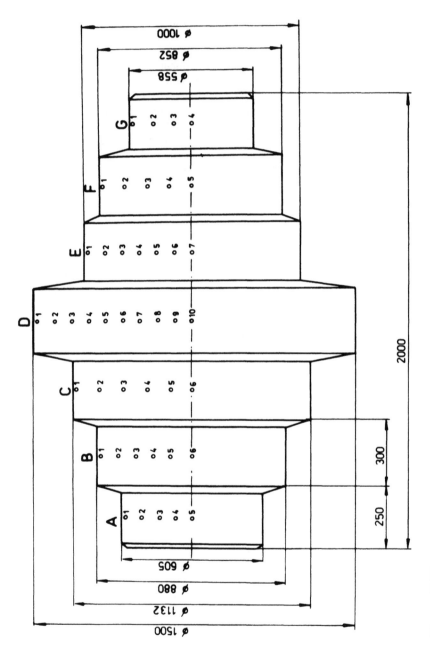

Figure 8.25 Locations on grooved roll diameter at which chemical composition was tested.

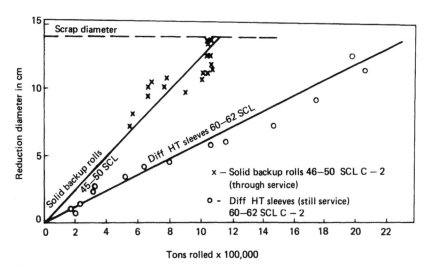

Figure 8.26 Comparison of performances between sleeved and solid backup rolls. (From AISE Year Book, p. 667, 1972.)

other. Setting a cross-angle between the upper pair of rolls and the lower one, the sectional shape of the rolled product can be controlled and the deformation of work rolls can be compensated. Double-layer centrifugally cast rolls exhibit outstanding performance. The outer layer of a roll consisting of white cast iron provides excellent wear resistance, whereas gray cast or nodular iron, of which the core is cast, gives necessary fatigue resistance. According to Grounes (1972), centrifugally cast double-layer rolls have a service life 30 to 50% longer than that of rolls made by conventional casting. The work surfaces of rolls cast of chromium cast iron consisting of 12 to 25% Cr have good wear resistance because of the high percentage content of hard-chromium carbides. Apart from that, this material is more resistant to spalling. Rolls made of nodular cast iron doped with molybdenum of centrifugal casting were used as work rolls of bloomer and slabber. Performance tests have proven the quality of these rolls. The resistance to thermal fatigue was attributed to the bainitic structure in the ferrite–pearlite matrix containing graphite.

As mentioned previously, worn-out rolls can be reclaimed by sleeving. An alternative method involves rebuilding the roll surface by submerged arc welding or electroslag welding (Foley et al., 1974). The service life of an overlayed roll is roughly equal to that of a new one. Trends in iron- and steelmaking technologies and resource saving compel more attention to be paid to the problem of the service life of rolls. The widely utilized continuous casting technology, which has revolutionized the steel industry, involves more severe working conditions for rolls. For example, water-cooled rolls used in these strip casters have to be reground for every 300 to 500 tons of aluminum alloy processed (Jaffrey et al., 1984). This new casting technology promotes the development of schedule-free rolling, which is connected with demands of better performance from rolls. There is also mutual dependence between the efficiency of computer control and roll quality.

The development of new roll materials and technologies is limited by insufficient knowledge of the rolling process. Some of the problems occurring in the field can be solved by processing data collected in the mill. An example of a measurable moving center (Mommertz et al., 1984) shows this means of cost improvement.

8.6 FORGING EQUIPMENT

In the case of two plastic working processes, forging and hot rolling, thermal fatigue can, and often does, seriously reduce the service life of work tools. In this section the factors affecting the durability of dies are reviewed.

Forging is historically the oldest method of working metals in the plastic state and is classified, by the geometry and the type of equipment used, into open-die and closed-die forging. The former is performed with flat or shaped dies, between which the forging stock is formed by successive closures of the dies. The linear and/or angular position of the forging can be changed in the course of the process, enabling complex shapes to be made with relatively simple dies. In closed-die forging a pair of forging dies in a single closure produces a forging close to the dimensions required in the final product, so little subsequent machining is required. Such dies frequently contain cavities of complex form and are expensive to manufacture. For this reason very large forgings are usually produced with open dies.

The compression of work tools in produced either by presses or forging hammers. Usually the upper tool is actuated while the lower remains motionless. Press forging is attractive because the dimensional accuracy of the product formed is greater than with hammer forging, as the separation between the two dies can be controlled more closely. The capacity of hydraulic presses reaches 10,000 tons and the design configurations can be divided into the two general groups: push-down presses and draw-down presses. In the former, the forging stock is shaped between a work tool attached to the moving crosshead, actuated by hydraulic cylinders which are fixed to the top entablature, and a lower work tool, which rests on the base plate. The disadvantages of this design are the large overall dimensions, the height being several meters, with the center of gravity located high above the fixed base, leading to sway of the press, which causes fretting of the hydraulic piping and reduces the forging accuracy. Presses of the draw-down type have the center of gravity below the floor, which provides stability of the construction and practically excludes the risk of hydraulic oil leakage and ignition. In this case, the force cylinders are located below the fixed base plate supporting the lower work tool and within a rigid frame consisting of lower and upper entablatures joined by two or four columns that move up or down within the fixed base plate, which also locates them laterally. The upper work tool is attached to the upper entablature and is drawn down when the lower entablature is forced down by the cylinder or cylinders.

Both forging hammer and press provide the same final effect—plastic deformation of forging stock—but they differ considerably in the dynamics of the process, as in the former case each successive blow lasts only for milliseconds, while in the latter case the deformation speed is far lower, usually a few millimeters per second, and deformation to the final shape is a continuous process occupying some seconds. Moreover, in the first instance the dynamic mechanical load on the work tools is far greater than in press forging, where transient forces are negligible.

In considering thermal loads it is necessary to distinguish between open- and closed-die forging. In open-die forging the duration of the complete forging cycle can even be some tens of minutes when a large rotor forging is being produced. Throughout practically the entire time the hot forging is in contact with the lower die and the upper die is exposed to radiant heating. In contrast, closed-die forging takes no more than several seconds. Dimensional accuracy and good surface finish, both matters of consideration in closed-die forging, demand an almost constant temperature of the forging stock, and this, as well as economic considerations, leads directly to short working cycles. Therefore, it can easily be seen that open dies are more heavily thermally loaded than closed dies.

The disadvantages of open-die forging in comparison with closed-die forging, longer cycle duration and hence higher thermal load on the dies, together with lower dimensional accuracy in the product, have previously been quite significant but are now being reduced by intensive development of the equipment for open-die forging and by a better understanding of metal flow during forging. As a result, open-die working is becoming competitive with closed-die working in areas for which it was previously unsuitable (Körbe et al., 1981). However, these authors confirmed that the state of knowledge regarding forging dynamics, particularly with respect to temperature distribution in the forging stock and the flow of metal, was still far from complete.

It is still true that open-die production technologies are worked out on the basis of small-scale experimental tests, and that the costs associated with introduction on a full industrial scale are only sometimes justified by the benefits obtained from eliminating expensive closed dies.

From this brief outline of forging processes it can be seen that dies are exposed to temperature variations that may cause thermal fatigue. The condition of the dies, which has a major effect on product quality, may deteriorate as a result of various other mechanisms, but thermal fatigue is a sufficiently serious factor to justify measures being taken to ameliorate the thermal condition of the die. Despite these, it is sometimes thermal fatigue cracking that ends a die's life.

8.6.1 Open Dies

From the earlier brief descriptions of the manner in which dies are used, it can be seen that open dies are exposed to dynamic mechanical loading as well as temperature variation. In the case of press forging, only the slowly varying component of the mechanical load is important, and for the lower die it can be expressed as the sum of the force exerted by the moving upper die and the weight of the forging stock, which may be as much as 200 tons. The necessity for precise control of forging stock positioning and the need to increase plant throughput has led to the development of overhead cranes, transfer cars, and manipulators. Overhead cranes, which some years ago were used as the main facility for transporting ingots to the forging shop, are now equipped with more accurate control systems and often work in association with manipulators to support the free end of the forging. Transfer cars are used to transport ingots from heating furnaces to the forging stand, to remove the forgings produced, and when equipped with turntables, to perform rapid rotation of the forging stock to facilitate the exit of forgings from both ends. The use of transfer cars increases forging shop throughput and, to a large degree, reduces the effect of ingot oxidation and cooldown on the quality of the product and the service life of the work tools. Modern, railbound forging manipulators are positioned to work integrally with the press, which is a great advantage over the overhead cranes and unconstrained mobile manipulators formerly used for the same purposes. The rail-mounted manipulator consists of two main parts: a peel assembly with grip mechanism, and a carriage running on rails. The motions of these two assemblies are independent of each other, which makes the manipulator flexible, providing accurate positioning of the forging stock as well as rotating and shifting it between successive closures of the press. In some instances, the grip mechanism is supported additionally by an overhead crane. To gain maximal profit from the flexibility of the complex comprising the press and railbound manipulator, microprocessor systems are incorporated in the control console. By the use of a railbound manipulator a dimensional accuracy of about 10 mm can be achieved when press forging a 180-cm-diameter bar between open dies. This is some three times better than can be obtained with the earlier and more orthodox use of overhead cranes (Hemingway et al., 1985). The result of these developments, which permit the use of larger billets and reduce idle time, is to increase the average temperature of the die bulk.

For hammer forging, the dynamic loads of the dies depend on the type of blow and, expressed in terms of peak acceleration, can reach some tens of g's. For this reason, the use of forging manipulators that maintain a constant grip on the forging stock is risky; only intermittent application of manipulators can be considered.

The magnitude of the thermal loads of open dies depends on the temperature at which forging is performed this being about 650 K for aluminum-based alloys and 1600 K for highly alloyed steels. In addition, the deformation heat evolved during forging modifies the thermal balance, adding an additional item. Heat is transferred to the environment and the forging equipment by radiation and conduction. The rate of heat transfer to the work tools is strongly dependent on the thickness and morphology of the oxides covering the forging stock and work tools, the differences in temperatures, and the compressive force. The existence of thermal contact between forging stock and work tools and radiative transfer to the environment produces large temperature differences within the forging stock, which can be illustrated with the results of the theoretical calculations of Braun-Angott and Berger (1981).

Considering a rectangular billet of side 360 mm and of such height that vertical transfer of heat could be neglected, they estimated that if the initial temperature throughout was 1470 K, the temperature difference between the midpoint of the interior and the external vertical edges opposite it would be 400 K after 3 min of

hammer forging. As a result of such temperature distributions, the temperature-dependent mechanical properties of various parts of the forging stock vary with time and with their position. This leads to uneven flow of the material in the course of forging, and consequent variation in the extent to which oxides are dislodged from the surface of the billet. This in turn results in the rate of heat transfer between billet and die varying in respect to time and locality.

Recapitulating the above and considering the repeated thermal and mechanical loads, the working conditions of open dies are characterized by cyclic mechanical and thermal stresses. For the upper die, however, shorter contact times with the forging stock reduce the severity of the thermal regime.

The surface temperatures of dies have been measured by optical pyrometry and also by the use of thermocouples. The former method indicates a maximum die surface temperature very close to that of the forging stock, whereas thermocouple measurements have indicated a maximum die surface temperature of 1120 K when forging a 1310 K billet. This disagreement is due to the thermal inertia of the hot junctions of the thermocouples and to the fact that the temperature measured is the average temperature of some thin subsurface layer adjacent to the surface of the die. Considering only the lower estimate of maximum surface temperature, it can be realized that thermal fatigue is a significant factor affecting the service life of heavily thermally loaded dies. For moderate thermal regimes (e.g., aluminum-alloy forging), mechanical fatigue and erosion are more serious factors. To reduce thermal shock, heavily loaded dies are preheated to the temperature of 420 to 520 K before forging is begun. At the end of the forging cycle the die temperature is usually above the original preheating temperature, so this somewhat troublesome operation does not have to be repeated unless there is a long interval before the arrival of the next billet. In the course of a number of successive forging cycles, the end-of-cycle die temperature at first increases until a balance is reached between heat gained from the workpiece and heat lost to the environment. As the temperature at which stabilization occurs might otherwise be excessive, dies are often brought down to the original preheat temperature at the end of each cycle by pressurized air or spraying with water—with or without a lubricant.

Of course, the temperatures developed in the forging dies during the production run can have a significant effect on the properties of a substantial part of the die material. The experimental results obtained by Szafraniec (1972) are helpful in quantifying this effect. The temperature distributions measured in the $1400 \times 500 \times 280$ mm lower die at points 35 mm below the working surface are depicted in Fig. 8.27 and 8.28. The temperature distributions were taken along die symmetry lines after forging out each of three successive 4.68-ton steel ingots initially at 1450 K in a press of 2000-ton capacity. The die material was low-alloy WNL steel (C = 0.55%, Mn = 0.7%, Cr = 0.7%, Mo = 0.25%, Ni = 1.6%) quenched and tempered at 793 K. The duration of each forging cycle was 60 min, which included a rest period of 15 min in which the completed forging was removed and a new billet brought in. The temperature distributions measured at various depths below the central region of the die are shown in Fig. 8.29. It is evident when comparing temperature distributions with the tempering temperature that the central region of

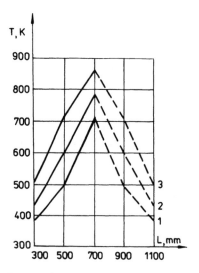

Figure 8.27 Temperature distributions along the longitudinal axis of the die for three successive billets.

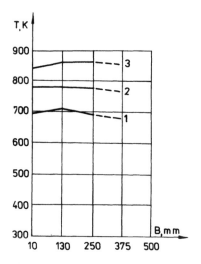

Figure 8.28 Temperature distributions along the transverse axis of the die for three successive billets.

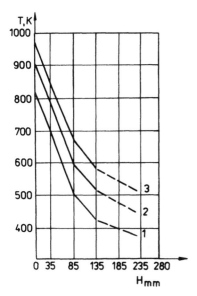

Figure 8.29 Temperature distribution in the interior of a die for three successive billets.

the die to a depth of approximately 50 mm undergoes structural transformations, as the temperatures developed during forging are in excess of the tempering temperature. Apart from other adverse effects, such transformations reduce the material's resistance to thermal fatigue when exposed to varying temperatures. The die material used in these experiments was a relatively low-alloy steel, but the problem is also encountered with high-alloy steels such as AISI H13, which is frequently employed for dies because of its higher yield point at elevated temperatures, at which its other mechanical properties are also better retained. With such steels transformation temperatures are still exceeded in the vicinity of the die surface, and structural transformations still occur, although less rapidly, than in lower-alloy steels at the same temperature. The last point is of some advantage, but the lower thermal conductivity of the high-alloy steels increases the magnitude of the thermal stresses induced by varying temperatures. Moreover, in today's forging practice with modern equipment, the die bulk temperatures tend to be higher than previously because of the reduction in idle time between completing one forging and beginning the next. The problem of adequate cooling of the dies during this briefer interval is increased by the trend toward larger presses and forgings, with a corresponding increase in the size and thermal mass of the dies themselves. In the case just described the dies were cooled only by application of pressurized air to their exteriors. Other methods of cooling are available, the most effective being to force water through passages in the die, situated below its working surfaces. However, improved cooling of the die increases the amount of heat extracted from the working stock, and when the forging cycle is of long duration, as in the example cited, the permissible degree of cooling is strictly limited. Thus thermal fatigue continues to be a factor of importance.

The large overall dimensions of the forging dies make their production as monolithic blocks of high-grade material expensive, and to obtain cost reduction the surfacing of low-grade steel with high-grade hot-work steel is often employed. However, the severe demands with respect to weld quality preclude this from being accepted as a general solution.

8.6.2 Failure of Open Dies

Field experience shows that the causes of the majority of die failures fall into two groups. The first group consists of metallurgical imperfections such as improper heat treatment, the presence of inclusions, and so on, and in the case of composite dies, errors in manufacture and assembly. These defects contribute to early cracking.

The second group comprises mechanical and thermal fatigue and erosion, which individually or in conjunction lead to die failure through wear (i.e., through loss of dimensional accuracy and surface quality).

Figure 8.30 illustrates the first case. The flat monolithic die failed after a short run and postmortem analysis revealed a fault in the die assembly (i.e., a misfit between the case of the die and the anvil stock). Its effect was intensified by the

Figure 8.30 View of a worn die.

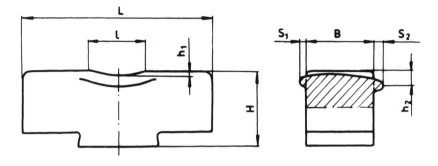

Open die dimensions mm			Press capacity	Permanent deformation mm				
L	B	H	t	l	h_1	h_2	S_1	S_2
600	240	400	800	150	5	20	10	16
1000	300	450	1000	240	7	45	16	28
1400	450	450	3000	300	12	95	27	35

Figure 8.31 View of scrapped press forging die.

large amount of inclusions in the die material, partly linked in bands that facilitated the initiation and subsequent progress of the crack. A similar effect, a through-body crack, can be produced, although more rarely, by low preheat temperature.

Wear of the forging die causes its retirement after some tens or hundreds of working hours, depending on the die material and working conditions. Field observations agree that the most damaged region of the die is its central portion, which results from performing most forging operations with centrally located forging stock. An example is shown in Fig. 8.31. The intuitive supposition that wear is more intensive for high-contact-pressure dies is well supported by the data presented. Considering the wear rate as, for example, a change in die height, measured in the central region per unit working time, it is found that in the early stage of service life the wear rate gradually increases, which is believed to be due to the effect of initial surface irregularities. Subsequently, the rate stabilizes at a certain level determined by the difference between fatigue processes, corrosion, spalling of corrosion products, formation of adhesive joints, and the resistive properties of the die material. In the final period of service life the wear rate increases as the die material deteriorates, due to thermally activated microstructural changes. Figures 8.32 and 8.33 show microstructures of die steel of WNL grade, hardened

Figure 8.32 Microstructure of the scrapped die, layer 40 mm below the initial surface (500×).

Figure 8.33 Microstructure of the scrapped die, layer 20 mm below the initial surface (500×).

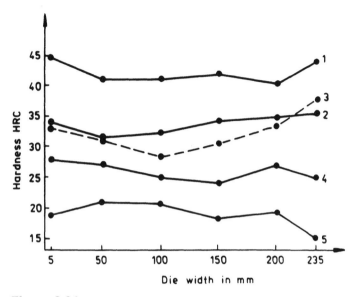

Figure 8.34 Results of hardness measurements on a scrapped forging die.

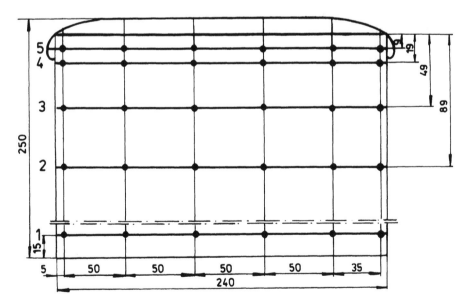

Figure 8.35 Locations of experimental points.

and tempered to 42 to 46 HRC. It can be seen that the martensitic structure developed by the initial heat treatment underwent diffusive transformation in the course of 100 h of use. For the layer 40 mm below the original die surface (Fig. 8.32) this is less evident than for the shallow 20-mm-deep layer of metal (Fig. 8.33). The consequence of microstructural changes is a deterioration in mechanical properties, evident from the hardness measurements shown in Fig. 8.34 (for the location of experimental points, see Fig. 8.35.

Dies are reclaimed when the forgings they produce no longer meet dimensional tolerances; this usually arises when the deformation of the central die region amounts to a few millimeters. Dies are sometimes reclaimed by milling and grinding to reproduce the original profile in the relatively sound material below the damaged surface. Alternatively, after an initial grinding/cleaning operation, the die may be built up by depositing weld metal, and finally machined to the required dimensions. In the course of its total service life a die may be reclaimed a number of times. In the case of dies containing passages for forced water cooling, the amount of metal that can safely be removed is less than with solid dies, and this reduces the number of reclamations possible. However, this is offset by the more efficient cooling, allowing longer runs between successive reclamations.

8.6.3   Closed Dies: Working Conditions

As with open dies, closed dies work under conditions of simultaneous thermal and mechanical fatigue. However, whereas it is accepted that open-die forgings require a considerable amount of machining to bring them to final dimensions, the use of more expensive closed dies is expected to produce more accurate forgings, such that the final machining is much less and may often be confined to those areas where the dimensional tolerances are least, the forging operation alone providing sufficient accuracy elsewhere. For this reason, erosion, which was mentioned in the preceding section, assumes greater importance in the case of closed dies, as it alone may often remove so much material that a closed die can no longer produce work of the accuracy required of it. As in open-die forging, the forging stock has a temperature, and thus mechanical properties, dependent on time and on the location of the point considered. Moreover, in the case of dies of a complicated shape, the magnitudes of the mechanical and thermal loads are far more variable in time than those for flat geometry. The rate of heat flow per unit area from the forging stock is greater in areas in contact with a die than elsewhere. In open-die forging with a flat die, contact is intermittent and limited to a relatively small area at each blow. In closed-die forging the area of contact is much greater, and hence the temperature of the forging stock falls more rapidly. In addition, the temperature falls faster in thin sections of the forging than in thicker sections of greater thermal capacity. Consequently, as a complex forging takes shape, the temperature distribution at the surface contacting the closed die varies with time and location. The complexity of the thermal and mechanical loads to which a closed die is exposed can be inferred from the map shown in Fig. 8.36. It can be seen that the most thermally loaded region of a die is

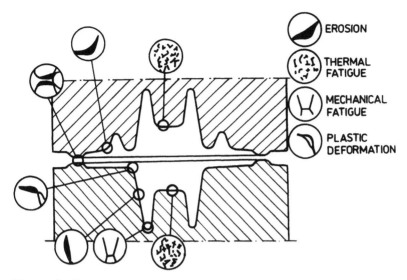

EROSION

THERMAL
FATIGUE

MECHANICAL
FATIGUE

PLASTIC
DEFORMATION

Figure 8.36 Different kinds of failures in forging dies and the locations at which they are likely to occur. (From Kannappon, 1969.)

its central part, which is subjected to the longest time in contact with the hot forging stock. Mechanical fatigue is a significant factor at the bottoms of cavities, where the radius of curvature is small and high mechanical stresses are produced by the wedge effect of forcing the forging stock into the cavities, an effect that is enhanced by the relatively thin sections of these portions of the forging, so that the temperature of the forging stock is lower and its bulk modulus higher in these regions than elsewhere. Erosion takes place where the flow of forging material is most vigorous. In the critical region of the die (i.e., in the flashland), erosion as well as plastic deformation takes place. In the case of improper flashland design, apart from a reduction of die durability, which is especially serious in hammer forging due to the buildup of excessive stresses, there can be poor filling of the die cavity.

Considering the working conditions of a die, it should be noted that during plastic forming the temperature of the forging stock falls at the expense of an increase in die temperature. The rate of heat transfer depends on the contact pressure between the die and forging stock, and this pressure is not uniform with respect to the die surface, and its distribution changes as the forging takes shape. Heat transfer is also a function of the rate at which forging stock flows across the die surface, this rate being a minimum at the center of the die. Hence the temperature increase in the die material is determined by several factors in addition to the initial temperature of the forging stock and the time for which a particular area of the die is in contact with it. In Fig. 8.36 it can be seen that thermal fatigue occurred in the central region of the die, indicating that the thermal regime was most severe in that area, and this may be regarded as a typical situation. However, all mechanisms that caused the types of damage observed in other parts of the die (e.g., erosion by forging stock material in regions of substantial flow) are accompanied by transient thermal stresses at the die surface.

Summarizing, there are mutually competing factors affecting die service life: thermal and mechanical fatigue, and erosion and mechanical stresses exceeding the yield point of the die material. Each of these factors can, by itself, cause die failure. However, different mechanisms dominate in various parts of the die giving patterns similar to that shown in Fig. 8.36. Slight deviation from the established forging practice or even a small error in die production or assembly can enhance the effects of some factors while making the others negligible.

Dean and Silva (1979) undertook experiments to estimate the temperatures developed in the die during forging. The important feature of their work is that the measurements were made on an instrumented, typical die set closed by commonly used forging equipment: a high-speed hammer with an impact speed of 9 m/s and an eccentric press rated at 48 strokes/min. The forging stock was a 38-mm die bar of 080 M40 (En8) steel sawn into billets 35 mm long. The temperature of the forged material was approximately 1310 K before forging was begun. Billets were heated by induction so that the thickness of the oxide layer was less than that which would be formed in fuel-fired furnaces. The average cycle duration for hammer forging was 8 s. The colloidal graphite lubricant was applied at a rate of 0.03 g per billet. The die set used in the experiment is shown in Fig. 8.37. Die inserts made of steel equivalent to AISI H13 were quenched and double-tempered to a hardness of 580 HV. Details of thermocouple locations are shown in Fig. 8.38. The temperature records are depicted in Fig. 8.39 for press forging and in Fig. 8.40 for hammer forging. The results correspond to those obtained with the first forging blow, when the die was initially at ambient temperature. It is seen from Fig. 3.39 that:

1. The temperature recorded from the fifth thermocouple increased before temperatures at other points started to rise, and this is due to the long contact time between the ejector and the billet.
2. A moderate increase in temperature at the first point, which shows that the oxide layer remained intact in this region.
3. Billet material flows more rapidly at the upper die than at the lower, which is in accordance with industrial observations. High temperature values demonstrate removal of the thermal barrier oxide layer, large compression forces, and a large amount of energy evolved during deformation.
4. The low-temperature values at points 3 and 7 are due to a reduction in the billet temperature through the prior central die region.

The results for hammer forging shown in Fig. 8.40 can be summarized as follows:

1. The rates of temperature rise are much greater than in the preceding case, due to the higher deformation speeds.
2. Temperatures developed at points 2 and 6 are much higher, which is caused by more complete removal of the oxide layer and by higher transient loads

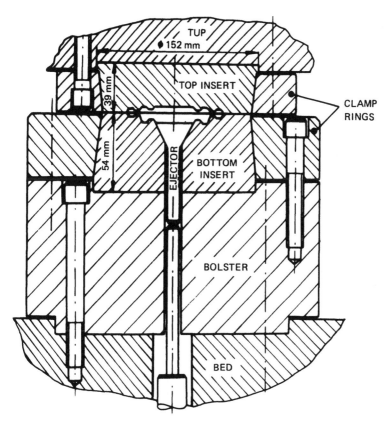

Figure 8.37 Die set. (From Dean and Silva, 1979.)

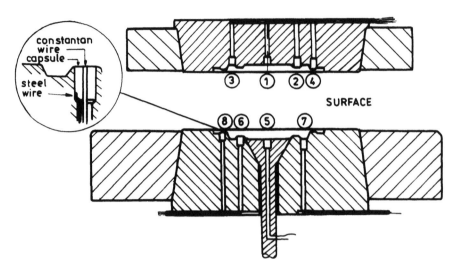

Figure 8.38   Thermocouple locations. (From Dean and Silva, 1979.)

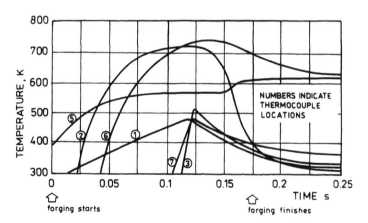

Figure 8.39   Temperature records in a pair of press forging dies. (From Dean and Silva, 1979.)

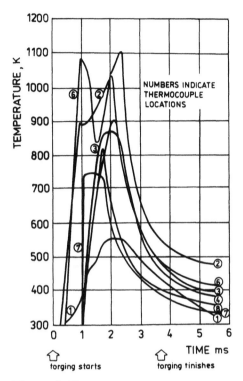

Figure 8.40 Surface transient temperatures for hammer forging dies. (From Dean and Silva, 1979.)

than those in press forging, accompanied by greater evolution of frictional heat. Saddlebacks on the temperature records seem to be due to a characteristic feature of hammer forging (i.e., rapidly decreasing deformation speed and abrupt increase of the load).

3. Regions 3 and 7 have a bigger thermal load than in the preceding case, which is again caused by the better thermal contact between the die and the billet and by the higher amount of energy evolved during forging. This is enhanced by the very short deformation time (ms) and the consequent high temperature of the billet when coming into contact with these regions of the die.

It follows from the pattern outlined above that the critical factors affecting the thermal load of the die are billet temperature, load and deformation speed, and the presence of an oxide layer. Additional factors, illustrated in Figs. 8.41 and 8.42, are lubrication and die temperature preceding the start of each forging operation. The intuitive expectation that lubrication reduces both friction and die temperature is supported by the results presented. Moreover, the distinct tendency to stabilize the die temperature after a certain number of forging cycles can be seen. This is due to

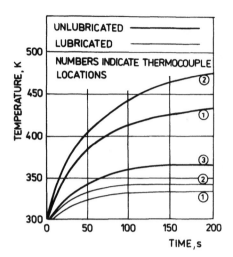

Figure 8.41 Surface temperatures developed in the upper die during press forging. (From Dean and Silva, 1979.)

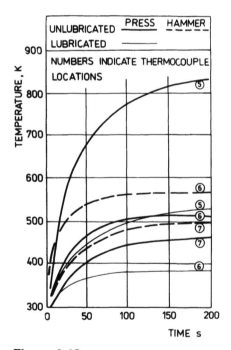

Figure 8.42 Surface temperatures developed in the bottom die during hammer forging and press forging runs. (From Dean and Silva, 1979.)

approaching a state of balance between the heat supplied and the heat extracted. Stabilizing of die temperature was also found by earlier authors (e.g., Beck, 1958; Demidov, 1966).

The effect of the oxide layer on the conditions of heat transfer mentioned in the preceding section needs some explanation and evaluation. This has been the subject of several investigations, including that of Kellow et al. (1985). The material used in their study was a typical forging steel O80M40 (En8) in the form of small billets that were heated in a muffle furnace at 1370 K in air. The surfaces of the cylindrical billets, 2.54 cm diameter and 2.54 cm long, reached furnace temperature in 6 min, at which time they were covered by only a thin film of oxide. On continued exposure to air at 1370 K, an oxide layer of appreciable thickness formed on the surface, at first rapidly and thereafter at a decreasing rate as the oxide already formed began to protect the metal below it from further oxidation. After 16 min, by which time the oxide layer had a thickness of 0.15 mm, the thickness was almost stationary and it appeared unlikely to exceed 0.2 mm even with prolonged exposure to air at 1370 K. This is an interesting and important result when we consider that the heating of large ingots, such as for turbine rotors, can take as much as several hours, during which potentially oxidizing conditions exist. The investigations revealed three types of oxides: a thin outer layer of hematite (Fe_2O_3), an intermediate layer of magnetite (Fe_3O_4), and an inner layer of wistite (FeO), the volume ratio of the oxide layers being 4:36:60. Wistite is unstable and transforms into α-iron and magnetite below 833 K. During rapid cooling, however, this transformation can be suppressed. The oxides have different heat capacities: that of Fe_3O_4 is 200 J/gmol, and for FeO and Fe_2O_3 the heat capacities (in J/gmol) are temperature dependent according to the formula $51.7 + 6.77 \times 10^{-3} \times T - 1.6 \times 10^5 \times T^{-2}$ and $133 + 7.35 \times 10^{-3}T$, respectively, where T is the absolute temperature.

Very few investigations of the thermal conductance of the oxides of iron have been made, but the following figures are quoted: fused and baked to a specific mass of 5.9 g/cm^3, the value is almost linearly temperature dependent and is 0.12 J/ms·K at 1473 K (Bidwell, 1917); for wistite of specific mass of 2.24 g/cm^3 the value is 0.17 J/ms·K (Mellor, 1934).

It is to be expected that the presence of layers of oxides on the surface of the forging stock, specifically the thickness and thermal conductivities of these layers, affects the rate of heat transfer from the stock to the die and hence the temperature attained by the die surface. In the study reported, the authors measured the surface temperature at the center of a flat die during the forging of small billets of the O80M40 (En8) steel which had been heated to 1370 K, as in the oxidation tests, and hence could be assumed to have surface oxide layers equivalent to the compacted thicknesses estimated in those tests (the oxide layer as formed may be somewhat porous but are reasonably assumed to become fully compacted immediately at the beginning of forging).

The thermo junction used to measure the temperature of the die surface was formed by an insulated Constantan wire running through the body of the die and joined to the 5% Cr die material by a sliver of the latter, which was formed during final grinding of the die. The junction was, accordingly, really at the surface and of low thermal inertia.

In press-forging 1370-K billets with a 0.125-mm oxide layer at a closing rate of 1 cm/s, the die temperature increased from ambient to 653 K in the course of the first second, the temperature rise thereafter being much slower. Using billets with an oxide layer of 0.08 mm, the corresponding temperature rise was to 693 K, and with an oxide thickness of 0.04 mm it was to 723 K. The authors quote from previous work a die surface temperature rise to 853 K in the same conditions but with oxide-free billets. It can thus be seen that the presence of even a thin layer of oxides on the forging stock reduced the surface temperature of the die very significantly and that the further reduction provided by thicker layers was much less than proportional to their thickness. The authors also calculated the die surface temperature rise by various methods, sometimes taking into account the specific heat as well as the thermal conductivity of the oxides. The calculated results agreed in form with those from experiment, but because of the inadequacy of the available data on the thermal properties of the oxides, the quantitative agreement was less satisfactory. The methods of calculation and qualitative agreement did, however, provide theoretical support for the intuitive expectation and experimental demonstration that the presence and low thermal conductivity of oxides on the forging stock reduces both the rate of change and maximum values of the die surface temperature. For hammer forging with an impact speed of 9 m/s, the die surface temperature developed after 1 ms was approximately 790 K for an oxide-free billet and only 470 K for billets covered with a 0.125-mm-thick layer of oxide. Summarizing and abridging the author's results, it can be stated that the influence of oxides is more pronounced for hammer forging where an oxide layer of 0.025 mm (which has practically no effect on the quality of the forging) reduces the die temperature by approximately 150 K. Increase in the thickness of oxide layer above 0.1 mm for press forging and 0.025 mm for hammer forging provides only an insignificant reduction in the die surface temperature.

Early in this section it was stated that thermal fatigue can cause die failure, and some areas in which it has been observed were illustrated by the work of Kannappen (1969). The factors that contribute to increasing the die temperature and those, such as lubrication and the presence of oxides on the forging stock, that reduce it have been noted. It can be seen that the range of temperature variation that may occur at the die surface can be sufficient to cause thermal fatigue there. It can also be seen that the somewhat less variable temperature in the metal below the die surface [e.g., in the range 340 to 830 K (after Dean and Silva, 1979)] can be sufficient to cause deterioration in the mechanical properties of the die material in the long term. Focusing on thermal fatigue and recalling the experimental data of Dean and Silva (1979), the following estimates of the effects likely to produce thermal fatigue can be made:

1. In hammer forging, after the production of a few forgings, regions 6 and 7 of the die are exposed to 563 $\leftrightarrow$ 1154 K and 498 $\leftrightarrow$ 874 K thermal cycling, respectively. In press forging the corresponding thermal cycles are 513 $\leftrightarrow$

874 K and 473 ↔ 634 K. On the basis of the work of Malm and Norström (1979), it can be said that the dies used in press forging will last a few times longer than those in hammer forging and that the thermal cycling in the region 7 will have a negligible effect on press forging, whereas in region 6 cracking will be a problem in hammer forging.

2. In press forging, the most thermally loaded element of the die is the ejector, encountering thermal cycles of 833 ↔ 1123 K. Because of thermal fatigue the service life of this element will be considerable shorter than that of the die insert.

However, die temperatures encountered in actual production use are the subject of disagreement, and according to some workers, they can even exceed 1270 K for heavily loaded dies. On the basis of such estimates, Izotov et al. (1982) undertook trials to study the influence of thermal cycling between surface temperatures of 853 and 1280 K on the microstructure of the 4H4WMFC die steel containing 0.38 to 0.48% C, 0.2 to 0.5% Mn, 0.6 to 0.9% Si, 3.5 to 4.1% Cr, 0.9 to 1.1% W, 0.7 to 0.9% V, and 1.2 to 1.3% Mo. The duration of the heating portion of the cycle was 2 s and that of cooling was 4 s. The difference in temperatures between the surface and the layer 2 mm below it was 300 K during heating but only some tens of degrees K during cooling. The microstructure developed by the thermal treatment in manufacturing the die contained lath martensite, $MC/VC/M_7C_3$ and M_6C carbides. The carbides of the former type were in the form of 20-Å-thick plates 100 to 150 Å long and the other carbides had a globular form 2000 to 3000 Å in diameter. In the thermally fatigued specimens, three zones were distinguished:

1. In the first zone the maximum temperature was in excess of A_{c3} (i.e., > 1183 K). Here it was found, after 50 to 100 thermal cycles, that partial transformation of martensite into austenite and M_6C carbides had occurred. Nonsoluble MC carbides underwent coagulation; further thermal cycling continued this process.

2. In the second zone maximal temperatures were between A_{c1} and A_{c3} (i.e., between 1103 and 1183 K). Here after 1000 thermal cycles, martensite crystals of high dislocation density and ferrite grains with dislocation polygons of 0.2 to 1.5 μm diameter were found. In addition, fine (ca. 1000 Å) globular carbides were observed at ferrite grain boundaries and also fine (ca. 200 Å) precipitates linked in chainlike patterns on ferrite–martensite boundaries.

3. The third zone was one in which the maximal temperature was below A_{c1}. Here recrystallization and precipitates of widely ranging diameter located at ferrite boundaries were found.

The pattern of changes described above shows that irreversible structural transformations reducing the mechanical properties had occurred. The seriousness of such structural changes and their dependence on the temperatures and stresses make it necessary to choose between the candidate materials for high-temperature duty on the basis of purposefully developed tests. Policinski (1979) employed a test rig in which thermal and mechanical loads could be exerted simultaneously. However, this method, although commonly used, provides a rather rough approximation to die surface working conditions. This method, developed by Tittagala and co-workers (1983), seems more realistic. The test apparatus was built on a conventional lathe, with the flanged disk of material being evaluated for use as a die positioned in the lathe chuck. A specimen of the anticipated forging stock was mounted on the tool slide together with an inductive heating coil by which it could be brought to the required temperature. Longitudinal movement of the slide enabled a specimen of hot forging stock to be brought in contact with the rotating specimen of die material with a predetermined force and for a predetermined time prior to withdrawal, in simulation of hot working. Test results were assessed by wear profiles. The great advantage of the method is the study of pairs of die candidate and forging stock materials. The variables in the test are temperatures (constant or varying), mechanical stresses, the rate of sliding speed, which resembles the metal flow in a die, and finally, lubrication.

8.6.4 Failure of Closed Dies

The experimental results that have been presented related to forging steel billets. Before attempting any discussion of the causes of closed-die failures and their avoidance, it should be stressed that the diversity of working conditions (e.g., forging temperatures being about 650 K for aluminum alloys, 1120 to 1420 K for titanium alloys, and even 1600 K for high-alloy steels), as well as the variety of die-cavity shapes, is such that the conclusions presented below are to be adapted by the reader to each particular situation, as it is virtually impossible to give a general solution or panacea for all problems with die performance. The examples quoted below illustrate that in industrial practice, evaluation of causes of die failure can differ considerably, depending on the type of duty.

A survey of dies withdrawn from use in the automotive industry showed that approximately 40% of dies had experienced normal wear (i.e., they were withdrawn when the forgings produced were outside the client's dimensional requirements), about 30% failed by through-thickness cracking, and the remaining 30% were retired for cracks that could shortly produce a breakup. Apart from that, a comparison of the die performances between upper and lower dies showed that the upper dies lasted 30 to 50% longer than lower ones. Press forging was more profitable than hammer forging with respect to durability.

The statistics for damage to dies and tooling in press forging of bearing rings on a Hatebur press showed that the dies used in preliminary operations were retired primarily because of erosion or plastic deformation, whereas dies used for final operations where limits on dimensional accuracy are tighter, were retired principally on the basis of thermal fatigue cracks, as was the case with ejectors and knives that experienced abrasive wear (Weronski, 1983). Examples of worn dies and tooling are shown in Figs. 8.43 to 8.47.

Aston and Muir (1969) employed statistical techniques to the effect of design and

Figure 8.43   View of scrapped Hatebur press die after 22,800 working cycles.

operational variables on the service life of hammer forging dies. The die material was mostly BS224—1938 steel No. 5, hardened and tempered: the records of a jobbing forge related to drop hammers with weights ranging from 1 to 5 tons and to board hammers. It was found that the chief factors affecting die life were, in order of importance: weight of forging, angle of die wall taper and fillet radius, initial area of forging—die contact, and hammer rating. The effect of these factors on the life of the die (in terms of number of forgings produced) was that die life decreased with increase in the weight of the forging, with hammer rating, and with increase in the angle of die wall taper, whereas it increased with increase in fillet radius and increase in the initial area of forging-die contact. These findings are in accord with the considerations discussed earlier, but it is worth noting that the statistical analysis made it possible to eliminate marginal effects and to arrange the remaining main factors in the order of importance given above. It can be expected, however, that in a more detailed and wider study, the list of factors would be amended and include forging temperature and die preheat temperature, factors of importance referred to in the preceding section.

   The ways of extending the life of closed dies can be divided into the two general groups: those that reduce the effect of the working conditions on the die and on methods of die repair. The first group contains the following: better match of die material to the known working conditions, redesign of the die cavity, improved

Figure 8.44   Crack pattern on the surface of a scraped die (85×).

Figure 8.45 View of a worn ejector.

treatment of the die material or die surface during manufacture, and applying cooling and lubrication.

Leonidow et al. (1983) demonstrated that the application of new highly alloyed steels can increase the durability of dies used to forge parts of agricultural machines by a factor of 1.5 to 3. The benefit was gained, however, by adding some expense to die production.

Szapovalov et al. (1981) showed that incorporating thermomechanical treatment with a degree of deformation of 20% into the process of producing dies from 3H3M3F steel (0.32% C, 0.39% Si, 0.25% Mn, 2.8% Cr, 2.7% Mo, 0.41% V) roughly doubled the service life. The hardness of the thermomechanically treated steel was slightly higher than that produced by the orthodox method of hardening followed by tempering; the greater durability was probably due to delaying thermal crack initiation as a result of a more homogeneous microstructure.

Laboratory tests performed on a rig simulating the forging of steel billets demonstrated the usefulness of diffusion treatment of the die surface. Specimens enriched in Cr, N, and C lasted from a few to several times longer than untreated ones (Gierzyńska-Dolna, 1985). There is also a report in the literature of the successful use of galvanic coatings (e.g., cobalt applied to the die surfaces) (Dennis and Still, 1975). The profit gained in this case is obviously due to the enhanced oxidation resistance.

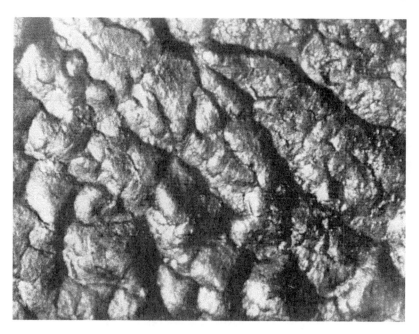

Figure 8.46 Surface of the scrapped ejector (34×).

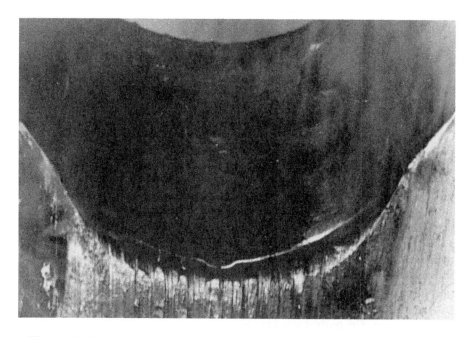

Figure 8.47 View of a worn shearing knife after 82,000 working cycles.

The application of a spray of water or water-based lubricant to the dies reduces both the maximum and mean temperatures they experience during the forging run; however, setting an excessive flow can suppress the effect of die preheat, making thermal fatigue more serious. Application of lubricants produces considerable benefits, but in choosing between potential lubricants a wide variety of facts, apart from product quality and die life. should be evaluated, including the level of smoke emission and its toxicity and the properties of scale produced. Replacement of one lubricant by another can necessitate altering the cooling and die-preheat systems (Lockwood et al., 1983). The restoration and repair of worn dies, constituting the second group of methods of increasing their service performance, includes:

1. Shot peening, where mechanical fatigue is the dominant factor. However, the state of stresses introduced gradually decreases with repeated forging cycles, and the die surface irregularities produced can disqualify this method in some instances. Apart from that, this treatment is useful only at an early stage of die service when there are no large cracks.
2. Restoration of mechanical properties by repeating the original heat treatment: hardening and tempering. This is capable of removing the effects of exposure to high temperature, and if repeated after a small fraction of the service life which would be anticipated in the absence of such treatment, it can, in some instances, extend the die life to several times that obtainable otherwise.
3. Welding of cracks followed by regrinding and proper heat treatment to remove welding stresses; this only partly restores the durability to the dies.

Although these methods are potentially of considerable economic benefit, they are not very widely used because they necessitate temporary withdrawal of the die from service use at a time when it is still in reasonably good condition. This may encounter psychological inhibitions and in any case requires careful organization of production.

8.7 DIESEL ENGINES

8.7.1 Working Conditions

Economic aspects and the specific demand for heavy field equipment causes more frequent application of diesel engines and promotes their development. This development leads to the involvement of higher working temperatures and is limited by the wear of diesel engine components. According to the available data concerning currently produced medium-speed diesel engines, the compression ratio reaches 25, the cylinder diameter 100 to 1000 mm, and the mean effective pressure reaches 3 MPa. Higher thermal and mechanical loads increase the possibility of cracking.

In diesel engines, fuel is injected into compressed air, which has a pressure of 3.5 to 7.0 MPa and a temperature of 500 to 750 K. The injected fuel vaporizes and mixes with air, forming a flammable mixture. Processes of fuel injection and combustion are still the object of basic investigations (e.g., Fujimoto and Sato, 1979;

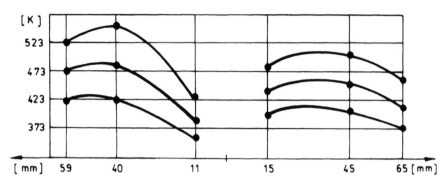

Figure 8.48 Typical temperature distribution across cylinder head of diesel engine. (From Slawinski, 1979.)

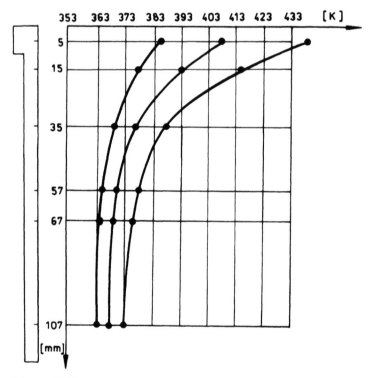

Figure 8.49 Temperature distribution in cast iron cylinder sleeve.

Fujimoto et al., 1981). However, the data obtained in these experiments cannot be used directly to optimize the design of engine components. More detailed information cannot be found in the literature at present.

Heat evolved during combustion is taken up by the engine block, pistons, and other movable components. Temperature distribution in the combustion chamber is not homogeneous, caused by an injection of fuel by a nozzle (or nozzles) of relatively small diameter, heterogeneity of the fuel spray, and different heat conductances of engine components. From results mentioned previously, it follows that bigger and faster droplets of fuel are in the central part of the spray. The degree of mixing between fuel and charge is also different in various parts of the spray. The combustion process is also complicated by the fact that some of the fuel is injected after ignition.

The measurement of the instantaneous temperature of the flame face of the diesel cylinder head is extremely difficult, as changes in temperature are very fast, thus cannot be measured by thermocouples or resistance sensors. A disagreement between experimental estimations of the average flame-face temperature given by different authors seems to be due to the type of engine tested and the experimental method employed in the reported study. Shalev and co-workers (1983) carried out experiments on an instrumented 6V diesel engine. The average flame-face temperature was estimated on 860 K by means of heat-transfer considerations based on temperature readings on wire thermocouples 3 mm distant from the surface. Results obtained by Britain (1964) point to 673 to 753 K. The typical temperature distribution across the diameter of the cylinder head is shown in Fig. 8.48. Temperature distribution was measured at three different levels of engine load. The diameter of the cylinder was 135 mm, the compression ratio 14:1, the piston travel 155 mm, and the number of valves 4. The observed asymmetry in temperature distribution was caused by nonhomogeneous combustion and heat abstraction. To complete the picture, the temperature distribution measured in the same experiment in the cast iron cylinder sleeve is shown in Fig. 8.49. It is worth noting that temperature profiles were taken by a thermocouple, the hot junction of which was performed by evaporation. The nickel–iron structure obtained was in direct contact with combustion gases and electrically insulated from the engine. Another measuring device employed in this study was a light sensor. The results obtained by means of it were in accord with those depicted in Figs. 8.48 and 8.49.

However, it follows from metallographic examinations carried out by Smith et al. (1971) that the temperature of the flame face of the head is in the range 573 to 953 K and in some cases can reach 1053 K. At such high temperatures, the material commonly used to produce cylinder heads, cast iron, is subject to severe corrosion in which the matrix and particular phases (e.g., graphite) are oxidized. It is noteworthy that at temperatures up to about 950 K, iron oxide and graphite can coexist in equilibrium in the atmosphere of the engine. In addition, at higher temperatures there is a reduction in iron oxide already formed by the adjacent graphite which accelerates corrosion, reducing the mechanical properties of the material. The conclusions of Smith et al. (1971) regarding maximal temperatures were drawn from the examination of oxide morphology. The thin subsurface layer of the flame face of the head is exposed to conditions of thermal fatigue, being heated by the flame and cooled by the colder layer of metal. However, the state of the cylinder head material considered as a whole can be referred to as creep in compression. Thermal fatigue also appears during startups and shutdowns. This can be explained in the following way:

1. During engine operation, thermal expansion of hotter layers of the cylinder head is restrained by the colder parts of metal adjacent to the coolant passages and by fixing to the cylinder block. It causes the flame face to creep in compression.
2. After shutdown, temperature distribution in a cylinder head becomes flat after a few minutes and the material, which has already yielded plastically, develops a tensile stress approaching several hundred megapascal.

Valves and valve seats are exposed to erosion by exhaust gases moving at subsonic velocities, corrosion, wear at the surfaces of valve–valve seat contact, and mechanical stress. They are also exposed to thermal fatigue, as they are rapidly heated by combustion gases and cooled through heat conduction.

The temperature of the postcombustion system is about 1023 K and at points is about 1223 K. This system is subject to mechanical loads caused by the motion of the equipment driven by the engine. An additional factor affecting its durability is thermal shock taking place while cooling by rainwater. Apart from corrosion caused by exhaust gases, there is corrosion enhanced by the chemicals frequently used as antifreeze agents on roads and this affects exhaust silencers. Damage to diesel engine components and methods of improvement are discussed in the following section.

8.7.2 Damage to Diesel Engines

The most expensive part of the diesel engine, which frequently fails prematurely, is the cylinder head. Cracks most often nucleate on the bridge between valve ports or between valve ports and the injector inlet, depending on the construction of the head. Normally, a group of fine cracks appears, and one or some of these cracks develop into a final fracture. However, postmortem metallographic examinations do not reveal any substantial change in the microstructure of the flame face of the head. It was found from the survey on a number of 6V diesel engines that 20% of failed cylinder heads were overheated (Shalev, 1981; Shohat et al., 1977).

It is difficult to change the temperature of the flame face considerably, although it follows from the preceding section that it would be profitable to maintain the temperature at the lowest possible level. The factor of prime importance in water cooling is the velocity of the water stream and the mass of water transported by it. A thin layer of water flowing along the cooled surface is found to have a slow speed, and this is dependent on the surface finish. It is obvious that the presence of this layer reduces heat conduction. Apart from that, the transmitted heat flux depends on the temperature of the cooled face. Supposing that the water flow is kept at a constant low rate, the dependence of the heat transfer rate on the temperature of the cooled face can be divided into the following (Fishenden and Saunders, 1950):

1. Initial, slow increase with increasing temperature of the cooled face. In this phase heat is transmitted by convection at a rate up to 500 W/m^2.
2. Rapid increase at temperatures exceeding 374 K with a maximum of about 1.1 MW/m^2 at temperatures of 390 to 420 K. This stage is referred to as nucleate boiling.
3. At higher temperatures, steam bubbles begin to join, and this reduces the contact of the cooled face with water.
4. At a temperature of 523 K, a continuous film of steam is formed which radically reduces heat transfer.

The last case mentioned can take place when an engine is switched off after working under high load and then restarted. In such a case the increase in the bulk temperature of the head can possibly cause the cylinder head to crack when the cooling system is defective.

Smith et al. (1971) have performed an experiment on the heat transfer of a specimen machined on both sides. The material examined was cast iron, commonly employed for cylinder heads, having a chemical coposition very close to that of normal carbon gray shown in Table 8.11 and a structure of fine flake graphite in a ferritic–pearlitic matrix. In this experiment, heat was conducted to one surface of the 8-mm-thick specimen through a graphite rod heated by an induction coil, and the other side was cooled by a controlled flow of cool water. The apparatus allowed the water flow to be controlled either parallel or directed at an angle to the cooled face.

The results obtained are shown in Fig. 8.50. The Y axis represents the heat flux transmitted through the specimen surface, and the X axis represents the temperatures of heated and cooled faces. The curves represent dependencies for four different water flows. Point A on the plot corresponds to the surface boiling in the form of fine bubbles. The change from parallel to directed flow, or the increase in flow rate, reduces boiling. Above point B, vigorous boiling affects water flow in the cooled region, but a change in the water flow alters temperatures on both sides of the specimen. It is seen that for a heat flux of about 2.5 MW/m^2, the difference in temperature of the cooled face is 105 K and that of the heated face is 65 K, depending on cooling conditions. From these facts it can be concluded that a distinct reduction in the temperature of the flame face can be achieved by high water flows and that in regions working under severe thermal load, directed flow can be used. Leaving pockets of low water flow should be avoided.

The second cause of head cracking is the creep phenomenon occurring in conditions in which the strain is restrained by the cooler part of the engine. In such circumstances, high tensile stress, approaching several hundred megapascal, will develop after some time upon cooling to room temperature. On the other hand, for an operating engine there is the difference in temperatures between the flame face and the cooled face, and this induces thermal stresses reaching up to 258 MPa. The chemical compositions and thermal conductivities of head materials are listed in Table 8.11, and compressive creep test results are given in Table 8.12. Alloying additives are used to improve creep performance, although they reduce thermal

Table 8.11 Thermal Conductivity of Cast Irons[a]

Grade of cast iron	Chemical composition (%)								Heat conductivity (W/mK)		
	C	Si	Mn	S	P	Ni	Cr	Mo	373 K	673 K	773 K
High-carbon gray	3.98	1.32	0.66	0.074	0.079	—	0.16	0.31	57.0	42.3	37.3
High-carbon gray	3.58	1.52	0.53	0.1	0.06	—	—	—	55.7	46.9	—
Normal carbon gray	3.18	2.08	0.6	0.112	0.097	—	0.14	—	49.0	40.2	36.9
Normal gray	2.92	1.75	0.73	0.126	0.1	—	—	—	35.6	—	—
Gray with additives	3.03	2.05	0.75	0.041	0.07	1.5	0.4	0.4	47.8	38.6	35.6
Nodular	3.64	2.20	0.23	0.010	0.021	0.15	0.05	—	37.3	34.8	34.0
Nodular	3.22	2.44	0.42	0.013	0.063	1.35	—	—	31.0	30.2	—
Nodular	3.15	2.27	0.32	0.013	0.06	—	—	Cu	26.8	—	—
Austenitic	2.4	1.8	1.15	—	—	13.7	3.4	0.4	33.9	31.4	—
Austenitic	3.0	1.0–2.8	1.15	—	—	13.5–17.5	1.75–2.5	5.5–7.5	39.8	—	—

Source: Mechanical Publications Ltd.
[a]Influence of graphite content and form is more appreciable than that of normal alloying additives.

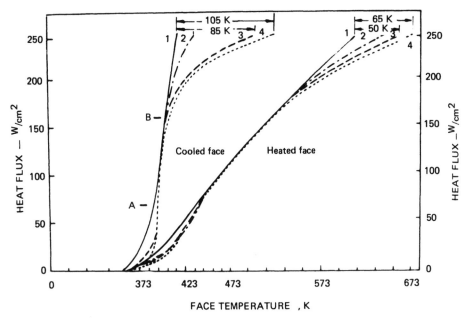

Figure 8.50   Effect of cooling conditions on face temperatures: 1, 7.3 dm³/min directed flow; 2, 2.3 dm³/min directed flow; 3, 7.3 dm³/min parallel flow; 4, 2.3 dm³/min parallel flow. (Reprinted with permission from Mechanical Publications Ltd.)

conductivity. It is clear that the factors mentioned above are dangerous to the head material.

It is advantageous to use high-carbon materials because of their high thermal conductivity and resistance to wear, but their machinability is low. Apart from that, the high value of Young's modulus causes an increase in thermal stresses induced by the thermal gradient existing between the cooled and fired faces of the head. It is seen from Table 8.12 that for gray cast iron and modified cast iron, the creep performances are greatly reduced at temperatures exceeding 673 K.

Cylinder sleeves are replaceable and their deterioration is mainly caused by the friction of piston rings as temperatures (see Fig. 8.49) are at a reasonably low level. The upper surface of the piston is in direct contact with combustion products and is exposed to high temperatures. However, the piston is also a replaceable part of the engine, and thermal gradients are usually small because of its thin-walled structure. The upper part is sometimes covered with protective material to avoid premature replacement. The deterioration of valve seats depends on the fuel. Trace elements such as vanadium and sulfur cause severe corrosion. Lead provides some lubrication on intake valve seats in spark-ignited engines and thus limits their wear, although it reduces considerably the durability of exhaust valves. Valve seat protection is usually achieved by a facing of cobalt-based alloy which mates up with

Table 8.12 Compressive Creep Results

Grade of cast iron	Chemical composition					Room-temperature properties		
	C	Si	Mn	S	P	Ultimate tensile strength (MPa)	Young's modulus (MPa)	Hardness (HB)
Hot-formed cast iron (N)	3.09	0.55	1.09	0.1	0.18	275	1.41×10^5	207
Modified cast iron (L)	3.30	1.18	1.50	0.067	0.345	291	1.45×10^5	234

Compressive Stresses (MPa) Necessary to Cause Proportional Strain in Defined Time

Grade of cast iron	Temperature (K)	Strain 0.1%			Strain 0.2%		
		100 h	1000 h	10,000 h	100 h	1000 h	10,000 h
Cast iron N	598	>243	194.6	88.2	>243	>243	167
	673	200.7	111	≈30	243	182.5	38
	748	76	45.6	≈30	111	60.8	≈30
Cast iron L	598	>228	>228	≈61	—	—	—
	400	226.6	164.2	≈30	228	205	88.2
	475	79.1	30	—	129	60.8	

Source: Mechanical Publications Ltd.

an insert made of cobalt-containing nickel-based alloy. It should be mentioned that any dimensional instability of the insert, or excessive wear, will accelerate the process of deterioration, as exhaust gas leakage rapidly oxidizes the valve seat and insert. The problem of developing new nickel-based alloys for such applications has been reported by Narasimhan and co-workers (1981).

An example of a solution to the problem of materials for postcombustion systems has been shown by Baudrocco and Corso (1979). The evaluation of candidate materials—ferritic stainless steel of AISI409 grade and austenitic steel of AISI304 grade—was based on their application for the hot zone of exhaust systems and casing of catalytic converters. The test performed included formability, thermal stress performance, corrosion, and oxidation resistances. It was proved that at temperatures lower than 1023 K, AISI409 steel is better than AISI304 steel. However, the latter can be used at higher temperatures because of the necessary mechanical properties that must be provided.

REFERENCES

Andrzejewski, C. (1980). *Proc. 2nd Symposium on Fabrication of Machine Parts*, Rzeszów, Poland.

Aston, J. L. and Muir, A. R. (1969). *J. Iron Steel Inst.* Feb.: 167.

Baudrocco, F. and Corso, S. (1979). *International Symposium on Automotive Technology and Automation*, Gratz, Austria.

Beck, G. (1958). *Stahl Eisen*, 78(22): 1556.

Bedford, A. J. (1974). *J. Aust. Inst. Met.*, 19: 61.

Berry, W. R. and Johnson, I. (1964). *J. Eng. Power*, 86: 361.

Bidwell, C. C. (1917). *Phys. Rev. A Phys. Soc.*, 10(11): 756.

Bradd, A. A. (1961). *Iron Steel Eng.*,38(2): 85.

Braun-Angott, P. and Berger, B. (1981). *Arch. Eisenhuettenw.*, 52(12): 465.

Brimacombe, J. K., Samarasekera, I. V., Walker, N., Bakshi, I., Bommaraju, R., Weinberg, F. and Hambolt, E. B. (1984). *ISS Trans.*, 5: 71.

Britain, E. W. M. (1964). Symposium on thermal loading diesel engines, *Proc. Inst. Mech. Eng.*, 179: 116.

Buchmayr, B., Danzer, R., Mitter, W. and Park, B. H. (1982). *Arch. Eisenhuettenw.*, 53: 35.

Burden, E. (1982). *Mod. Cast.*, (12): 20.

Chilton, J. M. and Roberts, M. J. (1981). *Iron Steel Eng.*, 58(1): 77.

Chladek, W. (1980). Influence of nickel–copper protective coatings on thermal fatigue resistance of cast iron. Ph.D. thesis, Silesian Technical University, Poland.

Chow, C. L. (1971). *J. Eng. Power*, 93(1): 13.

Cook, T. S., Pennick, H. G. and Wells, C. H. (1978). *Lifetime Prediction Analysis System in Steam Rotor Reliability*, Report NP 923, Electric Power Research Institute.

Danzer, R. and Sturm, F. (1982). *Arch. Eisenhuettenw.*, 53(6): 245.

Dean, T. A. and Silva, T. M. (1979). *J. Eng. Ind.*, 101(11): 385.

de Barbadillo, J. J. and Trozzi, C. J. (1981). *Iron Steel Eng.*, 54(1): 63.

Deilmann, W. and Schreiber, G. (1974). *Radex Rundsch.*, (1): 21.

Demidov, L. D. (1966). *Kuznechno-Shtampovochnoe Proizvod.*, (9): 14.

Droniuk, N. N., Gripatschevskij, A. N. and Oleksienko, A. J. (1982). *Isv. Vyssh. Ucheby. Zaved. Chern. Metall.*, (2): 73.

Fishenden, M. and Saunders, V. A. (1950). *An Introduction to Heat Transfer*, Clarendon Press, Oxford.

Fluck, D. E., Herchenroeder, R. B., Lai, G. Y. and Rothman, M. F. (1985). *Met. Prog.*, *128*(9): 35.

Foley, W. R., Pres, V. and Huber, W. R. (1974). *Iron Steel Eng.*, *15*(4): 72.

Fujimoto, H. and Sato, G. T. (1979). *Proc. 12th International Congress on Combustion Engines*, Tokyo.

Fujimoto, H., Sugihara, H., Tanabe, H. and Sato, G. T. (1981). *Proc. 14th International Congress on Combustion Engines*, Helsinki.

Gierzyńska-Dolna, M. (1985). *Friction, Lubrication and Wear in Plastic Working Processes*, WNT, Warsaw.

Greenberg, H. D., Wessel, H. T., Clark, W. G., Jr. and Pryle, W. H. (1969). Westinghouse Scientific Paper 69-ID 9-MEMTL-P2.

Grounes, M. (1972). *Iron Steel Eng.*, *49*(5): 73.

Harrison, J. M., Norton, J. F., Derricott, R. T. and Marriott, J. B. (1979). *Werkst. Korros.*, *30*: 785.

Hemingway, T., Oakley, J. W., Burt, G. and Evans, C. (1985). *10th International Forging Conference*, Sheffield.

Hohn, A. (1980). *Combustion*, (3): 10.

Hostetter, R. S. and Vyas, M. M. (1973). *Iron Steel Eng.*, *50*(10): 66.

Hoyle, R. and Mahabir, H. E. (1963). *The Engineer*, Aug. 30, *215*:353.

Izotov, W. I., Kotelnikov, G. A., Kulnitschev, G. P., Miftachov, R. G. and Tietiureva, T. W. (1982). *Met. Sci. Heat Treat.*, (9): 2.

Jaffrey, D., Dover, I. and Hamilton, L. (1984). *Met. Forum*, *7*: 67.

Jalovoj, N. I. (1973). *Tieplowyje processy pri obrabotkie mietallow i splawow dawlenijem*, Izdatelstvo Wysszaja Szkola, Moscow.

Jefimov, W. A. (1960). *Tieoreticzeskije csnowy razliwki stali*, Izddatelstvo Akademii Nauk USSR, Kiev.

Johnston, J. R., Weeton, J. W. and Signorelli, R. A. (1959). NASA Memo 4-7-59E.

Kannappan, A. (1969). *Metal Forming*, (12): 335.

Kellow, M. A., Dean, T. A. and Bannister, F. K. (1977). *Proc. 17th MTDR Conference*, p. 335.

Kobayashi, K. (1972). *Trans. Iron Steel Inst. Jpn.*, *12*: 172.

Körbe, H., Kopp, R., Schuler, G. and Stenzhorn, F. (1981). *Stahl Eisen*, *101*(9): 563.

Koul, A. K., Wallace, W. and Thamburaj, R. (1984). Report LTR, ST-1503, National Aeronautical Establishment, Canada.

Kumeno, K., Nishimura, M., Mitsuda, K. and Iwasaki, T. (1977). *J. Eng. Power*, *99*: 134.

Lamping, G. A. (1980). Recent experiences with steam turbine disc cracking, *Proc. 32nd Meeting of the Mechanical Failures Prevention Group*.

Leonidow, W. M., Nikitienko, E. W., Lopatkov, W. I. and Awtschuchow, I. A. (1983). *Kuznechno-Stampovocnoe Proizvodstvo*, (10): 17.

Lines, D. J. (1974). *Proc. Symposium on High-Temperature Materials in Gas Turbines*, Baden, Switzerland, Elsevier, Amsterdam, p. 155.

Linhardt, J. W. (1972). *Iron Steel Eng.*, *49*(12): 69.

Lockwood, F. E., Faunce, J. P. and Boswell, W. (1983). *Lubrication Engineering*, *39*(12): 753.

Lyle, F. F., McMinn, A. and Leverant, G. R. (1985). *Proc. Inst. Mech. Eng.*, *199*(A1): 59.

Maciejny, A. and Weronski, A. (1979). *Wiss. Z. Tech. Hochsch. Magdeburg*, *23*(7): 729.

Malm, S. and Norström, L. A. (1979). *Met. Sci.*, *13*(9): 544.

Maszkov, A. K. and Zawgorodniev, D. A. (1973). *Izv. Vyss. Uchebn. Zaved. Chern. Metall.*, (6): 114.

McCann, J., Boardmann, J. W. and Norton, J. F. (1973). *J. Iron Steel Inst.*, *211*(1): 66.

Mellor, J. W. (1934). *A Comprehensive Treatise on Theoretical Chemistry*, Vol. 13, *Fe*, Part 2.

Mirtich, M. J., Nieh, C. and Wallace, J. F. (1981). *Thin Solid Films*, *84*: 295.

Mommertz, K. H., Keck, R., Kessler, F., Sablotny, R. and Stelzer, R. (1984). *Stahl Eisen*, *104*(9): 453.

Nakajama, K., Omori, S., Kawamoto, T., Tsukamoto, H., Hatae, S., Hino, H. and Aratani, H. (1985). *Mitsubishi Heavy Ind. Tech. Rev.*, *22*: 143.

Narasimhan, S. L., Larson, J. M. and Whelman, E. P. (1981). *Wear*, *74*: 213.

Norton, J. F. and Strang, A. (1969). *J. Iron Steel Inst.*, *207*(2): 193.

Olejarski, B. (1986). Investigation of structural changes in low alloy Cr–Mo steel exposed to thermal cycling, Ph.D. thesis, Technical University of Lublin, Poland.

Pachowski, M. (1970). *Pr. Inst. Mech. Precyz.*, *18*(2): 33.

Policinski, J. (1979). Influence of load on thermal fatigue resistance of hot-work tool steels, Ph.D. thesis, Technical University of Czestochowa, Poland.

Rundell, G. (1982). *Met. Progr.*, *122*(8): 1.

Rundell, G. (1985a). *Met. Progr.*, *127*(4): 39.

Rundell, G. (1985b). *Met. Prog.*, *127*(5): 51.

Sawada, S., Ohnishi, T. and Kawaguchi, S. (1981). *Rotor Forgings for Turbines and Generators*, Workshop Proceedings WS 79-235 (Jaffee, R. I., ed.), Electric Power Research Institute.

Schmidt, W. (1979). *Neue Huette*, *24*(12): 479.

Shalev, M. (1981). Developing of cracks as a result of thermal stress in cylinder heads of internal combustion engines, M.Sc. thesis, Faculty of Mechanical Engineering, Technion, Haifa, Israel.

Shalev, M., Zvirin, Y. and Stotter, A. (1983). *Int. J. Mech. Sci.*, *25*: 471.

Shohat, U., Krasny, E. and Stotter, A. (1977). *Investigation of Cracks in G.M. 6V-53 Engine Cylinder Head*, Report 011-147, Israel Institute of Metals, Technion, Haifa, Israel.

Slawinski, Z. (1979). Investigation of local parameters of heat flux in a high-speed diesel engine cylinder head, Ph.D. thesis, Technical University of Leningrad.

Slusarjev, W. J. (1975). Influence of thermal treatment on physico-mechanical properties and durability of steel moulds fabricated by centrifugal casting and forging, Ph.D. thesis, Donieck Technical University, USSR.

Smith, L. W. L., Angus, H. T. and Lamb, A. D. (1971). *Proc. Inst. Mech. Eng.*, *185*: 807.

Somm, E. and Reinhardt, K. (1979). *Brown Boveri Rev.*, *66*(6): 365.

Spera, D. A. (1969). NASA TN D-5317.

Spera, D. A., Howes, M. A. H. and Bizon, P. T. (1971). NASA TMX-52975, p. 1.

Stachurski, W. (1972). *Zesz. Nauk. Akad. Gorn. Hutn.*, (29): 235.

Stevens, C. G. and Tidy, D. (1983). *J. Inst. Energy*, *55*(3): 12.

Szafraniec, B. (1972). Report 766, Instytut Metalurgii Zelaza im. St. Staszica, Poland.

Szapowalov, S. I., Alimov, W. I., Podliesnyj, L. S., Olifirenko, W. W. and Samojlov, W. A. (1981). *Steel*, (USSR), (4): 71.

Tittagala, S. R., Beeley, P. R. and Bramley, A. N. (1983). *Metallurgia*, *50*(10): 434.

Tlenkler, H. and Czech, H. (1962). *Berg Huettenmaenn. Monatsh.*, (9): 313.

Vingas, G. (1982). *Mod. Cast.*, (3): 43.

Viswanathan, R. and Jaffee, R. I. (1983). *J. Eng. Mater. Technol.*, *105*(10): 286.

Weronski, A. (1977). *Wiadomosci Hutnicze*, *2*: 46.

Weronski, A. (1979). *Arch. Hutn.*, *24*(1): 117.

Weronski, A. (1983). *Thermal Fatigue of Metals*, WNT, Warsaw.

Wigmore, G. (1979). Discussion on J. M. Hodge and I. L. Mogford, US experience of stress corrosion cracking in steam turbine discs, *Proc. Inst. Mech. Engineers*, *193*(11): 93.

9

Protective Coatings in High-Temperature Engineering

9.1 THE COATING FUNCTION

This chapter concerns the relatively thin protective coatings which today have an essential role in virtually all mechanical equipment that operates at high temperatures. A great variety of methods are used to form such coatings in or on the surface region of the component to be protected and an even great variety of materials are employed as coatings. The prime function of a coating may be to act as a thermal barrier or to reduce friction, stiction, and wear, but the most widespread use to date is to protect from oxidation, corrosion, and erosion, this being particularly important in the hottest areas of gas turbines.

As almost all cracks, whether from creep or mechanical and thermal fatigue, initiate at the surface, deterioration in the condition of the material in this region of a component will greatly reduce its serviceable life. If the alloy of which it is composed has inadequate resistance to chemical attack, such attack, which is generally selective, will create local stress raisers through pitting and may also modify the composition of the alloy in the surface region, impairing its mechanical properties just where they are particularly important. These factors contribute to the initiation and development of cracks. In addition, loss of metal by chemical attack (and erosion) increases the stress in the cross section, which remains to support the load. At high temperatures creep is accelerated and time to failure, either by exceeding deformation tolerances or by creep rupture, is reduced. The potential significance of this is apparent when we note that the use of high-strength alloys has resulted in the wall thickness of an aircraft turbine blade being, in some cases, scarcely 1 mm initially.

There are other mechanisms by which environmental attack may accelerate failure. Solid oxides formed within a crack generally have a greater specific volume than the metal which was oxidized and may thus wedge the sides apart, extending the crack and exposing fresh material for oxidation. Other oxides and corrosion

310

products which are molten at the operating temperature cause decohesion where they penetrate grain boundaries. In all, the possible interactions between metal alloys and chemically aggressive high-temperature environments are numerous and often complex, depending on the chemical composition of the alloys and environments, the temperature and pressure, and also the relative velocity of the metal surface and the medium adjacent to it. The theoretical and empirical information which exists has enabled reasonably adequate and even good solutions to be found for most practical problems, but more remains to be discovered. Currently, there are few coatings that do everything required of them and none that last indefinitely.

Of course, coatings, which may be described as protective, are also widely used in circumstances where temperatures are not necessarily very high or variable, and even if they are, the purposes of the coatings do not include protecting the substrate from corrosion-accelerated thermal fatigue failure. They are used to improve the performance of skis on snow and of tools to cut metal at high speeds. In a helium-cooled nuclear reactor, coatings may be employed not to prevent oxidation, but to prevent the serious increase in friction and scuffing that may occur when metal parts in sliding contact lack the oxide layer which they surfaces would possess in a more normal environment. The subject of coatings could fill many books. The present survey is confined to a rather brief outline of those coatings and associated processes which include in their functions the mitigation of thermal shock or, as is more common, protection from chemical and erosive attack at high temperatures.

The many methods used to provide such coatings fall into two groups. *Overlay coatings*, as the name suggests, are applied to the exterior of the component, the method of application or some subsequent treatment ensuring an adequate bond. *Diffusion coatings* are obtained by modifying the composition of the surface region of the component by diffusing an element, or possibly elements, into it, and thus the processes used somewhat resemble nitriding and carburizing. More exactly, the diffusion processes commonly used involve the outward diffusion of a substrate element to combine or alloy with an element that is being supplied to the free surface. In such cases a resultant protective layer exists above the original level of the substrate surface (as would an overlay coating), while a further zone enriched with the protective elements may also exist below it and also a region somewhat depleted of the alloy constituent, which diffused outward.

In theory and largely in practice the two methods may be combined in various ways: an initially overlaid coating may be diffused into the substrate by subsequent heat treatment or itself improved by diffusion treatment from an external source. Overlays may be put down on coatings formed by diffusion to provide a multilayer coating, and so on. The division into two groups is a convenience for descriptive purposes rather than an absolute technical separation. Coatings of either group offer the same range of advantages, although one group or combination may provide the best advantage in a particular situation. Similarly, to do their job all coatings must satisfy certain requirements. We may therefore look at these common factors before looking at the nature of different coatings and processes.

9.2 ADVANTAGES OBTAINED
FROM PROTECTIVE COATINGS

Before considering specific coatings, we may note the advantages that may be obtained from any coating which is capable of performing the functions indicated above in the environment in which it is used: prolonged life and increased reliability and extended life and reduced maintenance. There are special cases (e.g., in rocket engineering) where a vital component would not have any life at all but for the thermal protection provided by a refractory coating. However, this section relates to situations in which unprotected metal components might be used, and sometimes have been used, but improvement in service life and reliability is desirable. In this instance the advantage provided by the coating is not that the equipment performs better but that it performs longer. This can contribute to safety and, safety being adequately secured, to profit.

The economic value of a capital asset such as a gas turboalternator is reduced by damage to the surface of vanes, nozzles, and especially blades exposed to combustion products at the hot end. Expenditure on fuel, being less than the value of the electrical energy derived from it, provides profit. Withdrawal from service at intervals of a few thousand hours to inspect and, where necessary, replace components interrupts the profitable use of both turbine and alternator and also adds costs for labor and components. Hence any processing of blades and so on which assures a longer life is profitable: Intervals between inspections can be increased when warranted by confidence in the components lives and, in any case, the costs incurred by component replacement are reduced.

Aircraft engines provide another instance where increase in component life, once established, provides advantages additional to those of reduced maintenance costs. The loss of revenue while an airliner is out of service for engine change or overhaul is substantial. Gas turbines provide convenient examples, but there are obviously many other areas in which prolonging component life with protective coatings may be of direct and indirect financial advantage.

We must always remember that oxidation and corrosion are not the only mechanisms of failure that affect real-life load-bearing components exposed to high temperatures. Although these mechanisms are interactive at high temperatures, they stem from individual factors one or more of which will always be present.

In systems that operate at high temperatures, thermal fatigue will always make some contribution to failure. For any such equipment, each startup and shutdown will provide a major cycle, and during operation there will generally be significant thermal cycling, as temperatures change with power output or with change in other parameters. Although base-load nuclear power stations do not really run quite continuously at constant load and temperature, they do take an increasing share of the constant part of the network load so that it is now common for even very large fossil-fueled generators to be run up only for peak loads, a mode of operation that provides one or more major thermal cycles every 24 h. Engines in airline service experience a major cycle, on average, every 2 or 3 h. Solar generators in any probable location on earth will have at least a daily cycle, however clear the sky.

There is generally a variable element in the mechanical forces to which the parts of mechanical forces to which an element is subjected, and even if it is small, some mechanical fatigue must be expected at high temperatures where metals do not usually display the infinite endurance to moderate cyclic stresses which they may have possessed at lower temperatures.

Finally, load-bearing metal components will always be subject to creep at high temperatures and will eventually exceed dimensional tolerances, or rupture.

Thus even a perfect protective coating (it has yet to be found) cannot extend the high- temperature working life of a component indefinitely. However, it can extend working life, often by a large factor, by preventing oxidation and corrosion accelerating failure; and this is very valuable.

9.2.1 Reduction in Material Cost

The alloys used at high temperatures are expensive, partly because they must almost always have good corrosion resistance as well as excellent mechanical properties. Some years ago it was said, in reference to the use of uncoated blades in industrial gas turbines, that "approximately 20 to 30% of the total gas turbine cost is spent on blades, about half of that cost being accounted for in corrosion protection, i.e., about 15% altogether" (Endres, 1974). It might therefore appear that great savings could be achieved if the task of corrosion protection was transferred from the base material to a protective coating. However, in the gas turbine field the development of protective coatings has not been exploited to enable cheaper base materials to provide the same life under the same conditions as before. It has been more advantageous to use the best available superalloys together with protective coatings to permit working at higher temperatures, use of cheaper fuels which have more corrosive combustion products, or to extend working life to the maximum. Nevertheless, where reduction in material cost is more important than intensification of working conditions, the use of protective coatings may enable it to be obtained. Thus experiment has shown that in the proposed use of molten drawsalt ($NaNO_3$–KNO_3) at 873 K for the heat transfer and storage medium in a solar concentrator, the required corrosion resistance can be obtained by the use of an aluminum coating on a relatively low-alloy steel at a combined cost some 50% below that of the alloy required if no coating were used (Carling et al., 1983). Although in this instance the thermal fatigue and creep resistance of the cheaper combination had still to be fully established. The example indicates the considerable cost saving potential of protective coatings in hostile high-temperature environments.

9.2.2 Increasing Performance

The efficiency with which thermal energy is converted to mechanical increases with the temperature at which the process begins and also with reduction in the temperature at which it ends. Practical considerations as well as the temperature of the surroundings set limits to reducing the end temperature so that thereafter, increasing efficiency demands increasing the initial temperature. The extent to which this can be done depends on the materials available for constructing the heat engine. Some years ago in the case of gas turbines, increasing the inlet temperature from 1170 to 1370 K increased thermal efficiency by some 6% and power output by around 20%, despite an increase in losses due to bleeding more air from the compressor to cool the blades. The increase in power output, of course, arises from a combination of the increased heat input to the turbine together with its conversion at the higher efficiency. Naturally, the trend to higher inlet temperatures continued and still does. This example merely serves to indicate how increased inlet temperatures improve fuel consumption and, to a useful degree, power-to-weight ratio. The trade-off among fuel economy, size, material costs, and maintenance intervals varies with the application, so that while inlet temperatures have risen more slowly in the industrial turbine area, in high-performance aircraft engines they passed 1500 K more than a decade ago. Increase in temperature is made possible by protective coatings which were first used in aircraft turbines some 30 years ago.

At high temperatures oxidation and corrosion are accelerated, and at very high temperatures even very low levels of corrosive substances, sometimes only in the ppm range, can cause rapid damage. Initially, coatings were used to extend the life of components made of superalloys which themselves possessed very good corro- sion resistance: the protection, useful at moderate temperatures, became almost essential as they rose. With further rise in temperature the coating became quite essential because new superalloys formulated to have the necessary strength, creep and thermal fatigue resistance in very high temperature usage had less corrosion resistance than the preceding generation. The increased responsibility of the coat- ing, together with the intensification of chemical attack on it, due to the very high temperature, has led to much development and the appearance of a second genera- tion of coatings, often of a complex character. As none of these are perfect, and as temperatures continue to rise, research and development continue.

It should be remembered that while blade cooling and sometimes vane cooling is employed at the hot end of the turbine and component temperature is appreciably lower than inlet gas temperature, the coating's task is still arduous. Cooling is essential to maintain adequate mechanical properties of the component material, but does less to assist the coating, which is some way from the cooling air inside the blade or vane and directly exposed to the combustion gas.

9.3 NECESSARY PROPERTIES OF A COATING

These may be listed individually, although they are not entirely independent of each other.

1. Ability to form and maintain a continuous layer highly resistant to oxidation, corrosion, and in some cases, erosion
2. Good adhesion to the substrate
3. Ductility

4. Compatibility with the substrate material
5. Low interdiffusion with the substrate material
6. Resistance to thermal and mechanical fatigue
7. Practicability
8. Material and processing costs no greater than necessary

Next we discuss each of the foregoing properties in detail.

1. The first requirement is satisfied by metallic coatings if they form and maintain a continuous film of oxide, since it is the metal oxide that is chemically resistant. Ceramic coatings have inherent chemical resistance but must provide a dense fissure-free layer. Used as thermal barriers, they must have low thermal conductivity.

2. Good adhesion to the substrate is more readily obtained with metallic coatings, but the protective oxide formed on the outer surface of the coating must also adhere well. Some loss of oxide by spalling during thermal cycles has to be accepted and is made good automatically by the environment oxidizing the freshly exposed surfaces of the coating, but excessive spalling causes premature exhaustion of the coating and exposure of the substrate. Good adhesion to the substrate is essential; adhesion between oxide and coating must also be as good as possible under thermal cycling and may be improved by additives.

3. Ductility is required to accommodate differences between the thermal expansion and deflection under load of the coating and the substrate and also creep in the load-bearing substrate. Although the coating does its work at high temperatures, it has to get up there and from time to time get back, and ductile–brittle transitions en route have to be taken into account; this is a factor that sets some limits on coating composition. Ceramics having little ductility are usually applied in a mix with metal constituents. Bonding pure ceramic to a metal substrate may require one or more intervening layers of suitably chosen metals. Ductility in a coating also accommodates any difference between the specific volume of the oxide formed on its surface and that of the coating element from which it was formed.

4. Compatibility of coating and substrate includes matching thermal expansion sufficiently for such ductility as the coating possesses to accommodate the cumulative difference in strain throughout the temperature range. Compatibility of chemical composition is obviously a decisive factor for diffusion coatings. They and overlay coatings bonded by diffusion must not contain elements that would weaken the substrate, insofar as some interdiffusion may occur in the course of use, this also applies to all overlays. Chemical bonds have specific requirements; mechanically bonded overlays have the least restraints on chemical compatibility.

5. The requirement of low interdiffusion has to be interpreted with regard to the type of coating, its composition, and that of the substrate together with the service environment. Obviously, high rates of interdiffusion in service will generally result in premature loss of protection: either because the whole coating, or an element vital to its function, has largely disappeared from the surface by diffusion into the substrate or because elements that reduce the corrosion resistance of the coating

have entered it by diffusion from the substrate. In either case the mechanical properties of the substrate are also likely to have been impaired to some extent. Nevertheless, with many types of coating some degree of interdiffusion is inevitable and is beneficial by providing an intermetallic bond between coating and substrate. Moreover, moderate outward diffusion of a substrate element such as chromium can extend the life of a coating that was deficient in it. Regarding inevitability, differences in composition of coating and substrate imply concentration gradients which must result in diffusion at high temperatures except where the physical structure of the medium precludes it. The diffusion is also influenced by temperature gradients and stress fields, both of which are normally present. More particularly, in processes that are widely used, the protective coating is produced either by diffusing a suitable element into the surface zone of the substrate or by the combination of an element diffusing outward from the substrate, with one being supplied to the surface. The essence of these processes is fairly rapid movement of one or the other constituent by diffusion at temperatures which are not vastly different from that used in forming the coating. However, in the absence of a continued supply of the element provided during processing, the concentration gradients do not encourage rapid diffusion, and in practice this has much less effect on the life of the coating than the oxidation, corrosion, and thermal cycling to which it is exposed.

Concerning the other effects of interdiffusion, we have noted that moderate outward diffusion of chromium, which will generally occur when a chromiumless coating is formed on a chromium-rich substrate, can be beneficial. We shall shortly reach an example in which tungsten diffusing from a coating into the substrate reduced its resistance to thermal fatigue so that it cracked under thermal cycling. In summary, coating compositions and processes must be chosen to limit interdiffusion but with particular regard to the effects of specific elements.

6. Although protective coatings are not intended to, and generally do not, take an appreciable share of the main mechanical load on the substrate, they are fully exposed to thermal variations, and being bonded to the substrate, follow the fluctuating strains which are the substrate's response to varying loads. Coatings must therefore be resistant to thermal and mechanical fatigue.

7. Practicability is a reminder of fairly obvious considerations; for example, processes in which the substrate is heated enough to change its microstructure can only be used if the microstructure can be restored by subsequent heat treatment with the coating in situ. Substrates that have been cold worked (e.g., to improve resistance to radiation damage) must not be strongly heated. Processing in vacuum chambers is unsuitable for large components. Line-of-sight processes such as spraying run into difficulty if the interior of a hollow component is to be coated.

8. That costs should be no more than necessary is also obvious. High temperatures, an aggressive atmosphere, and long life may justify the use of precious metals and complex and expensive processing (e.g., for first-stage blades in a turbine running on low-grade fuel); cheaper and simpler methods may suffice elsewhere.

9.4 EVALUATION OF COATINGS FOR A CENTRIFUGAL CASTING MOLD

Before proceeding to methods of coating, we may illustrate some of what has been said by a straightforward example. Weronski (1980) investigated the use of coatings to extend the life of the class of centrifugal mold described in Chapter 7. The substrate material was 20H2M steel, generally used for such molds in Poland. A number of coatings were applied and tested; the behavior of 20-µm thick coatings of Ni, Cr, and Ni-W deposited galvanically provide our example. From Chapter 7 it can be seen that 20H2M steel is virtually devoid of Ni and contains only a trace of W. Tests consisted of repeated thermal cycling between 293 and 973 K in air. The criteria used to compare performance were (1) number of cycles to the appearance of the first crack, and (2) the total length of cracks per unit area after prolonged cycling. Preliminary tests were made by the turning ring method (see Chapter 5) using batches of samples in the form of rings 60 mm in diameter.

The principal tests were made with samples in the form of rings of internal diameter 120 mm and external diameter 180 mm by the method described in Chapter 5, that is, with periodic induction heating of the inner surface and continuous water cooling of the outer surface of the cylinder in close simulation of casting cycles. From the results shown in Fig. 9.1 it can be seen that the unprotected samples failed first. Those plated with Ni-W failed almost as rapidly, while the Ni coating

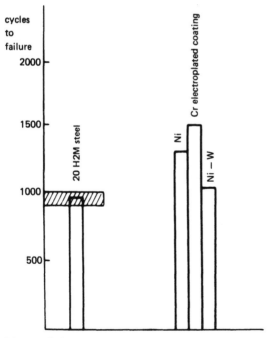

Figure 9.1 Influence of protective coatings on mold durability. (From Weronski, 1980.)

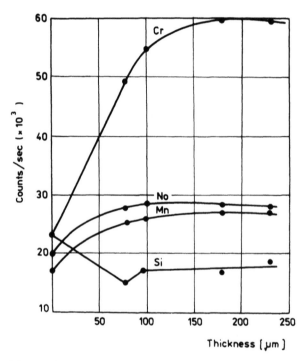

Figure 9.2 Variations of elemental concentrations versus depth for uncoated specimen after 1100 thermal cycles.

had extended sample life by about 30%, and Cr-coated samples lasted 50% longer than those without any coating. The surface of the unprotected samples showed signs of oxidation and some pitting. In continued testing some, though not all, of the cracks had started at the pits. Metallography and x-ray fluorescence analysis were used to clarify the behavior of the materials. Figures 9.2 and 9.3 show the results of the fluorescence analysis from the surface to a depth of 250 μm in an uncoated sample and in a Ni/W-coated sample (for clarity, elements that did not significantly affect the results were not plotted). Microphotographs of the structure to a greater depth are shown in Figs. 9.4 and 9.5. In the uncoated specimen, depletion of Cr can be observed up to 150 μm below the surface and that of Mn and Mo to a depth of 100 μm. The loss of Cr will have been due to its oxidation at the surface exposed to the air, loss of the oxide (Cr_2O_3) by spalling in the course of the thermal cycling, followed by new formation and subsequent loss of oxide, repeated a number of times. Loss of Mo and Mn may have resulted from similar processes occurring after the amount of Cr available at the surface had become too low to provide effective protection with Cr_2O_3.

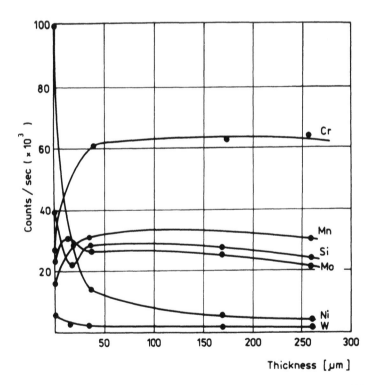

Figure 9.3 Variation of elemental concentration versus depth for Ni/W-coated specimen after 1040 thermal cycles.

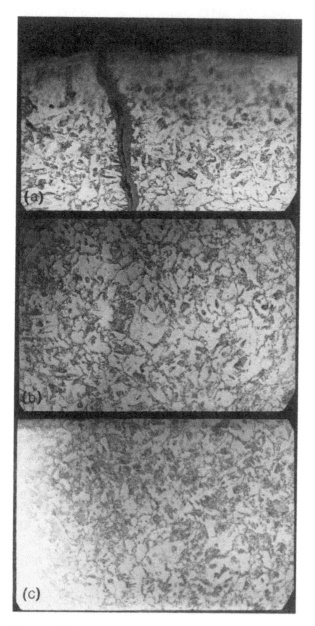

Figure 9.4 Subsurface layer of uncoated 20H2M steel after 1100 cycles of thermal fatigue (125×). (a) Subsurface layer; (b) zone 0.5 mm distant from the surface; (c) zone 10 mm distant from the surface.

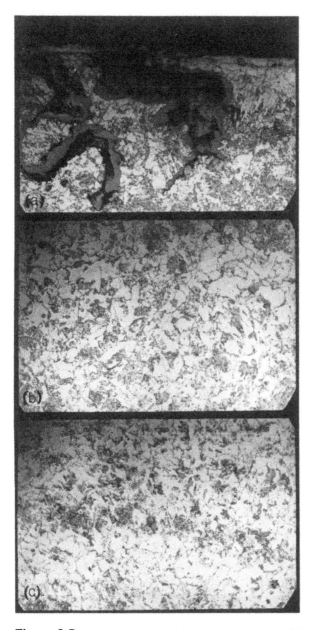

Figure 9.5 Subsurface layer of 20H2M steel coated with Ni-W after 1040 cycles of thermal fatigue. (a) Subsurface layer; (b) zone 0.5 mm distant from the surface; (c) zone 10 mm distant from the surface.

The specimen coated with Ni-W is particularly interesting because while it illustrates in some measure the protection a coating can provide, it also shows that excessive interdiffusion has occurred. An ideal case would show a surface zone composed entirely of the coating materials, in this instance Ni and W, followed by a fairly narrow interdiffusion zone of mixed composition, serving to provide an intermetallic bond, and below this the substrate material of unchanged composition. The actual specimen shows that although the surface zone is enriched with Ni and W, of which the original 20-μm coating was entirely composed, it now contains substantial amounts of other elements as a result of diffusion from the substrate, while Ni and W from the coating have diffused quite deeply into the substrate. The overlarge interdiffusion zone extends all the way to the surface and Ni and W have entered the substrate to depths of 50 and 100 μm, respectively, below the surface. On the credit side, the coating has greatly reduced the loss of alloying elements from the substrate. Cr and Mn remain at their normal substrate levels to within 30 μm from the surface, in comparison with 150 μm and 100 μm, respectively, in the uncoated specimen. It is not certain that there has been any actual loss of these elements; the fall in their concentration near the surface may be due to reduced diffusion rates in the Ni/W-enriched region. In the coated specimen the level of Mo is normal, or slightly above normal, right up to the surface, whereas in the uncoated specimen Mo depletion could be observed to a depth of 100 μm.

Under thermal cycling this coating exhibited good resistance to oxidation and pitting, and crack initiation did not occur at the surface. Nevertheless, despite this, and the better retention of substrate composition, the lives of the Ni/W-coated specimens were very little longer than those of the uncoated ones.

This can be explained by the inward diffusion of W. It is known that W has an adverse effect on the thermal fatigue resistance of alloys such as 20H2M, and the tests were of course characterized by severe and prolonged thermal cycling. Microscopy did in fact show that the cracks had initiated *below* the coating and in the region where substrate composition was modified by the addition of Ni and W (see Fig. 9.5).

The specimens with simple coatings of Ni or Cr lasted longer than uncoated or Ni/W-coated specimens. Both the Ni and the Cr coatings reduced the loss of substrate alloy constituents to much the same extent as the Ni-W coating, and they preserved the condition of the exposed surface through the same number of cycles as the Ni-W coating. What they did *not* do was infuse W into the substrate. When they eventually failed, all cracks initiated at the surface in the case of Ni coatings and frequently at the surface and only sometimes in the substrate close to the surface, in the case of Cr coatings. These results confirm that impairment of the thermal fatigue resistance of the substrate by inward diffusion of W from the Ni-W coating was responsible for the failure of specimens. This account, though much abridged, serves to show how coatings can extend component life, why low interdiffusion between coating and substrate is a requirement and why the composition of the coating must be compatible with that of the substrate.

9.5   COATING METHODS

9.5.1   Diffusion Coatings

These coatings are widely used with nickel and cobalt superalloys and have been used successfully with other materials. The object is to provide an outer layer that is richer than the underlying high-strength alloy in protective elements such as Al, Cr, or Si. Protection is provided by the oxide of the protective element, which is formed at the surface exposed to the environment. All environments in which these coatings are used in practice contain free oxygen. In gas turbines, which are currently the major area in which this class of coating is employed, excess combustion air is always present. The coating, being rich in the protective element, re-forms the oxide layer where it has been lost by spalling during thermal cycles and can continue to do this until, eventually, the amount of protective element remaining is too small to maintain a continuous oxide film. The substrate with a smaller proportion of effectively protective elements, would, if similarly exposed, be unable to maintain a protective film for so long a period.

The aluminizing of components made of a nickel-based superalloy illustrates the process. A typical composition of such an alloy might be around 65 wt% Ni and only some 5 to 10 wt% Al, the balance being Co, Cr, Mo, Mn, and so on. The aim may be to produce a nickel aluminide coating with a thickness of some 50 μm and an aluminum content of 30 wt%. (A thicker coating and higher aluminum content are possible but tend to brittleness and an increased brittle–ductile transition temperature, so these figures are fairly representative.)

The coating is produced by heating the components and a source of aluminum in an inert or reducing atmosphere in the presence of a small amount of a halide. The halide provides movement of aluminum from source to components in the form of a monohalide gas which decomposes at the surfaces of the components and deposits aluminum on them. This greatly accelerates the processing and is, for practical purposes, an essential part of it. The halide may be provided in various forms, for example, ammonium fluoride or calcium chloride, either mixed with the source material or contained separately within the treatment chamber. Diffusion takes place for a number of hours during which the ensemble of source, components, and so on, is held at a specific temperature in the range 1170 to 1470 K. Diffusion can proceed in one or other of two ways, inward diffusion of aluminum or outward diffusion of nickel, the activity of the source determining which mechanism occurs. If the source activity is high, the Ni_2Al_3 phase is formed initially at the surface, and thereafter diffusion is inward, as aluminum can move in Ni_2Al_3 and nickel cannot. Thus the coating is formed within the upper region of the substrate. The aluminum diffusing into the substrate combines with the nickel within it to form more Ni_2Al_3 so that the layer formed, at this stage, is Ni_2Al_3 together with the substrate alloy constituents other than nickel. Heat treatment at about 1370 K is required after the diffusion process to convert the brittle and low-melting Ni_2Al_3 to NiAl. During the heat treatment aluminum moves from the Ni_2Al_3 to form, with nickel from the substrate, a layer of NiAl in which aluminum cannot move and nickel can.

Conversion of the Ni_2O_3 above this layer is completed by outward diffusion of nickel from the substrate, leaving behind it a zone of nickel depletion. As the majority of elements are not soluble in NiAl, the substrate alloy constituents other than nickel form precipitates within the coating and to some degree in the nickel-depleted zone immediately below it.

The outward diffusion process is obtained by reducing the activity of the source of aluminum. Whereas for inward diffusion the "high-activity" source might contain pure aluminum with some inert dilutent, or an alloy of aluminum with another metal, such as chromium or nickel, the proportion of aluminum being fairly high, for outward diffusion the "low-activity" source contains aluminum only in the form of an alloy in which it is a relatively minor constituent. In this case the phase formed at the substrate surface when the process begins is NiAl, in which nickel can move and aluminum cannot, so the coating builds up by outward diffusion of nickel from the substrate, via such a layer of NiAl as exists above it, to form more NiAl, with the aluminum arriving at the surface of the NiAl already formed. Thus the coating is formed primarily above the substrate surface, and this external zone does not contain precipitates in the form of intermetallic compounds and carbides of elements contained in the substrate. Such elements may diffuse into the external zone but not in such concentrations as to form precipitates. Below the external zone, precipitates are formed as in the case of inward diffusion coatings.

The general structures of the two types of diffusion coatings are illustrated in Fig. 9.6. The cement particles shown in the external zone of the outward diffusion coating (Fig. 9.6b) are inclusions due to contact between the component and the source material—"cement"—during the cementation process described below. In a noncontact process, also described, they would not occur. The low-activity outward diffusion process is usually preferred to the high-activity inward diffusion process because the intermetallic compounds and carbides of substrate elements precipitated in the outer zone with the latter process are prone to corrosion and provide sites for local attacks which can undermine the coating's protective func-

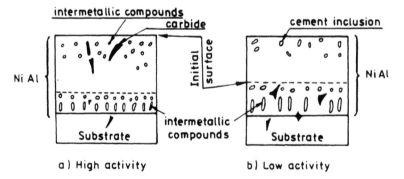

Figure 9.6 Comparison of the structures of two Ni/Al-type coatings. (After Gauje and Morbioli, 1983.)

tion. In addition, with inward diffusion a second course of heat treatment is necessary to convert the Ni_2Al_3 formed initially to NiAl.

9.5.2 Diffusion Coating Methods

The three methods generally used are pack cementation, the slurry method, and the vapor-phase method, all of which work in the manner outlined above. Regardless of the method, careful preparation of the surfaces to be coated is essential. Diffusion coatings do not depend on a mechanical bond for adhesion, and surface roughening is not required, but cleaning and degreasing with acetone, trichlorethylene vapor, or similar agents is necessary.

Pack Cementation

For pack cementation the components are treated in a sealed or semisealed box in which they are surrounded by, and in close contact with, an appropriate compound in the form of a powder. A nonoxidizing atmosphere, frequently hydrogen, is provided. The box is heated in a furnace for a number of hours during which diffusion takes place. After the pack has cooled, the components are taken out and dusted to remove particles of the cementation compound adhering to their surfaces. Further cleaning may be performed by barreling or vapor blasting with a mild and noncontaminating abrasive such as alumina or fine glass beads.

The precise formulations of the compounds used in the cementation process are seldom publicized by their developers, but for simple aluminization of a nickel-based alloy they are probably much of a kind. Dupre et al. (1981) have described NiAl coatings produced by outward diffusion of Ni from nickel-based superalloys by use of a low-activity pack composed of alumina, an 85:15 Cr/Al alloy, and ammonium chloride, all in the powder form. The alumina dilutent was the predominant component; the ammonium chloride, present as a small addition, provided the halide activator to facilitate transport of the aluminum. A total coating thickness of about 60 μm was produced by 16 h of cementation at 1320 K.

The outer zone of the coating contained inclusions of cement particles; this is typical of a low-activity cementation process in which outward diffusion from the substrate causes part of the coating to be formed in space previously occupied by the cementation compound. The inner zone, below the original surface of the substrate, was less rich in aluminum and contained precipitates of substrate alloy constituents which were not wholly soluble in NiAl.

As the cement particles included in the external zone are generally insert (e.g., Al_2O_3) or protective (Cr derived from the 85:15 Cr/Al particles in the case just described), their presence is not necessarily harmful and may sometimes be beneficial. Particles of chromium may provide a local source from which this element, diffusing in the course of high-temperature service life, helps maintain the surface film of Al_2O_3 when the level of aluminum in the coating has become too low to maintain it in the absence of a small amount of chromium. In some temperature ranges chromium improves resistance to sulfidation. However, cement inclusions are at best a very haphazard means of including beneficial elements and they very often provide sites for accelerated local corrosion. More systematic processes are available to optimizing the composition of a coating. Disadvantages of the process include: explosive and toxic hazards in handling the finely divided dry powders used for the cementation pack.

A relatively long process time is due to the thermal mass of the pack and the inclusion of materials such as alumina, which are poor thermal conductors. Because the slow rates of heating and cooling may not match the requirements of the substrate material, further heat treatment may be required to restore its microstructure. The temperature within the pack is not uniform throughout the cycle and the coating thickness depends on the position of a component within the pack, leading to a variation on the order of ±20% within a batch. Despite these disadvantages, the pack cementation process, which can handle a fair variety of sizes and shapes without special jigs or supports, has achieved quite wide acceptance.

Slurry Method

The slurry method avoids some of the disadvantages of pack cementation by applying the same class of ingredients in the form of a slurry. The medium, often organic, contains a small amount of binder. Components are coated with the slurry by dipping, spraying, or brushing, and then oven dried. This process is usually repeated two or three times to obtain a coating of sufficient thickness. The components are then placed in a furnace and heated to an appropriate temperature (typically, 1270 to 1370 K) for sufficient time for the desired amount of diffusion to take place. The atmosphere in the furnace is inert or may contain a halide to provide transport of the aluminum from the dried slurry to the actual surfaces of the components if the slurry contained no halide. After cooling, the components are cleaned, as in the case of pack cementation.

In comparison with pack cementation the slurry method confines the hazards of dry powder usage to the preparation of the slurry. Further advantages claimed arise from the reduced thermal inertia and greater dispersion of the load, allowing more rapid and better controlled heating and cooling with more uniform temperature distribution within the batch being treated. These in turn provide a shorter cycle time with better matching to the heat treatment requirements of the substrate material together with less variation in coating thickness throughout the batch.

Vapor-Phase Deposition

The vapor-phase method to be described here is that pioneered and extensively used by the French enterprise SNECMA and is chosen for comparison with pack cementation and slurry methods, as all three are widely used for the same purpose of protecting parts of gas turbines. Of course, even in the first two methods the coating element is transported t the substrate in the vapor phase, as a monohalide gas, but this is more evident in the SNECMA process, where the source and the substrate are not in direct contact.

In this process the components to be coated are, after degreasing, supported adjacent to, but not in contact with, perforated baskets containing the source material. For aluminizing by outward diffusion the low-activity source is a 70:30

Cr/Al alloy in the form of 1- to 8-mm granules (i.e., not in the form of powder and without additions such as the alumina used in other processes). The halide source, ammonium fluoride, is separately accommodated nearby. Processing is performed in an atmosphere of argon. The layout of the components and sources in a cylindrical box is shown diagrammatically in Fig. 9.7. The covers shown on the top right of the diagram are to protect turbine blade roots from aluminizing. For heating, a muffle and furnace are lowered over the loaded boxes. To save time the furnace may be preheated to some 1070 to 1170 K before being lowered into position. The temperature is then raised for the diffusion process, the precise time and temperature required depending on the substrate alloy and desired thickness of coating (e.g., 3 h at 1420 K) produce a 60-μm coating on an In 100 substrate with properties shown in Fig. 9.8. It can be seen that the coating is entirely free of cement particles, the chromium present near the surface being that which has diffused from the substrate and the precipitates in the internal zone being substrate constituent displaced from solution by the formation of nickel aluminide.

At the end of the diffusion cycle the furnace may be raised, leaving the muffle in place to preserve the inert atmosphere until the temperature has fallen to some 470 K. Used in this way the process can provide heating and cooling rates satisfying

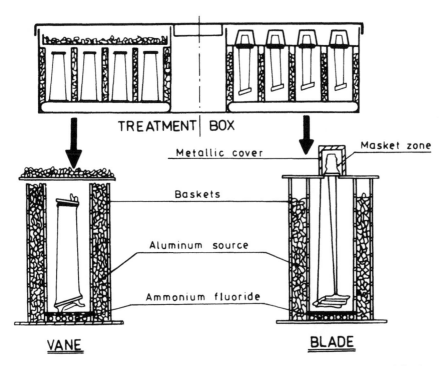

Figure 9.7 Schematic view of components and tooling for vapor-phase aluminization. (From Honnorat and Morbioli, 1984.)

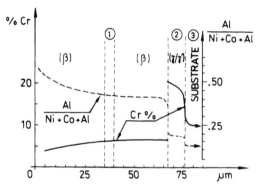

Figure 9.8 Elemental concentrations for vapor-phase aluminized IN-100 specimen versus depth. (From Honnorat and Marbioli, 1984.)

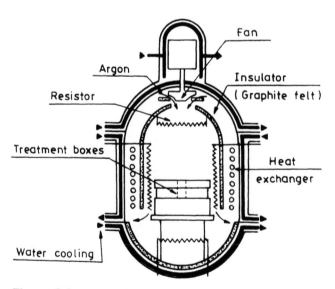

Figure 9.9 Cold wall furnace. (From Honnorat and Marbioli, 1984.)

the requirements in respect to microstructure of many alloys. If very high cooling rates are required, they can be attained by use of a cold wall furnace with fan-assisted circulation of overpressure argon during the cooling phase. This is shown in Fig. 9.9.

Apart from the flexibility and control of heating and cooling rates together with a relatively short overall cycle time, the process has other advantages. The use of a granular material rather than powders for the source and an argon atmosphere rather than the hydrogen frequently required in pack cementation makes the process less hazardous. The coating if free of cement inclusions.

Coatings are of uniform thickness (e.g., ±5 μm) within each batch and from batch to batch because the low thermal inertia of the system permits good control and uniform distribution of the temperature. Coatings are smooth; it is claimed that coatings with a roughness of only some 1.5 μm are obtained on substrates that had been finish machined to a roughness of 3 μm. Coated parts do not require cleaning or other treatment after the vapor-phase diffusion process.

Despite their differences, all three methods are used on a large scale to produce aluminide coatings on nickel- and cobalt-based alloys and also to renew coatings approaching exhaustion when the underlying component is still undamaged and has creep life in hand. For renewal it is usual to strip off the remains of the old coating by chemical means before replacing it by the process used initially.

The same processes may be used for chromizing, the source being pure chromium or an alloy of iron and chromium. This is frequently followed by aluminizing the combination, being of most value when the substrate alloy is low in chromium. It will be remembered that chromium prolongs the period for which a given initial quantity of aluminum can maintain a protective layer of Al_2O_3 and also that Cr_2O_3 has better resistance than Al_2O_3 to sulfide attack at temperatures below 1120 K, particularly if NaCl is present to form Na_2SO_4.

A further method of protection involving diffusion consists of electroplating the substrate with the protective element, which is then diffused into it by heat treatment. This rather simple and direct method has been used successfully by Brown Boveri (Felix, 1977) for components of industrial gas turbines. Although in this case chromizing by the methods described earlier could presumably have produced a similar result, prior deposition of a particular element or elements galvanically or by sputtering can be a valuable adjunct to the diffusion processes. Thus coatings with enhanced corrosion resistance have been produced by electrodepositing a thin layer of platinum (about 10 μm), followed by aluminizing by a conventional diffusion process. The platinum and aluminum interdiffuse during this stage, and subsequent heat treatment can produce a variety of structures, of which the most useful is that in which the outer zone is rich in platinum and aluminum and the lower zone is nickel aluminide.

9.5.3 Overlay Coatings

Although diffusion coatings are relatively cheap and simple to produce and are still enjoying a successful career, they have serious disadvantages and limitations. Their production inevitably involves modifying the composition of the substrate to a depth at least equal to the thickness of the coating. The mechanical properties of the modified zone are much inferior to those of the substrate and contribute little or nothing to load bearing, so that some increase in substrate thickness may be required at the design stage to compensate. As the diffusion coating is partly formed from substrate material, its optimization is limited by what is available in the substrate.

Finally, inasmuch as the methods used chiefly to produce diffusion coatings depend on vapor-phase transport from source to substrate, they cannot in practice be used to provide a coating containing materials with very low vapor pressure, such as yttrium and ceramic substances.

Overlay coatings offer much greater freedom in regard to their composition and should have minimal effect on substrate composition and strength if adhesion is provided by a mechanical bond and only a small effect if limited interdiffusion is used to secure an intermetallic bond.

Overlay coatings can be produced by a variety of processes, including electrophoresis, plasma arc spraying, electron-beam evaporation, and ion plating, the last two methods becoming commonly employed.

Plasma Arc Spraying

In this the coating materials are transported to the substrate by a high-velocity stream of gas which has been heated by a sustained plasma discharge in a dc arc. The arrangement of an arc plasma gun is shown schematically in Fig. 9.10. The carrier gas is inert, typically argon, or sometimes contains some hydrogen to render it reducing. The coating materials in the form of a fine powder are introduced to the main stream by a subsidiary flow of carrier gas. The particles are small, less than 60 μm, and are melted in the high-temperature gas stream before reaching the substrate, where molten particles form splats that fill in irregularities in the pre-

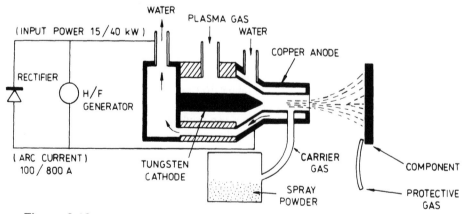

Figure 9.10 Arc plasma gun. (Reprinted with permission from Springer Verlag GmbH.)

roughened substrate surface to provide a mechanical bond, and as the coating builds up, to densify it. Provided that the compositions of substrate and coating are compatible, an interdiffusion bond can be produced by heating the component during spraying or by subsequent heat treatment. Although this results in modification of the substrate in the interdiffusion zone, the volume of material affected is less than when coatings are produced entirely by diffusion.

The temperatures produced in the plasma gun are very high, in excess of 15,000 K, so that there is no limit to the range of materials that can be applied. In particular, it is possible to include yttrium in the coating by this method, and coating powders frequently include about 1% of yttrium to improve oxide adhesion. The majority of coatings are entirely metallic and comprised of a major component which provides plasticity, together with chromium, aluminum, yttrium, and so on, for protection. Such coatings are frequently designated MABC, where M is Ni, Co, and Fe according to the nature of the substrate and ABC represents the protective elements. A typical formulation or a Co-Cr-Al-Y powder might be 60:25:14:1 by weight, and in the absence of oxidation, the coating produced would consist of these elements in much the same proportions. Spraying metallic compositions is usually performed in vacuum or with an inert gas shield to avoid oxide inclusions in the coating. The former requires the process to be carried out in a chamber with airlocks for insertion and removal of the components, together with a system of remote, and preferably automated, control of the movements of the components and the gun during spraying to provide a uniform coating of the desired thickness over the entire area to be protected. Relative movement of component and gun is necessary to secure full coverage, as spraying is essentially a line-of-sight process.

Refractory oxides used to provide thermal barrier coatings can also be applied by the arc plasma gun. Despite their high melting points (zirconia 2953 K) the particles of powder melt readily in the hot gas stream and splat out and bond at the substrate in the same way as metal particles. With these materials spraying can, of course, be performed in a normal atmosphere of air either manually, relying on the sprayer's skill for a good coating, or with automatic control.

While the plasma arc spray has the potential to produce excellent coatings, it is not altogether easy to achieve them. Preparation of the substrate includes cleaning, roughening by such means as vapor blasting with an inert abrasive (e.g., Al_2O_3), and very thorough final cleaning to remove abrasive residues from the surface to ensure good adhesion of the coating. Spray parameters must be carefully controlled to minimize the porosity and surface roughness, which tend to characterize the coatings produced by a process that is essentially the agglomeration of minute particles. Porosity is reduced insofar as the particles are fine, are completely melted during their transport in the hot gas, and impinge on the substrate with sufficient velocity, but even so, further treatment after spraying is required with metallic coatings intended to provide corrosion resistance.

Shot peening with glass beads is frequently employed to densify the coating and tumbling in an inert abrasive such as Al_2O_3 to produce a good surface finish. Aluminizing or chromizing by pack cementation is sometimes used after arc plasma coating as a means of consolidating the coating.

More elaborate postcoating treatments that have been explored (Burman, 1984) include electron-beam remelting, laser glazing, and hot isostatic pressing. The latter process was performed by placing the coated component in a glass vessel which was then evacuated, placed in a press, heated to 1270 K, and subjected to a pressure of 250 MPa for 1 h. It was then removed from the press and the glass peeled off. This process, applied to an initially porous and rough coating of plasma-sprayed Fe-Cr-Al-Y, converted it to a smooth dense coating in uniformly close contact with the substrate, as evidenced by the interdiffusion between coating and substrate. There was no evidence of diffusion from the glass envelope into the coating. Despite its complexity, or cost, hot isostatic pressing was considered to be one of the most promising treatments for plasma-sprayed coatings intended to provide protection from chemical attack.

The very high temperature of the arc plasma spray enables it to be used to produce thermal barrier coatings of refractory oxides. With these materials, oxidation is of course no problem, and some degree of porosity can improve their performance in the role of thermal barriers. The very low thermal conductivity of zirconia (2 $Wm^{-1} K^{-1}$) make it valuable for this purpose. The partially stabilized form is normally used, applied directly to the substrate or on top of an intermediate layer.

Electron-Beam Evaporation

In this process, overlay coatings are produced by heating the coating material in vacuum by means of a beam of high-energy electrons sufficient to vaporize it. The vapors then condense on the substrate, which is itself heated and positioned above the pool of molten coating material. Figure 9.11 illustrates the process and equip-

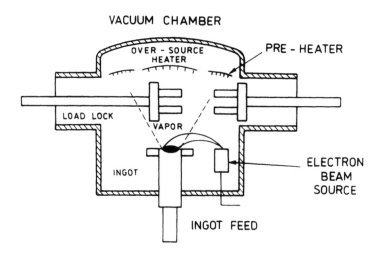

Figure 9.11 Electron-beam evaporation. (After Restall, 1979.)

ment schematically. Components to be coated are preheated and maintained at a high temperature to provide sufficient bond. They are rotated mainly to secure a uniform coating, as otherwise the surface nearest the source would have the thickest coating and shield other surfaces from the vapor. Rotation and accompanying translation also reduce the effect of varying composition within the vapor cloud. After cooling in vacuum the components may be further processed by peening with glass beads and heat treatment to improve the structure of the coating.

Very high temperatures can be produced by electron-beam heating and refractory oxides can be vaporized, but it is somewhat difficult to maintain the desired composition in the vapor cloud when materials with widely different vapor pressures at the same temperature are being evaporated from a common source. Very satisfactory coatings of the M-Cr-Al-Y type have been produced by this process, but it is currently expensive.

Ion Plating

The several processes known as ion plating are variants of the sputtering technique in which atoms are detached from the surface of a target composed of the coating material as a result of momentum exchange with ions that strike it. Bombardment is almost invariably with argon ionized at low pressure by collision in an electron stream and a number of sources or ion guns are known. Material leaving the target travels to the substrate to be coated, which is maintained at a negative potential relative to the source. The rate at which material is removed from a source depends on the magnitude of the ion flux and the energy of the ions striking it and also on the binding energy of the material. Thus with an alloy source there is initially some preferential removal of the element with the lowest binding energy, but this is immediately countered by depletion of this element at the surface, and removal proceeds similarly with the other elements in turn until the outermost atomic layers of the source have been removed, after which the composition of the material removed is constant and the same as that of the source from which layer after layer is removed. Thus the difficulty in controlling the composition of the coating, which is found with electron-beam evaporation, does not arise with ion beam plating, in which the steady state is reached almost instantaneously. This extends the range of materials that can be deposited by this method. The process has the further advantage that it is possible to accomplish a final and very thorough cleaning of the substrate immediately prior to coating by subjecting it to ion-beam bombardment after its insertion in the coating chamber simply by setting the applied voltages so that the argon ions are directed to it. In this way, by mild ion milling all contaminants on or in the immediate surface are removed.

The type of equipment that may be used for ion beam plating is shown schematically in Fig. 9.12. The process in its present state of development is rather slow and expensive, primarily because the number of components that can be treated at one time is small. It has, however, been used to produce excellent coatings.

A coating different in composition and method of application from those already mentioned is described in a paper by Tamarin and Dodonova (1977). It was developed with the aim of protecting turbine blades from sulfides and vanadium

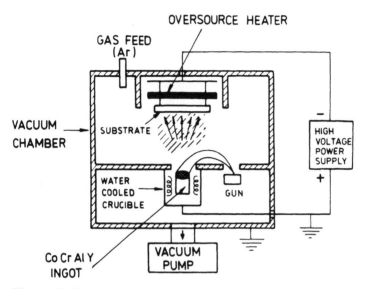

Figure 9.12 Schematic illustration of ion plating process. (From Nakamori et al., 1983.)

oxides in the combustion products of aviation fuels. The substrate material was a Ni-based alloy of 9.5 to 12% Cr, moderate additions of W, Mo, Co, and Ti, with 5 to 6% Al. It was aluminized and then given a second treatment by coating it with a slurry composed of MgO, Al_2O_3, and SiO_2, with a small addition of powdered NiAl. The latter was incorporated to improve thermal fatigue resistance, as NiAl was a major phase in the surface zone of the aluminized substrate. The coated samples were then heated in a muffle furnace at 1450 to 1470 K for 3 to 5 min. The thickness of the coating produced in this way was 60 to 80 μm.

Thermal fatigue resistance was evaluated by heating wedge-shaped samples to 1370 K in a flame of burning aviation fuel followed by cooling to 470 K in a blast of compressed air, the total cycle time being 3 min. The criterion of failure was the appearance of a surface crack or spalling. Aluminized samples survived 80 cycles, and aluminized and coated samples survived 100 cycles.

Corrosion resistance was tested by rotating samples immersed in a molten mixture consisting of 85% V_2O_5 and 15% $NaSO_4$ at 973 K for 24 h. After this test aluminized samples appeared to be badly corroded and microscopic examination showed a surface of ragged oxides with oxide penetration deep into the material. In much of the aluminized zone the content of Al was reduced to 15%, with 28% Al retained in the lower zone of NiAl.

The aluminized samples with the additional coating looked good, and examination showed an unbroken layer composed of NiAl and spinels of $MgOAl_2O_3$ type and glass.

In further tests under in-service conditions, aluminized blades showed signs of

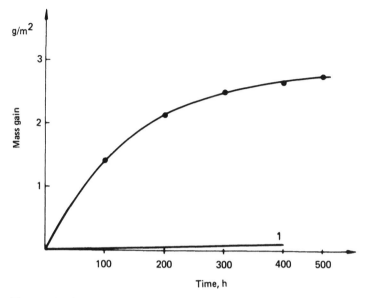

Figure 9.13 Corrosion resistance of an aluminized specimen (1) and of an aluminized and coated specimen (2).

corrosion after 100 to 300 h, and those with the additional coating were entirely unaffected. Figure 9.13 shows the relative performance in terms of weight change per unit area, which indicates the low interaction between the glass coating and environment.

9.6 TESTING COATINGS

While the use of sundry forms of protective coatings has proceeded successfully on a large scale for many years, their development continues intensively. It is aimed at increasing the life of components operating in existing conditions and at making it possible to use metal components in still more arduous conditions. It is a complex task partly because frequently the materials to be protected are themselves being developed simultaneously, and mechanisms of high-temperature corrosion are still being investigated, so that the coating designer's data base is subject to change. The most serious obstacle to progress is the length of time required to fully establish the merits of combinations of coating formulation, method of application, and base material. It is usual to screen the possible combinations by exposure to exceptionally severe thermal cycling and to abnormally high concentrations of corrosive chemicals and to restrict prolonged testing under more realistic conditions to those which best survived the accelerated life tests. This brute-force method of selection is sometimes misleading, but its use seems inevitable because of the cost and delay

Table 9.1 Protective Coatings Used in the Laboratory and Operation Tests

Type of coating	Procedure	Chemical composition (%)							
		Si	Al	Cr	Mo	Ni	B	Co	Y
Chromized	Diffusion			45–80					
Aluminized	Diffusion		30						
Chrome–aluminized	Diffusion		30	15–35		25–37			
L DC 2	Diffusion	Duplex coating: Pt-Al							
Elcoat 360	Diffusion	Duplex coating: Ti-Si							
Ni-Cr-Al-Y (ATD 1)	Plasma spray (vacuum)		11	38		Balance			0.25
Co-Cr-Al-Y (ATD 2)	Plasma spray (vacuum)		12.5	25				Balance	0.35
Ni-Co-Cr-Al-Y (LN 20)	Plasma spray (atmosphere)		7.5	29		41		22	0.5
Co-Cr-Al-Y	Plasma spray (atmosphere)		13	23				Balance	0.6
Ni-Cr-Si-Mo-B	Plasma spray (atmosphere)	4		16	2	70	3		

Source: Thien, 1984a.

involved in a genuine life test. Progress already achieved means that the lives now being sought are typically thousands of hours for aircraft engine components and many tens of thousands of hours for land-based gas turbines. The equipment for laboratory testing under conditions simulating the thermal cycles, gas velocity, temperature, and composition is expensive and scarce. Simpler apparatus, such as that described in Chapter 5 is employed more, and once the coating and base have established a measure of reliability, testing actual equipment in service is begun. This has the advantage of complete realism, but with in-service testing the opportunities of inspecting the components are limited. Even in laboratory testing it is not possible to observe a component closely, continuously, and directly, while it is in the test rig and at high temperature, but each removal for examination adds a thermal cycle to the test and may itself be the cause of any crack that is found. To detect the onset of cracking due to extension alone, as for example in a hot tensile test simulating creep of substrate, use may be made of the sound (acoustic emission) that accompanies the occurrence of the crack, this being conveyed by a suitable acoustic conductor to a transducer in a cool zone for amplification, discrimination, and display, as described in Chapter 5.

The problem of durability is acute in the case of land-based gas turbines used for power generation and marine turbines. For economic reasons such equipment runs on relatively low-grade fuel and must operate continuously for very long periods. The development and proving of suitable protective coatings for the blades of these turbines is a lengthy and expensive procedure involving prolonged testing. An example is provided by the work of this nature which has been carried on continuously since the early 1970s at Kraftwerk Union Ag (Mühlheim, Germany). A valuable account of some of these investigations may be found in (Thien et al., 1984a,b), and the second of these papers is discussed briefly here.

The coatings used to provide protection and the methods used to deposit them are shown in Table 9.1. Laboratory tests were conducted, at temperatures corresponding to service use, in an atmosphere simulating the turbine gas and consisting of air, synthetic ash, and 0.03% SO_2/SO_3 by volume. The specimens were in the form of disks 15 mm in diameter and 3 mm thick. The condition of the surface was monitored almost continuously, the appearance of the first macrocrack being used as the criterion of failure.

The protective coatings were also tested on turbine blades in actual service use, where they were exposed to real turbine gases and thermal and mechanical stress. In these cases their surfaces were inspected at intervals of 5000 h during normal routine maintenance.

The results of both laboratory and operational service tests are shown in Fig. 9.14, the base material in each case being the alloy Udimet 520, having the composition shown. Since defects that arose in service use could only be discovered and recorded during the 5000 h maintenance intervals, although they may have occurred beforehand, the agreement between laboratory and in-service tests is probably even better than it appears. It can be seen that the chromized coating obtained by diffusion gave the best protection. The coating of ATD1 deposited by plasma spray in vacuum showed durability comparable with or better than the diffusion coatings other than

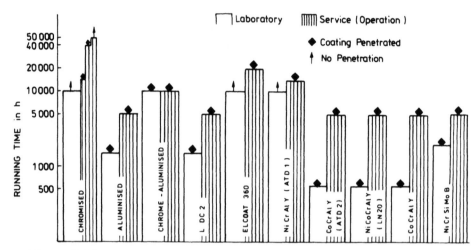

Figure 9.14   Corrosion resistance of various coating systems on Udimet 520 (57% Ni, 19% Cr, 12% Co, 6% Mo, 1% W, 2% Al, 3% Ti, 0.05% C, 0.005% B). (From Thien et al., 1984a.)

the chromized coating, showing that chemical composition and not method of application alone determined a coating's durability. However, all the coatings deposited by plasma spray in the atmosphere failed very quickly. It was found to be due to the formation of oxides during deposition; these constituted sites for the initiation of subsequent corrosion.

The reader will have realized that Fig. 9.14 presents only one aspect of a very wide investigation which included studying the vital mechanical properties (including creep resistance) of a number of alloys potentially suitable for the turbine blades, together with a detailed examination of the effects of heat treatment. Their resistance to corrosion at high temperatures was also studied, and although the alloy adopted was better overall than the others investigated, it was nevertheless necessary to develop a "protective coating/base material combination" to provide the required durability in service.

REFERENCES

Burman, C. (1984). Thesis, Institute of Technology, Linkoping, Sweden.

Carling, R. W., Bradshaw, R. W. and Mar, R. W. (1983). *J. Mater. Energy Syst.*, **4**: 229.

Dupre, B., Steinmetz, P. and Morbioli, R. (1981). *Proc. International Conference on High Temperature Corrosion*, San Diego, Calif.

Endres, W. (1974). *Proc. Symposium on High-Temperature Materials in Gas Turbines*, Baden, p. 1.

Felix, P. C. (1977). *Brown Boveri Rev.*, 40.

Gauje, G. and Morbioli, R. (1983). *Proc. AIME Symposium on High-Temperature Protective Coatings*, p. 13.

Honnorat, Y. and Marbioli, R. (1984). *Proc. Conference on Coatings for Heat Engines,* Acquafredda di Maratea.

Nakamori, M., Harada, Y. and Fukue, I. (1983). *Proc. AIME Symposium on High-Temperature Protective Coatings,* Atlanta, p. 175.

Restall, J. E. (1979). *Metallurgia, 46*: 676.

Tamarin, J. A. and Dodonova, R. N. (1977). *Zashch. Met., 13*: 114.

Thien, V., Schmitz, F., Slotty, W. and Voss, W. (1984a). *Scanning Electron Microsc., 4*:1629.

Thien, V., Schmitz, F., Slotty, W. and Voss, W. (1984b). *Z. Anal. Chem., 319*: 646.

Weronski, A. (1980). *Improvement of Resistance of Steel Moulds to Thermal Fatigue,* Report.

10

Matters for Consideration in Designing Equipment Exposed to Thermal Fatigue

10.1   INTRODUCTION

Thermal fatigue is only one element in the complex of problems that may face a design engineer, but it is of increasing importance. Constantly growing demands for new or better products frequently lead to intensified thermal regimes in industrial processes, while the trend to their integration and operation on a continuous basis vastly increases the cost of interrupting production by breakdown or the need to replace components. This also applies in the case of the very large individual units that are now constructed in the interest of technological efficiency—blast furnaces with a volume of $5000 \, m^3$, convertors handling 400 to 600 tons of metal at one time, ladles and mixer-type ladles with a capacity of, say, 500 tons, and giant presses for hot forging. The highly efficient continuous casting process combines the features of great size and integrated processing. In such equipment the scale factor itself, as well as the required operating conditions, generally contribute to the thermal fatigue problem; the temperature differences across thick sections create intense stresses. Any item that operates at a high temperature may be subjected to significant temperature variation in use and major temperature variation at startup and shutdown. Shutdown, particularly emergency shutdown, usually produces the most rapid temperature changes and the most dangerous stresses. Useful machines and structures are never built simply to get hot and cold. The elements exposed to thermal fatigue must support loads and transmit forces that may be constant or variable and in many instances are also exposed to corrosion and erosion. Hence machinery and structures subjected to elevated and varying temperatures are exposed to many mechanisms of failure, among which pure thermal fatigue may have a dominant or subsidiary, but still significant role. Methods of estimating the durability of components when mechanical fatigue, creep, and thermal fatigue occur simultaneously were discussed in Chapter 6.

In the general case, other potential causes of failure will also be present, ranging from excessive stress due to unnecessary constraints on expansion to simple underdesign. Diagnosis is complex and time consuming so that failures in the field are frequently rectified on an ad hoc basis and only if the part repaired fails repeatedly is consideration given to the overall design. A vast amount of data from such failures still awaits scientific description and explanation.

The material in this chapter, based on laboratory data and industrial experience, is intended as an *aide-mémoire* for the designer and user. The specific examples and case histories described will serve a more general purpose if they highlight the importance of awareness of thermal phenomena when examining any initial design concept from every aspect, with an eye alert for the unexpected contingency and the obvious flaw. The purpose of the specific examples and case histories described is to highlight the general need for awareness at the design stage of all the thermal phenomena that may be encountered when the design is put to use. They also show that the measures that may be taken to mitigate the effects of thermal fatigue are frequently different from, and sometimes contrary to, those that would be appropriate in dealing with mechanical fatigue or for the purpose of increasing the strength and rigidity of a component and may sometimes constrain design into directions contrary to conventional practice.

Design of a structural member subject to a constant or varying mechanical load provides an instance of such a situation. In the absence of temperature variation, the greater the cross section, the lower the stress and the resultant elastic and plastic deformation, susceptibility to mechanical fatigue, and in the case of high-temperature operation, to creep. However, if the temperature of the environment is subject to variation, increasing the cross section is not the royal route to safety. In this case the greater the cross section, the greater the lag between the temperature changes in the interior of the item and those at its surface, and the stresses resulting from differences in thermal expansion may be sufficient to cause failure even in the absence of an external load. The risk is obviously greater with materials of low thermal conductivity and components of complicated form and varying thickness. The likelihood and number of imperfections in the material increases with cross section and inclusions of slag, oxides, sulfides, and silicates; the presence of blowholes, discontinuities of the structure, and the existence of metastable phases are particularly important when thermal fatigue may occur. There are various ways around the impasse arising from the need to increase the strength of a component and the dangers of merely increasing its cross section; several are illustrated in examples later in this chapter. However, for cases where failure in the field rather than design calculations suggest that a component is too weak, it is worth considering that it may merely be too thick.

The process of reaching an optimum design consists of a small number of phases that interact. Factors that may seem relatively unimportant in one phase can have an unexpectedly large effect on another. The designer must pay great attention to all possible phenomena and exercise great caution in dealing with unfamiliar phenomena or in deciding to neglect any factor.

Design of a component starts with an awareness of:

1. Complete data on the environment in which the component will work and its variation during startup and shutdown.

2. The stiffness required of the component, the structure in which it works, and the tolerances allowed.
3. The properties of the proposed material and the changes in those properties throughout the range of temperatures, which include startup, working conditions, and shutdown.
4. The fact that all adverse thermal effects can be diminished by a reduction in temperature.

The importance of this information and the numerous ways in which its parts may interact are rather apparent: for example, knowledge of the thermal environment may necessitate adopting a material other than that originally proposed, while consideration of the raw environment and the properties of economically available materials may dictate protecting the component from part of the thermal regime.

10.2 WORKING CONDITIONS

Working conditions are often difficult to define at the design stage as, for example, technological developments and competitive factors may result in the raw materials or final product of a process plant being other than originally envisaged, with some accompanying change in the mode of operation. Economic factors may lead to operation being more intermittent than had been anticipated. This can be very harmful: It has been found many times that the thermal gradients created when shutting down are more detrimental than long periods of continuous running.

Insofar as it is under control of the designer, intermittent operation should be avoided and integrated plants and production lines designed to provide the most stable operating conditions possible. If a component must be subjected to discontinuous use, it should be designed with an extra factor of safety and mounted so as to facilitate inspection and placement.

A reduction in the maximum temperature attained by a component extends its life both by reducing the intensity of the thermally activated processes contributing to failure and by limiting, in one direction, the excursions in temperature and hence the thermally induced stresses within it during use, startup, and shutdown. Reducing heat flow into the component and increasing heat flow from it serve this purpose.

Heat flow to a component may be reduced by application of protective coatings to its surface by covering the surface with ceramic plates, or by interposing not-load-bearing and usually thin shields between the component and the heat source to intercept radiation or deflect the flow of hot gases. The method most commonly used in industry is the application of a layer or layers of refractory material (e.g., firebricks). Water or water-based coolants are cheap and generally used for the removal of heat, with air and radiation cooling, cooling with liquid metals, fuel or oxidants employed in special cases where water cooling is impracticable.

Where the temperature of the heat source or the heat flux is variable, it is possible and generally advisable to control the rate of cooling in order to reduce the fluctuations in component temperatures. Some examples of measures taken to

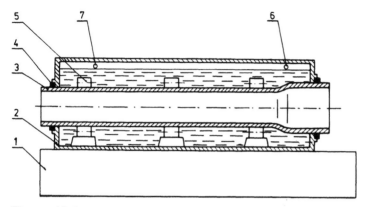

Figure 10.1 Schematic diagram of mold cooling system. 1, Base; 2, casing; 3, steel mold; 4, seal; 5, supporting roll; 6, water outlet; 7, water inlet.

reduce the temperatures at the surface and within a component are illustrated in Figs. 10.1 to 10.5.

Figure 10.1 shows a typical steel mold used to produce centrifugally cast pipes in cast iron, as described in Chapter 7. In this process the flow of cooling water should be controlled so that its temperature is substantially constant and uniform within the cooling box. In Chapter 8 it was shown that a relatively small change in coolant temperature can have a very large effect on heat transfer. In the centrifugal casting machine excessive water temperature reduces mold life, impairs the quality of the castings, and with water of industrial quality, leads to the buildup of mineral deposits from the water on the surfaces to be cooled and a further reduction in cooling efficiency. However, too low a temperature results in a chilled casting, which is unsatisfactory. These considerations lead to the employment of a simple servo-temperature sensor, electronic amplifier, and motor-controlled value to keep the water temperature in a range that is typically 320 ± 10 K. The precise set point is found by checking the quality of the castings produced at production startup, neglecting the first few, whose quality merely reflects transient conditions due to the initially cold mold. The procedure is discussed more fully in the senior author's earlier book (Weronski, 1983).

Figure 10.2 shows a cupola furnace for melting cast iron in the foundry. The shell is water cooled, the flow being controlled in accordance with the stage reached in the production of the melt. Although such furnaces were formerly constructed without water cooling, experience has proved its worth, particularly on the smaller furnaces and in the rehabilitation of older furnaces prone to shell cracking. It is increasingly used in new furnaces, partly because it also prolongs the life of the refractories.

Figure 10.3 shows the application of water cooling to a blast furnace, where it serves a dual purpose. In normal operation the cooling system is used to reduce and stabilize the shell temperature. During startup of the furnace the temperature of the

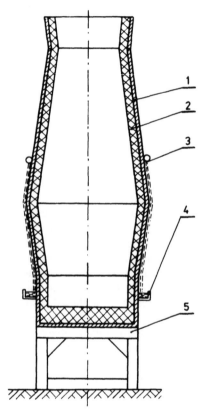

Figure 10.2 Schematic diagram of cupola furnace water cooling system. 1, Shell; 2, refractory material; 3, water nozzles; 4, gutter; 5, frame.

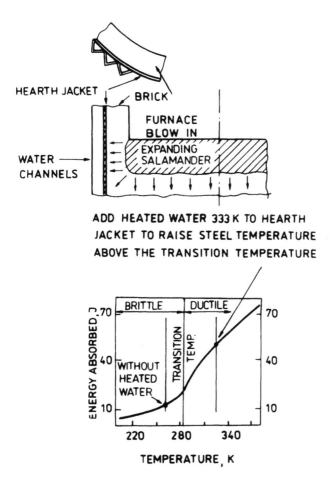

Figure 10.3 Schematic diagram of blast furnace water cooling and temperature dependence of impact strength of shell steel. (From Wei, 1980.)

salamander rises much more rapidly than that of the refractory-lined shell surrounding it, and the stress due to the difference in their expansions can cause the shell to crack. This situation becomes critical in winter, when the temperature of the shell is below the brittle-to-ductile transformation temperature. The problem is solved by passing steam or heated water through the channels provided for cooling and thus raising the temperature of the shell above the transition temperature.

Figure 10.4 shows, schematically, a furnace used for annealing large components. The doors at either end can be raised to allow entry or removal of the component and are lowered when it is in place. While the doors are down, their refractory linings are heated by the furnace gases, some of the heat passing through them to the steel shell, which provides support for the door linings. During periods of loading or unloading the furnace, the doors are in the upper position and the refractory linings and their shell cool off. Thus, in the absence of the water-cooling arrangements shown, the steel shell of the doors is subjected to thermal cycling.

Prior to applying water cooling, thermal fatigue cracking of the door shells necessitated frequent repair by welding, with consequent interruption of production; even so, door life was undesirably short. This problem is eliminated by the very simple water cooling system shown. Automatic flow control serves to keep the steel door shell at a moderate and substantially constant temperature throughout the operating cycles. This also improves the life of the refractory linings of the doors.

It will be appreciated that when a door is being raised, its refractory lining, which may then be at a high temperature, is close to the outer surface of the end of the main steel shell of the furnace itself. The resultant radiant heating of the shell in this

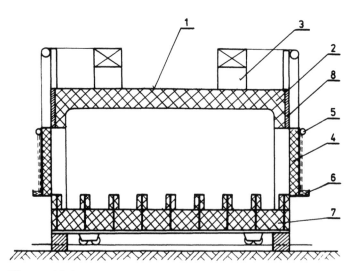

Figure 10.4 Door water cooling system of large annealing furnace. 1, Shell; 2, refractory layer; 3, door power drive; 4, door; 5, water nozzles; 6, gutter; 7, trolley; 8, asbestos layer.

region might be so rapid as to cause thermal shock and would certainly cause local heating and expansion, undesirable in themselves, together with thermal cycling of the shell in this region, as it, of course, cools once the door is lowered and clear of it. As this exposure to radiant heating is of short duration, it is rendered harmless by the interposition of a simple and relatively thin (2 to 3 cm) asbestos heat shield, as shown in the figure.

Figure 10.5 illustrates the cooling of steel ingot molds by air or water mist. Controlled cooling provides the maximum solidification rate appropriate for the chemical composition and purpose of the ingot, and this reduces the time in which the mold material is exposed to the deleterious effects of high temperature and speeds up the production cycle. Cooling is particularly beneficial when killed steel is made as the addition of aluminum to the ladle increases the temperature of the metal poured into the mold and agitates it, which increases the heat transfer. Further cooling of the mold after removal of the ingot also reduces its exposure to high temperature and hastens its availability for reuse.

Cooling is not always used when casting small ingots (e.g., less than 10 tons) but

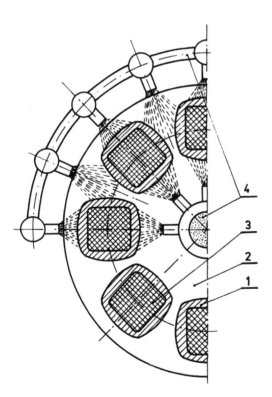

Figure 10.5 Cooling of ingots and ingot molds. 1, Ingot mold; 2, turntable; 3, ingot; 4, manifold with cooling medium.

can still be used to hasten cooling of the emptied mold. The magnitude of the stress cycle to which the mold is subjected during casting is not increased by cooling during solidification, as its maximum is attained when the molten metal first enters the mold.

The effect of the cross section of an ingot mold on its durability was investigated by Deilmann and Schreiber (1974), who found that with the traditional shape of walls (shown on the left of Fig. 10.6), numerous cracks occurred, especially in the corners, cracking being accelerated when killed steels were cast. Local concentration of stresses due to the nonuniform thickness of the walls was identified as a cause of cracking, and molds redesigned to the form shown on the right of Fig. 10.6 were more durable.

Although in the case of ingot molds, cracking was reduced by a small increase in thickness at the corners, this was merely to avoid stress concentrations by maintaining uniformity of cross section. As said earlier, thermal fatigue problems are seldom solved simply by increasing the thickness of the section involved. The next few examples demonstrate improvement in thermal fatigue life by the removal of metal that was not needed for load bearing and created unnecessary restraints on thermal expansion and contraction.

For the first example of the benefits obtainable in this way, we return to the mold for centrifugally cast pipes. In Chapter 7 we saw that the stresses which lead to cracking in the surface zone of the mold arise from the temperature differences within the material, which are the result of delay in the propagation of temperature changes across the thickness of the mold. Continuation of this analysis led to the improved design, shown in Fig. 10.7, in which the thickness of the mold has been reduced by over 40%. The reduction in thickness desirable with respect to thermal fatigue is limited by mechanical factors, particularly the forces the mold must sustain when centrifuging the liquid metal and the need for stability when getting up to speed, running, and slowing down. To satisfy the requirements, the mold is

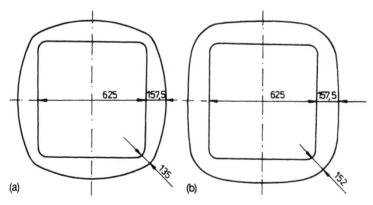

Figure 10.6 Redesign of ingot mold corner: (a) old design; (b) new design.

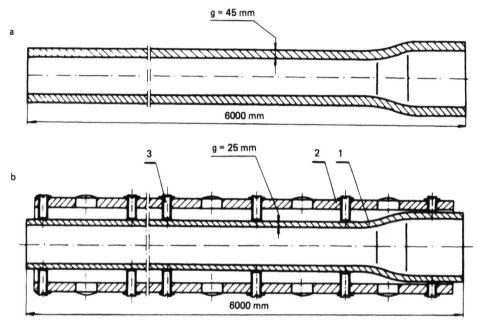

Figure 10.7 Steel mold for centrifugal casting: (a) orthodox design; (b) redesigned mold. 1, Thin-walled mold; 2, casing; 3, supporting element. (From Weronski, 1976.)

supported at intervals along its length by a water-filled casing which revolves with it within the main water box. This imparts sufficient rigidity to ensure that the deflection of the tapered mold under load does not exceed 0.5 mm at any point along its 6-m length. The taper, to facilitate extraction of castings, has some effect on buckling strength and is kept to 1.5 to 2 mm total change in diameter, which is sufficient for the purpose of extraction. Orifices in the supporting casing connect the inner water jacket with the water box proper and are so shaped and positioned as to provide optimum circulation. To avoid chilled castings, the water temperature is set to approximately 10 K higher than would be used with a 45-mm-thick mold. The reduction in stress with the thinner wall more than offsets the effect of the higher temperature at the outer surface. In service, molds of the new design have outlasted the old pattern by a factor of 2 (i.e., 44% off the thickness has added 100% to the life).

A problem that has frequently been encountered is the cracking of steel rolls used to transport items to, through, and from furnaces. On hardening lines sheets that may be only 2 or 3 cm thick are passed through tubular furnaces on steel rolls 15 cm in diameter. Cracking often results from the surface of the roll coming into contact with the item transported, which is at a different temperature, as when a cool sheet comes onto hot rolls inside the furnace or a hot sheet is carried away on cool rolls. With solid rolls the large mass of metal comprising the core is slow to follow the temperature change at the surface in contact with the load, and it resists dimensional change in the surface region, so that high stresses are induced there. Repetition of this process leads to cracking.

This problem has been solved by using hollow, tubular, rolls, which is practicable, as load on the rolls is not great. The sheets have merely to be supported and moved along. It has been found in practice that where the solid rolls lasted only a few months, the hollow rolls last two or three times as long and are also cheaper to make.

A somewhat similar problem and solution can be found in the case of the rolls in the secondary cooling zone of a continuous casting plant (see Fig. 10.8). Gochfeld and Tscherniavski (1979) state that solid cylindrical rolls failed by thermal cracking after 2 to 4 months, corresponding to handling some 1000 heatings. The severe stresses that caused the cracking arose partly from expansion of the surfaces of the rolls in contact with the hot slab being restrained by the cooler interior portions of the rolls, which expanded less; and partly by differences in temperature along the length of the roll. The end sections of the rolls are not in contact with the hot slab and hence are cooler than the main central sections of the roll, which do press on it. Thus the circumferential expansion of the hot central sector of the roll is not matched by the expansion at the cooler ends, the stress being concentrated where these parts abut (i.e., where the edges of the slab are in contact with the rolls). The stress in this region of the roll surface would not be reduced by removing metal

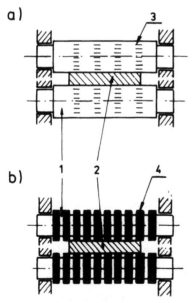

Figure 10.8 Rolls in slab transportation system of continuous caster: (a) traditional design; (b) redesigned system with rings shrunk onto rolls. 1, Roll; 2, slab; 3, crack locations; 4, ring.

Figure 10.9   Cross section of typical wagon wheel showing plate contour. (From Haley et al., 1980.)

from the center of the roll, as the temperature gradient causing the stress is longitudinal rather than radial. Moreover, hollow rolls would not be sufficiently strong in this application, in which a thick and heavy slab is being plastically deformed.

The solution is found by using composite rolls of the form shown in the figure, the working surface being provided by rings that are a shrink fit on a solid core. This reduces the restraints and hence the stresses by eliminating rigid metallic connection, longitudinally and radially, between large volumes of metal at different temperatures. Rolls constructed in this manner outlasted solid rolls by a factor of 2. To avoid slippage and fretting corrosion, the shrink fit of the rings to the roll must be carefully toleranced; an alternative that is used successfully is to bond the rings to the roll with high-temperature epoxy resin.

The cast steel wheels on which railway wagons run and by which they are braked provide another example in which thermal cracking is avoided by minimizing restraints on dimensional change during heating and cooling. The necessary compliance is provided where it is needed by keeping the thickness of metal in that region to the minimum consistent with safety and by giving it a contour that permits deflection in the required direction, as shown in Fig. 10.9. A wheel is a single casting comprised of the tread, rim, and flange at the circumference joined to the hub by the somewhat thinner section of metal, which is known as the plate. During braking, friction with the brake shoe raises the temperature of the tread and rim and their expansion is restrained by the plate, which despite some heat flow into it, is at a much lower temperature. The hub, of course, remains cool. In these circumstances the stress in the tread and rim is compressive and, provided that the temperature and stress are moderate and plastic flow does not take place, the wheel will return substantially to its original condition after cooling. That is, it will have experienced a stress cycle but will not contain residual stresses or strains of thermal origin. However, if braking is heavy and prolonged, as it may be in descending a long incline, the rim temperature may rise considerably, to perhaps 770 to 870 K, and the compressive stress is very great, so that some creep or plastic flow may occur. When this has happened, the rim in subsequent cooling will contract toward a diameter somewhat less than it had originally, but as the plate prevents this, the rim will be subjected to a tensile stress that may produce cracking. The crack will then be extended by the tensile stress in the course of further thermal cycles until failure occurs. If cracking does not occur initially, it may occur after a number of cycles, in the course of which the tensile stress builds up. This pessimistic picture assumes that the plate imposes excessive restraint on the expansion and contraction of the rim, and such failures do occur, although fortunately with modern wheels systematically checked in service, they are rare. The point is that an overly strong plate is a contribution to failure.

The reduction in thickness in order to avoid excessive rigidity is subject to some limitations, particularly in the case of castings, as the thinner the cross section, the more serious is the effect of flaws and inclusions, but compliance is determined by shape as well as by size.

Figure 10.10 illustrates this, a slight straightening of the plate permitting an increase in wheel diameter, and vice versa. In a recent study of improved design of railcar wheels, Haley et al. (1980), using computer modeling and full-scale laboratory testing, showed that optimizing the wheel geometry was the most fruitful of the various means they considered for reducing the stresses due to temperature changes in the braking cycle. Their old and new designs are shown in Fig. 10.10, the new and more compliant design having a greatly improved life under severe testing.

Figure 10.11 shows the overhead crane used in a steelmaking plant. The winch trolley runs on rails supported by a girder frame that spans the foundry and itself moves on rails, enabling it to traverse the length of the building. The crane is used to move large ladles of molten metal, and the girders supporting the trolley and its load are I sections fabricated from thick sheet steel. As originally made, they were conventional in form, the top and bottom flanges of the I being parallel to each other. With this form of construction the lower flange of each beam is in tension due to the weight being supported. When a full ladle is raised, the lower flanges of the beams immediately above it are heated by intense radiation from its contents. When the ladle is moved laterally by the trolley, or lowered, the section of the flanges that had been heated cools down. In the course of service serious cracks developed on the lower flanges of the I beams, sometimes penetrating the web. Despite frequent repairs the beams had to be replaced every year. The mechanism causing the cracks

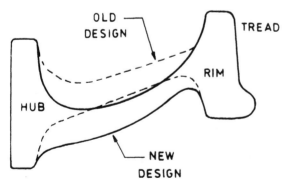

Figure 10.10   Old and new design of railway wagon wheels. (From Haley et al., 1980.)

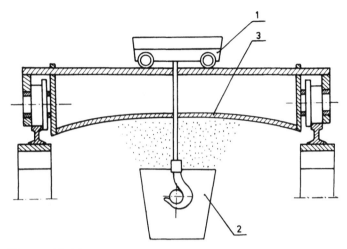

Figure 10.11   Redesigned overhead crane. 1, Winch trolley; 2, ladle; 3, I beam.

appears to have been the familiar one in which during heating, the temperature of the surface nearest the heat source rises more rapidly than that of the metal behind it. Hence the surface zone is in compression as its expansion is restricted by the initially cooler metal below it. The metal in the interior does, however, become hot as a result of heat flow through the surface, and it, too, expands. When the heat source is removed, the surface zone cools rapidly and is subject to tensile stress as its contraction is prevented by the still hot and expanded metal below.

Thus, during cooling, the stresses due to the mechanical load and to the thermal load were both tensile and their combined magnitude was sufficient to cause cracks after a small number of cycles, the tensile stress in subsequent cycles causing them to extend rapidly. The author's solution was to fabricate the I beams in the form shown in Fig. 10.11. Prior to assembly, the lower flange was cold rolled to the form of a shallow arc and was then welded to the bottom edge of the web, which had been cut to the corresponding profile. With this construction the lower surface of the bottom flange was in compression, which could not be relieved by flattening under load because attachment to the web prevented longitudinal extension, which also prevented the assembly run. During the heating portion of the thermal cycle the compressive stress in the flange was increased by the thermally induced stress, but as compressive stress is far less dangerous than tensile stress with regard to cracking, this caused no damage. In the cooling phase the thermally induced tensile stress was opposed by the compressive stress in the flange. As a result, no cracks have appeared in the course of several years.

In major items of equipment, troubles seldom come singly or from a single cause. The high cost of such equipment and the need to interrupt production as little as possible usually precludes a complete rebuild to an improved design and specification; adequate performance has to be secured by specific remedies applied where essential and as opportunity allows. This is illustrated by the case of the cement kiln shown in Fig. 10.12. It consists of an inclined tube (1) 170 m long and 5 m in diameter made of medium-carbon steel with a wall thickness of 6 cm. The tube is supported by a small number of rings (2) that rest on rollers, allowing it to be rotated by the driving motor and gears (5) at a rate of approximately one revolution every 2 min. The wet slurry (essentially, limestone, clay, and water) enters from the hopper (3), partially filling the tube to a depth of about 1.6 m, and as it proceeds down the tube is dried and then strongly heated by combustion gases from the pulverized coal burner (4), below which the cement is discharged.

The temperature distribution along the kiln is shown in Fig. 10.13, the upper curve being the temperature of the combustion gases and the lower curve the temperature of the charge, which is, of course, initially wet slurry and farther along the kiln a hot dry powder. The first 40 m of the tube or shell is unlined; the remainder is lined with 15 cm of firebrick. In the unlined portion, stout metal chains are strung between attachment points on the inner surface of the tube. At one stage in the rotation of the kiln, a chain will hang in a catenary exposed to the combustion gas and on further rotation of the kiln will be dragged through the slurry, or if someway down the kiln, through the clinker formed as the slurry dries. The chains serve several purposes: As heat exchangers they transfer heat from the hot gas to the wet slurry, hastening its drying before emerging from it to be reheated in the combustion gas. They also stir the slurry, and in the hotter zone, pulverize the dried slurry. They are thus exposed to temperature variation, to a complex of varying mechanical stresses, and to abrasion.

At the time when the author was invited to propose remedies, the problems were:

1. Chains broke and had to be replaced after 3 to 6 months of service.
2. Cracks developed in the tube in the sections between the rings and had to be repaired by welding.
3. Tube cracking also occurred in the vicinity of the rings, and in this region the fireclay lining frequently became detached from the shell and broken.

The latter was the most serious, as repairing the lining is an expensive process that interrupts production for a considerable period. Prompt repair is, of course, required, as detachment of the lining leads to serious overheating of the shell in the vicinity when it is exposed to the hot gas, followed by a greater-than-normal fall in

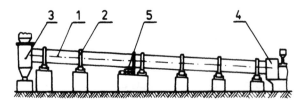

Figure 10.12 Schematic diagram of rotary cement kiln. 1, Shell; 2, supporting rings; 3, charging machinery; 4, heating stove; 5, driving mechanism.

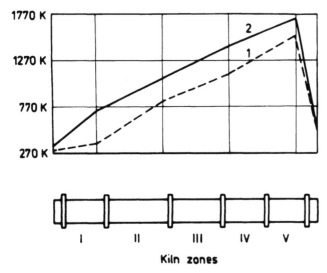

Figure 10.13 Temperature distribution along the kiln. 1, Temperature of charge; 2, gas temperature.

temperature when rotation of the kiln submerges the overheated area in the cement powder, which shields it from the gas. These factors accelerate cracking to unacceptable proportions.

The chains had been constructed of alternate links of malleable cast iron joined by welded links of low-carbon steel. The ultimate tensile strength of the malleable cast iron is 200 N/mm^2. The stresses in the material of the chain depend on many factors, with random characteristics (e.g., the temperature and mechanical properties of the portion of slurry or clinker that a link encounters at a particular moment). Hence they must be treated statistically. Some maximum values of the stresses of different origins can be estimated for applied links that due to the chain's weight and adhesion of dry slurry on links, 15 N/mm^2; that due to dynamic drag in the slurry, 73 N/mm^2; and that due to temperature gradients in the hot zone, 40 N/mm^2. Fortunately, these stresses do not all add up simultaneously, but the probable variation in net stress is a sufficient proportion of the ultimate tensile stress to account for the failures, which of course occurred in a somewhat corrosive environment.

Clearly, the functions of the chains left no scope for ameliorating their working conditions, and this was a case where better materials were essential. Welded chains of H 13 JS steel were specified for the hot zone. This Cr-Ni heat-resisting steel has high tensile strength at elevated temperatures. Its specification does not include thermal fatigue resistance, but laboratory tests showed that this, too, was good. For the cooler zone, welded chains of the less-expensive low-alloy 25 HGHM steel were specified. Chains of this C 0.25%, Cr 1.1%, Ni 15%, Mo 0.2% material were hardened and tempered and before installation were strained to introduce work hardening. These changes in specification increase the life of the chains by a factor of 6.

Shell cracking and lining detachment in the vicinity of the rings was more difficult to correct. The shell was not welded to the rings, which was probably good, as it avoided a restraining connection between items, which at startup would certainly have been at quite different temperatures, the shell heating first and more rapidly. The shell was also not a very close fit within the ring. The result is shown (with some exaggeration for clarity) in Fig. 10.14. Under its own weight and that of its contents, the currently lower portion of the shell deformed to make contact with the ring over the arm diameter, which was approximately 210°. As the ring rotated, successive portions of its circumference were distorted in this way, the previously distorted section resuming a substantially circular form, as the deflection was within elastic limits. The distortion described probably contributed to lining detachment, but there was also a further more localized and acute distortion. To permit rotation the ring is supported on rollers, with which it is (nominally) in line contact. Under the high pressure the ring distorts locally, and as the shell is in contact with the ring in this region, it too is distorted. As before, the distortion affects successive portions of ring and shell as the kiln rotates. This distortion was particularly damaging.

The remedy that was quickest to apply was to weld short lengths of stiffening bars around the circumference of the shell at each place where it was supported by a ring, the internal diameter of the ring being increased suitably to accommodate them. This reduced the trouble but did not eliminate it entirely. The next step was to replace the rings with new ones cast in the form shown in Fig. 10.15. It can be seen that the gaps between the projections of the shell now constitute slots engaged

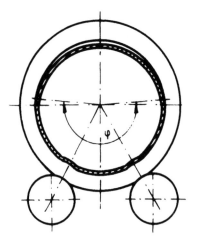

Figure 10.14   Schematic diagram of ring and shell deformation-deflection exaggerated for clarity. (From Weronski and Kielbinski, 1983.)

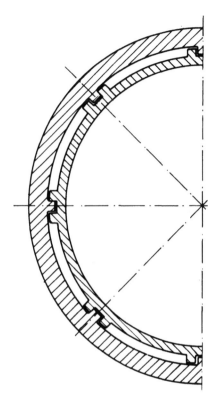

Figure 10.15 Schematic diagram of redesigned shell supporting system

by splines on the interior of the ring. (For clarity only a few projections and splines are shown in the figure. In fact, there are 64 projections uniformly distributed around the shell and engaging corresponding splines on the ring.)

With this system, for which a patent has been obtained (Weronski et al., 1985), the shell is supported at a number of places around its circumference, and sagging distortion is effectively eliminated. In addition, the shell itself does not come in direct contact with the inner surface of the ring. This, together with the radial clearance between the splines on the ring and the bottom of the slots on the shell, ensures that the slight deflection of the ring when passing over a supporting roller is not imparted directly to the shell; the friction between the sides of the projections and slots is not sufficient to transfer a force of damaging magnitude. Frictional forces were carefully assessed during the design, and while sufficient to stabilize the configuration in motion, do not create excessive restraint on differences in thermal expansion of shell and ring. The new construction put an end to shell cracking and lining detachment in the vicinity of rings. The lining in these regions now lasts as long as the lining elsewhere in the kiln.

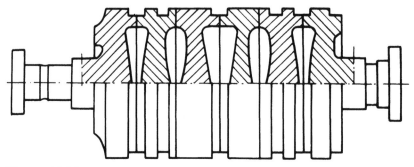

Figure 10.16   Welded turbine rotor. (From Hohn, 1980.)

Cracks do still appear in the shell. In the area midway between supporting rings the stress due to the weight of tube and contents is greatest and of course as the kiln rotates, the stress in any element of the shell alternates between compressive and tensile. The temperature of the shell is comparatively high, generally in the range 423 to 623 K, depending on the position, and it varies, particularly at the unlined input end. In this region the upper part of the shell is exposed directly to the combustion gas and may reach an instantaneous temperature of 670 K before rotation brings it into contact with cold slurry (see Fig. 10.13). Thus there are many factors contributing to shell cracking and no single, inexpensive way of eliminating them. Fortunately, these cracks occur with low frequency and can be repaired quickly and cheaply by welding, which is therefore the most economical way of combating them. Repairs are made promptly before a crack weakens the shell sufficiently to cause cumulative damage. The support rings are subjected to high stresses from their mechanical loads and also thermally induced stress, the interior of the ring being at about the same temperature as the shell and the exterior at a much lower temperature, depending in part on the weather. The thermally induced stress has been calculated as 90 MPa, but in extreme conditions and at startup can be twice as great. Ring life is adequate, but they ultimately suffer from spalling on the outer surface, and improvement could justify some extra expenditure.

As they become due for renewal, the two in the hottest zone will be replaced by rings provided with air cooling and the remainder by rings incorporating channels containing a suitable liquid (e.g., quenching oil) for the purpose of reducing temperature differences within each ring rather than cooling them.

In the case of large components there is sometimes a choice between integral and composite construction: that is, making the item as a single large casting or forging, or assembling it from a number of smaller parts. Both methods have technical and economic advantages that must be evaluated for each instance. As an example, Fig. 10.16 shows the composite construction of the rotor for a large steam turbine as built by Brown Boveri. Other manufacturers also use this method, but many prefer to make the rotor as a single forging, which is the method used in Poland. The author has no desire to adjudicate on the construction of steam turbines but simply uses this example to illustrate the types of advantages that may be obtained from composite construction when the different portions of a large component are exposed to different temperatures and temperature variations during operation— and frequently to still larger differences at startup and shutdown.

First, composite construction allows the use of materials best suited to the conditions in which each subunit works. In the case of the rotor, composite construction allows the heat treatment of each section to be optimized for its load and temperature. Also, it is generally easier to carry out heat treatment well on parts of moderate size rather than on very large ones. Finally, the risks of inclusions and flaws is less with the smaller components, their inspection is easier, and the total mass of the rotating component is reduced. Advantages of this sort are always worth considering whenever portions of a large component are exposed to different thermal environments. Composite construction does not have to be restricted to the protective coating of turbine blades described in Chapter 9.

10.3 MATERIAL PROPERTIES

A preliminary choice of material can be made on the basis of the anticipated thermal cycle and a knowledge of the magnitude and distribution of the stresses due to mechanical loading. The properties of materials change with temperature and in some cases with rate of change of temperature. Figure 10.17 shows some of these changes for the case of carbon steel, and similar data for the material being considered, including any variation in its specific heat and thermal conductivity,

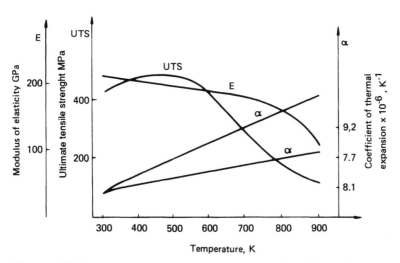

Figure 10.17 Dependence of mechanical properties of medium-carbon structural steel on temperature. The upper curve relates to a rapid temperature change; the lower curve, to a slow change.

must be taken into account in the next and subsequent stages of estimating temperature distribution and resultant thermally induced stresses. This may lead to an amended choice of material and recalculation until a satisfactory design and set of specifications are obtained. Formerly this was a matter of trial and error or successive approximations guided by intuition, but it can be carried out more precisely and formally by, for example, the method of finite differences, for which computer programs are available. This permits taking account of more variables, in a reasonable time, provided that he physical data are available. It must be accompanied by intelligent human monitoring, assessment, and where necessary, intervention. A factor that must not be overlooked is the fact that changes in the structure and hence properties of a material may take place over a long period of time. The process of thermal fatigue itself can cause such things as changes in the size of carbides, grain coarsening, embrittlement, and phase transitions, and thus deterioration in properties.

REFERENCES

Deilmann, W. and Schreiber, G. (1974). *Radex Rundsch.*, *1*: 21.

Gochfeld, D. A. and Tscherniavski, O. F. (1979). *Resistance of Structures to Repeated Loading*, Maszinostroienie, Moscow, p. 344.

Haley, M. R., Larson, H. R. and Kleeschulte, D. G. (1980). *J. Eng. Mater. Technol.*, *102*: 26.

Hohn, A. (1980). *Combustion*, (3): 10.

Wei, M. L. (1980). *Iron Steel Eng.*, *57*(2): 46.

Weronski, A. (1976). *Zesz. Nauk. Politech. Swietokrzy.*, *M11*: 1.

Weronski, A. (1983). *Zmeczenie cieplne metali*, WNT Warsaw, Poland.

Weronski, A. and Kielbinski, J. (1983). *Przegl. Mech.*, *9*: 11.

Weronski, A., Milanowski, W., Kielbinski, J., Frackiewicz, H. and Skrzydlo, K. (1985). Polish patent 128571.

Index

9 780367 402952